W0050412

Respiratory
Pharmacology and
Pharmacotherapy

Series Editors

Dr David Raeburn
Discovery Biology
Rhône-Poulenc Rorer Ltd
Dagenham Research Centre
Dagenham
Essex RM10 7XS
England

Dr Mark A Giembycz
Department of Thoracic Medicine
National Heart and Lung Institute
Imperial College of Scie ce, Technology and Medicine
Lor on SW3 6LY
Engla d

Pulmonary Actions of the Endothelins

Edited by
R. G. Goldie
D. W. P. Hay

Springer Basel AG

Editors

Prof Roy G Goldie
Faculty of Medicine & Dentistry
Department of Pharmacology
University of Western Australia
Perth Nedlands, WA, 6907
Australia

Dr Douglas W P Hay
Department of Pulmonary Pharmacology
SmithKline Beecham Pharmaceuticals
King of Prussia
Pennsylvania 19406–0939,
USA

Library of Congress Cataloging-in-Publication Data

Pulmonary actions of the endothelins / edited by R G Goldie, D W P Hay
     p   cm – (Respiratory pharmacology and pharmacotherapy)
     Includes bibliographical references and index

     1 Endothelins–Physiological effect   2 Endothelins–
     Pathophysiology   3 Respiratory organs–Pathophysiology
     4 Inflamation–Mediators   5 Asthma–Pathophysiology   I Goldie,
     R G (Roy G) II Hay, Douglas W P, 1956– III Series
        [DNLM 1 Endothelins–pharmacology   2 Lung–drug effects
     3 Bronchi–drug effects   4 Muscle, Smooth–drug effects
     5 Respiratory Tract Diseases–physiopathology   QV 150 P981 1999]
     QP552 E54P85   1999
     616 2' 407 – dc21
     DNLM/DLC
     for Library of Congress   99–11900
                                                                         CIP

Die Deutsche Bibliothek – CIP-Einheitsaufnahme

Pulmonary actions of the endothelins / ed by R G Goldie , D W
P Hay - Basel , Boston , Berlin Birkhauser, 1999
     (Respiratory pharmacology and pharmacotherapy)

     ISBN 978-3-0348-9786-0   ISBN 978-3-0348-8821-9 (eBook)
     ISBN 10.1007/978-3-0348-8821-9

Printed on acid-free paper produced from chlorine-free pulp TCF ∞

© 1999 Springer Basel AG
Originally published by Birkhauser Verlag, PO Box 133, CH- 4010 Basel, Switzerland 1999
Printed on acid-free paper produced from chlorine-free pulp TCF ∞

9 8 7 6 5 4 3 2 1

Contents

List of Contributors VII

1. History of Endothelin
   Katsutoshi Goto   1

2. Endothelin Receptors and Ligands
   Timothy D. Warner   21

3. Cellular Localisation of the Endothelin System in the Lung
   Nicholas W. Morrell, Carlos Orte, John Wharton and Julia Polak .   49

4. Activity and Distribution of Endothelin-converting Enzyme
   in the Lung
   *Pedro D'Orléans-Juste, Jean-Philippe Gratton, Ghassan Bkaily
   and Adel Giaid*   63

5. ET Receptor-Linked Signal Transduction Processes
   in the Airway Wall
   Peter J. Henry   83

6. *In Vitro* Effects of the Endothelins on Airway
   and Vascular Smooth Muscle Tone
   Claude Bertrand, Emmanuel Naline and Charles Advenier . . .   107

7. *In Vivo* Effects of the Endothelins in the Lung
   William M. Abraham   125

8. Endothelin in Lung Development and Tissue Growth
   Vera P. Krymskaya and Reynold A. Panettieri, Jr.   143

9. Endothelin and the Airway Epithelium
   *Joaquim Mullol, James N. Baraniuk, Cesar Picado
   and James A. Shelhamer*   155

10. Endothelin as a Proinflammatory Mediator
    Janos G. Filep and Douglas W.P. Hay   177

11. Endothelin-induced Neuromodulation
    Karen O. McKay   197

12. Role of Endothelin in Pulmonary Hypertension
    Bernadette Raffestin, Saadia Eddahibi and Serge Adnot   213

13. Endothelin as a Putative Mediator in Asthma
    and Other Respiratory Diseases
    Anthony E. Redington   231

Index   265

List of Contributors

William M. Abraham, Department of Research, Mount Sinai Medical Centre, 4300 Alton Road, Miami Beach, FL 33140, USA

Serge Adnot, Département de physiologie et INSERM U492, Hôpital Henri Mondor, F-94010 Créteil, France

Charles Advenier, Faculté de Médicine Paris-Ouest, Laboratoire de Pharmacologie, 15 rue de l'Ecole de Médicine, F-75270 Paris Cedex 06, France

James N. Baraniuk, Division of Reumathology, Allergy and Immunology, Department of Medicine, Georgetown University Medical Center, 3800 Reservoir Road NW, Washington DC 2007, USA

Claude Bertrand, Roche Bioscience, Inflammatory Diseases Unit, 3401 Hillview Avenue, Palo Alto, CA 94304–1397, USA

Ghassan Bkaily, Department of Anatomy and Cell Biology, Medical School, University of Sherbrooke (Québec) J1H 5N4, Canada

Pedro D'Orléans-Juste, Department of Pharmacology, Medical School, Université de Sherbrooke, Sherbrooke (Québec) J1H 5N4, Canada

Saadia Eddahibi, Département de physiologie et INSERM U492, Hôpital Henri Mondor, F-94010 Créteil, France

Janos G. Filep, Research Center, Maisonneuve-Rosemont Hospital, University of Montréal, 5415 boulevard de l'Assomption, Montréal, Québec, Canada H1T 2M4

Adel Giaid, Department of Pathology, Montréal General Hospital, 1650 Cedar Avenue, Montréal (Québec) H3G 1A4, Canada

Katsutoshi Goto, Department of Pharmacology, Institute of Basic Medical Sciences, University of Tsukuba, Tsukuba, Ibaraki 305–8575, Japan

Jean-Philippe Gratton, Department of Pharmacology, Medical School, Université de Sherbrooke, Sherbrooke (Québec) J1H 5N4, Canada

Douglas W.P. Hay, Department of Pulmonary Pharmacology, SmithKline Beecham Pharmaceuticals, King of Prussia, Pennsylvania 19406–0939, USA

Peter J. Henry, Department of Pharmacology, University of Western Australia, Stirling Hwy, Nedlands 6907, Western Australia, Australia

Vera P. Krymskaya, Pulmonary and Critical Care Division, Department of Medicine, University of Pennsylvania School of Medicine, Room 805 East Gates Building, 3400 Spruce Street, Philadelphia, PA 19104, USA

Karen O. McKay, Department of Respiratory Medicine, Royal Alexandra Hospital for Children, Westmead 2145, NSW, Australia

Nicholas W. Morrell, Section on Clinical Pharmacology, Divison of Medicine, Imperial College School of Medicine, Hammersmith Hospital, Du Cane Road, London W12 0NN, UK

Joaquim Mullol, Fundació Clínic per a la Recerca Biomèdica, Institut d'Investigacions Biomèdiques "August Pi i Sunyer", Hospital Clínic i Universitari Villarroel 170, E-08036 Barcelona, Catalonia, Spain

Emmanuel Naline, Faculté de Médicine Paris-Ouest, Laboratoire de Pharmacologie, 15 rue de l'Ecole de Médicine, F-75270 Paris Cedex 06, France

Carlos Orte, Department of Histochemistry, Imperial College School of Medicine, Hammersmith Campus, Du Cane Road, London W12 0NN, UK

Cesar Picado, Servei de Pneumologia i Al·lèrgia Respiratòria, Hospital Clínic i Universitari, Departament de Medicina, Universitat de Barcelona, Villarroel 170, E-08036 Barcelona, Catalonia, Spain

Julia M. Polak, Department of Histochemistry, Imperial College School of Medicine, Hammersmith Campus, Du Cane Road, London W12 0NN, UK

Reynold A. Panettieri, Jr., Pulmonary and Critical Care Division, Department of Medicine, University of Pennsylvania School of Medicine, Room 805 East Gates Building, 3400 Spruce Street, Philadelphia, PA 19104, USA

Bernadette Raffestin, Hôpital Ambroise Paré, 9 av Charles de Gaulle, F-92104 Boulogne, France

Anthony E. Redington, Department of Respiratory Medicine, 2nd Floor, Thomas Guy House, Guy's Hospital, London SE1 9RT, UK

James H. Shelhamer, Critical Care Medicine Department, Clinical Center, National Institutes of Health, 10 Center Drive Bethesda, MD 20892, USA

Timothy D. Warner, St. Bartholomew's and the Royal London School of Dentistry and Medicine, Charterhouse Square, London ECIM 6BQ, UK

John Wharton, Department of Histochemistry, Imperial College School of Medicine, Hammersmith Campus, Du Cane Road, London W12 0NN, UK

Preface

The biology of the family of endogenous autocrine peptides known as the endothelins (ETs) has been a source of intense study for researchers since 1988, following the identification of ET-1 as the previously described endothelium-derived contractile factor. Initial interest focussed on the actions of the ETs in the cardiovascular system, and this remains the primary source of published work involving these peptides. Importantly, evidence is mounting for a significant mediator role for ET-1 in cardiovascular diseases including systemic hypertension and congestive heart failure. However, it was also recognized early on that the ETs exert an array of powerful actions in the respiratory tract, many of which prompted speculation concerning mediator roles for ET-1 in several lung diseases including asthma and pulmonary hypertension.

In recent years the evidence for the involvement of ET-1 in lung diseases has become more compelling. There is now real hope and anticipation that novel ET receptor antagonists may eventually be used therapeutically against asthma and other major lung pathologies. It is this atmosphere of excitement created by new discoveries and the increasing scientific rationale for a significant impact of the ETs in pulmonary diseases that prompted the creation of this book, the first on this topic. Most of the information presented relates to ET-1, which is the most widely studied member of the ET family.

We were very privileged to have been able to recruit many of the world's leading pulmonary researchers in putting this volume together. We thank them sincerely for their contributions to this project. We have attempted to combine chapters describing the fundamental scientific foundations of the lung biology of the ETs with contributions which place this work in potential clinical contexts. Thus, the book moves from the opening chapter describing the history of ET research, to contemporary reviews of studies describing what is now established concerning the identities and functions of ET receptors, signal transduction processes and the sites at which the ETs are synthesized, released and inactivated in the lung. Subsequent chapters deal with the activities of these peptides in *in vitro* and *in vivo* systems, and analysis of their impact as physiological mediators of lung development and tissue growth. Evaluations of the potential of the ETs to induce abnormal airway responses via actions on the epithelium, neuronal tissue and on pathways leading to airway inflammation then set the stage for consideration of their roles as mediatiors and/or promotors of lung disease, with particular emphasis on pulmonary hypertension and asthma.

We believe that this volume represents a very contemporary evaluation of ET lung biology and the potential pathophysiological role of the ETs, in particular ET-1, in lung diseases, and trust that it will serve as a useful reference for the professional researcher, teacher and medical specialist.

Dedicated to my parents, Gordon "Bill" Goldie (1914–1994) and Sylvia Goldie and to my wife Rhonda and daughters Lauren and Susanne.

R. G. Goldie

This book is dedicated to Susanne and my family for their love, support and patience, especially during my education and career years. They have been an inspiration and pillars of strength.

D. W. P. Hay

Pulmonary Actions of the Endothelins
ed. by R. G. Goldie and D. W. P. Hay
© 1999 Birkhäuser Verlag Basel/Switzerland

CHAPTER 1
History of Endothelin

Katsutoshi Goto

Department of Pharmacology, Institute of Basic Medical Sciences, University of Tsukuba, Tsukuba, Ibaraki 305, Japan

1   Introduction
2   Endothelium-dependent Vascular Relaxation and Contraction
3   Isolation, Purification and Characterization of Endothelin
4   Endothelin Family Peptides
5   Regulation of Endothelin Biosynthesis
5.1 Regulation of Gene Expression
5.2 Endothelin Converting Enzyme
5.3 Release and Inactivation of Endothelin-1
6   Endothelin Receptors and Intracellular Signal Transduction
7   Measurement of Plasma and Tissue Endothelin-1
8   Endothelin Receptor Antagonists
9   Pathophysiological Significance of Endogenous Endothelin-1
10   Production of Endothelin Gene-targeted Animals
11   Conclusions and Perspectives
     References

1. Introduction

It had long been considered that noradrenaline released from autonomic sympathetic nerves and classical humoral factors, including renin-angiotensin, vasopressin and adrenaline, regulate vascular tonus by directly contracting vascular smooth muscle. In other words, peripheral vascular resistance is maintained mostly by vasocontraction produced by these factors, and vasorelaxation is passively induced when their influence diminishes. Within the past two decades, however, the existence of definite active vasorelaxation mechanisms has been uncovered, e.g., peptidergic [1] and nitric oxidergic [2] nerves, vasorelaxing substances including atrial natriuretic peptides (ANP) [3], adrenomedullin [4], and endothelium-derived relaxing factors (EDRF) [5]. Vascular endothelium consists of only one layer of endothelial cells on the inner surface of the blood vessel and plays an important role as a barrier against blood cells directly contacting vascular smooth muscle cells, hence preventing blood coagulation. In addition to this influence, vascular endothelial cells have been revealed to regulate vascular tonus by producing and releasing potent vasoactive substances, endothelium-derived contracting factors (EDCF), as well as EDRF. A number of vasoactive substances exert not only short-term contracting or relaxing actions but also long-term hypertrophic or hyperplastic actions on cardio-vascular tissues. Thus, cardiovascular structure and function appear to be controlled by a large number of endogenous vasoactive substances in a complicated manner.

2. Endothelium-dependent Vascular Relaxation and Contraction

An endothelium-derived vasoactive substance, prostacyclin, was first described by Moncada et al. in 1976 [6], but vascular endothelium was actually first recognized as an important functional unit involved in the regulation of vascular smooth muscle tonus after endothelium-dependent vasorelaxation was discovered by Furchgott and Zawadzki in 1980 [5]. It was hypothesized that, when stimulated by vasoactive substances such as acetylcholine, bradykinin, etc., endothelial cells secrete endothelium-de-rived relaxing factors (EDRF), causing relaxation of underlying vascular smooth muscle cells. Because EDRF was chemically labile, it took nearly 7 years since this initial observation to identify one EDRF as nitric oxide or a closely related substance [7, 8].

Vasocontraction dependent on or enhanced by intact endothelium has also been observed in response to various chemical and physical stimuli such as noradrenaline [9], thrombin [9], hypoxia [10], increased transmural pressure [11] and mechanical stretch [12]. These observations lead to the speculation that endothelial cells might release certain vasocontracting substance(s), endothelium-derived contracting factors (EDCF). Initially, one EDCF was suggested to be thromboxane A_2 or a related substance(s) by Fujiwara et al. [13] or to be superoxide anion by Katusic et al. [14].

In 1985, Hickey et al. attempted to test the biological activity of the culture medium of bovine aortic endothelial cells on isolated porcine coronary arteries, and found that the culture supernatant unexpectedly contained peptidergic factor(s) which triggered a slowly developing and long-lasting contraction of the coronary arteries [15]. At that time, they might have anti-cipated that cultured endothelial cells may release EDRF into the culture medium and that, upon application, the medium might cause relaxation of vascular smooth muscle preparations. Because of the extremely labile nature of EDRF, no vasorelaxing activity was detected when the culture supernatant was applied to the isolated arterial preparations and, instead, vasocontracting activity was detected. Based on analysis by means of gel chromatography, they suggested that this particular EDCF is a peptide with molecular mass of ca. 8500 Da. At about the same time, Gillespie et al. also detected vasocontracting activity in the culture supernatant of porcine aortic endothelial cells [16]. They subsequently reported that the activity was increased when the cultured endothelial cells were stimulated by thrombin, and suggested that this EDCF is a peptide with molecular mass of approximately 3000 Da [17].

As described above, EDCF seemed to consist of a variety of substances belonging to different chemical categories, including arachidonic acid metabolites, superoxide anion, and peptides. Nevertheless, only one peptide, endothelin, has been chemically identified among EDCFs thus far investigated.

3. Isolation, Purification and Characterization of Endothelin

In 1987, we performed similar experiments to those of Hickey et al. and confirmed that the supernatant from confluent monolayer cultures of porcine aortic endothelial cells contained a slowly developing and long-lasting vasocontracting factor(s), peptide in nature, since the vasocontracting activity was abolished by pretreatment of the conditioned medium with trypsin (1 µg/ml at 37°C for 2 h). In contrast, unconditioned medium or medium conditioned with human fibroblast IMR-90 cells had no activity. Furthermore, the activity was also detected in serum-free conditioned medium, and no appreciable change in activity was observed even after long-term (4–5 weeks) maintenance of the endothelial cell culture in serum-free conditions. These observations clearly indicated that the vasocontracting factor in the supernatant was not a derivative or degradation product of a serum component. The successful attempt at serum-free maintenance and detection of vasocontracting activity prompted us to isolate and purify the active peptide in the supernatant because of the absence of interference with proteins and/or peptides in the serum itself.

Endothelial cells isolated from porcine thoracic aortas and grown to a confluent monolayer were maintained in serum-free minimum essential medium. Medium was changed every 5 days and the conditioned medium was pooled and stored at −20°C. The pooled conditioned medium (approximately 10 l) was first centrifuged at 1000 g for 20 min, and subsequently, the supernatant was desalted and concentrated. The concentrated medium was loaded onto an anion-exchange column and eluted by applying a linear gradient of NaCl. The vasocontracting activity of the eluent was assayed by adding a small amount of each fraction directly into a muscle chamber where a helical strip of porcine coronary artery with the intima denuded was suspended, and the active fraction was collected. The active fraction was subjected to reverse-phase high performance liquid chromatography (HPLC) and elution with a linear gradient of acetonitrile, and the vasocontracting activity of each fraction was similarly assayed. A second trial of reverse-phase HPLC enabled us to purify the active component. Approximately 3 nmol of final product was obtained, which was just enough to perform subsequent amino acid sequence analysis.

The purified peptide was subjected to analysis of amino acid sequence by means of an automated gas-phase peptide sequencer and carboxy-terminal analysis by hydrazinolysis (Edman's reaction). As a result, the peptide was revealed to be comprised of 21 amino acid residues with free amino- and carboxy-termini. At first, the carboxy-terminal amino acid (trypto-phan) was not detected, and a 20-amino acid peptide synthesized artificial-ly did not show any biological activity. It was soon recognized that the structure of tryptophan (indole ring) is easily degraded during Edman's reaction. The peptide of 21-amino acid residues subsequently synthesized did exhibit an identical vasocontracting activity to that of the natural peptide. The four cysteine residues at the amino-terminal portion were found to form two intrachain disulfide bonds, the topological analysis of which pro-vided another interesting finding. Since the amino-terminal alignment of this peptide is quite similar to that of several α-scorpion toxins, i.e., neuro-toxins (*Buthus epeus* neurotoxin M 10, *Androctonus australis Hector* neu-rotoxin III, etc.), and because their intrachain disulfide bonds usually intersect, we first hypothesized that the disulfide bonds of this peptide might also intersect ($1-11$ and $3-15$), and the submitted manuscript con-tained this intersecting structural formula. According to more detailed NMR analysis of the three-dimensional structure performed at the laboratories of Takeda Chemical Ind. Ltd. and Peptide Inst. (Osaka, Japan), the arrange-ment of disulfide bonds was finally settled to be a coaxial form ($1-15$ and $3-11$), and the figure in the manuscript was corrected at the stage of galley proof. Because it was originally discovered from the culture supernatant of endothelial cells, the peptide was termed "endothelin" [18].

Based on the amino acid residues 7--20 of endothelin, Yanagisawa syn-thesized several kinds of oligonucleotide according to the mammalian codon usage statistics. Using one of the synthesized oligonucleotides as a single "optimal" DNA probe, he screened $\sim 2 \times 10^6$ clones from aλgt10 cDNA library constructed for porcine aortic endothelial cell mRNA and soon identified 38 hybrydization-positive clones. Four of these clones were subjected to further characterization, and finally, a complete nucleotide sequence of porcine preproendothelin cDNA and the deduced amino acid sequence were determined.

The 203-residue porcine preproendothelin contains 19 residues of de-duced N-terminal amino acid sequence, being characteristic of a secretory signal sequence, that is, a hydrophobic core followed by residues with small polar side-chains. As anticipated, paired basic amino-acid residues Lys^{51}-Arg^{52}, which are recognized by processing endopeptidases, directly precede the endotelin sequence, but no dibasic pair is found thereafter until Arg^{92}-Arg^{93}. This indicated that mature endothelin may be generated via unusual proteolytic processing between Try^{73} and Val^{74} in a 39-amino acid residue intermediate, or so-called "big-endothelin," presumably by an endopeptidase exhibiting a chymotrypsin-like specificity, which was puta-tively termed "endothelin-converting enzyme (ECE)." This presumptive pathway of endothelin biosynthesis in endothelial cells was confirmed later. The human preproendothelin cDNA was soon cloned, and its sequence of

212-amino-acid residues containing big-endothelin of 38-amino-acid residues was determined [19].

In parallel with molecular biological studies, analysis of the pharmacological characteristics of synthesized endothelin was also carried out. *In vitro* studies with isolated porcine coronary artery revealed that endothelin exerted an extremely potent and long-lasting contracting action; the contracted tissue did not relax even after repeated washing out for several hours, without irreversible tissue damage. Furthermore, when a small amount of endothelin, e.g., 1 nmol, was intravenously injected as a bolus in the anesthetized rat, after transient hypotension, a sustained elevation in blood pressure for more than 1 h was evident [18]. This was an extraordinary finding, since we had never observed such a long-lasting response to a substance of endogenous origin.

It took less than 10 months (March to November, 1987) to complete most of the studies relating to isolation, purification and characterization of endothelin. The rapid achievement of these investigations was the fruit of good collaboration of specialists in molecular biology (M. Yanagisawa), peptide chemistry (S. Kimura) and pharmacology (K. Goto), with the contribution of M. Yanagisawa being particularly notable. One more important point is that the methods and know-how for identifying bioactive peptides were already established at that time, since S. Kimura and K. Goto had been searching for neuropeptides, and had isolated a peptide from porcine spinal cord exhibiting myocardium-stimulating and vasorelaxing activities and determined its amino-acid sequence [20]. Unfortunately, however, the structure of the peptide was identical to that of calcitonin gene-related peptide (CGRP) and its existence had been predicted from the analysis of calcitonin gene by Amara et al. (1982) [21].

4. Endothelin Family Peptides

In our search for the possible existence of endothelin-related peptides(s) at the DNA level, we soon encountered three distinct human genes for putative endothelin precursors [22]. Southern blot analysis of human genomic DNA under low hybridization stringency with a 42-mer synthetic oligonucleotide probe corresponding to amino acid residues 7–20 of endothelin, demonstrated that three different restriction fragments were always detected irrespective of the restriction endonucleases used. The nucleotide sequences encoding amino acid residues of the three endothelins are highly conserved among the three genes, with 77–82% of the nucleotide residues being identical. In contrast, the nucleotide sequences upstream from the mature peptides are very poorly conserved. These observations suggest that, although the three genes are evolutionarily relatively distant from each other, the genes evolved from a common ancestral gene under strong pressure to preserve mature endothelin sequences. The three peptides were designated endothelin-1, endothelin-2 and endothelin-3. Endothelin-1 is the original product, i.e., the peptide corresponding to that detected in the culture medium of porcine aortic endothelial cells. Although vascular endothelial cells are the major source of endothelin-1, Northern blot analysis revealed that the genes encoding the three endothelin isopeptides are expressed with different patterns in a wide variety of cell types including cardiac myocytes, vascular smooth muscle cells, renal tubular epithelial cells, glomerular mesangial cells, glial cells, pituitary cells, macrophages and mast cells, suggesting that the peptides may participate independently in complex regulatory mechanisms in various organs (Fig. 1).

The number of endothelin-related genes was also demonstrated to be conserved in several mammalian species examined, including human, pig and rat. The sequence of human endothelin-3 was identical to a sequence of the peptide previously found in the rat genome, which is therefore now considered to be the rat endothelin-3 gene [23]. Further, four cardiotoxic peptides, sarafotoxins (S6a-S6d), which are highly homologous to endothelins, were isolated from the snake venom of *Atractaspis engaddensis* [24]. Dr. C. Takasaki from Tohoku University was fascinated to find that sarafotoxins, the structure of which he determined, were so similar to endothelin [25]. Endothelin-like immunoreactivity was also found in several

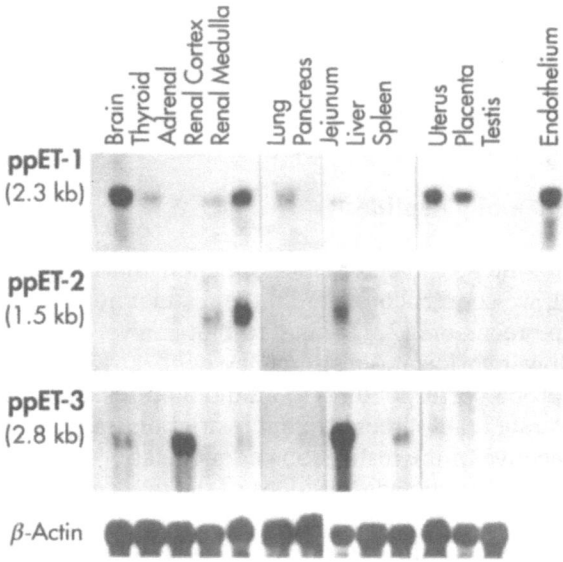

Figure 1. Northern blot analysis of mRNA of endothelin-1, -2 and -3, and *-actin. Poly(A)RNA was extracted from various human tissues, and 8 µg of each RNA was subjected to electrophoresis and subsequent Northern blot analysis using each cDNA as the probe. (In the case of endothelial cells, total RNA was employed.)

species of invertebrates [26] and fish [27], indicating that endothelins found in humans seem to have a long evolutionary history.

According to the predicted amino acid sequence, 21-residue endothelin-1, endothelin-2 and endothelin-3 were synthesized by solid-phase chemical method, and their biological activities were assayed for contractile responses of porcine coronary artery strips *in vitro* and for pressor responses *in vivo* in anesthetized chemically denervated rats [22]. *In vitro*, the three synthetic peptides exhibited strong and long-lasting vasoconstrictor activity, although the effects were quantitatively different among the isopeptides. In terms of molar potency, endothelin-1 was most potent, whereas the maximum contractile response was greater for endothelin-2, and endothelin-3 was the least potent in either sense. The three peptides also produced strong pressor responses in anesthetized rats *in vivo*; typical tracings for the changes of blood pressure in response to a bolus injection of each peptide are shown in Figure 2. A transient depressor response lasting 1–2 min always preceded the long-lasting increase in blood pressure. The peak pressor effect was markedly smaller for endothelin-3, but not significantly different between endothelin-1 and endothelin-2. In contrast, the initial depressor effect was considerably greater for endothelin-3. The pressor effect was longest for endothelin-2 and shortest for endothelin-3. Since

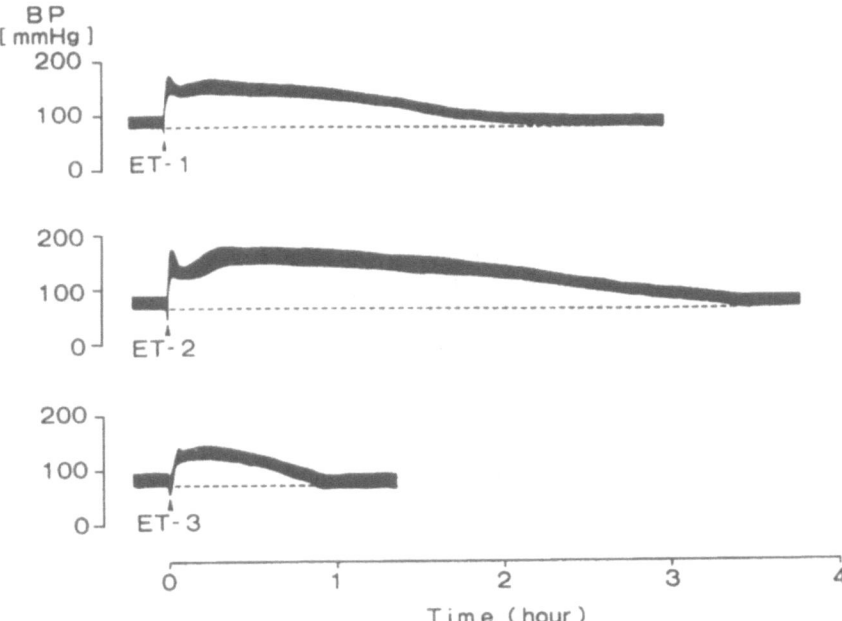

Figure 2. Typical recordings of pressor responses of anesthetized chemically denervated rats to intravenous injection of bolus of endothelin-1, -2 or -3 (1 nmol/kg). BP: blood pressure.

endothelin-1 is an isopeptide which was discovered first and appears to be concerned with various cardiovascular diseases, a large amount of information concerning the effects of endothelin-1 of either exogenous or endogenous origin has accumulated. However, there is still a paucity of comparable information on endothelin-2 and endothelin-3.

5. Regulation of Endothelin Biosynthesis

5.1. Regulation of Gene Expression

The induction of endothelin-1 mRNA and/or peptide by various chemical and mechanical stimuli has been studied mainly in cultured endothelial cells. The growing list of physiological stimuli that can increase endothelin-1 production includes thrombin [28], transforming growth factor-β (TGF-β) [29], tumor necrosis factor-α (TNF-α) [30], interleukin-1 (IL-1) [31], insulin, and vasoactive substances [32, 33] such as norepinephrine, angiotensin II, vasopressin, bradykinin, oxidized or acetylated low density lipoprotein (LDL) [34], and fluid dynamic shear stress. Thrombin, angiotensin II, and vasopressin stimulate phospholipase C activity in endothelial cells, leading to the formation of the second messengers inositol 1,4,5-triphosphate which mobilizes Ca^{2+} from intracellular storage sites and 1,2-diacylglycerol which stimulates protein kinase C. Indeed, the expression of endothelin-1 mRNA and/or production of peptide are also induced by Ca^{2+} ionophores and phorbol esters [35, 36]. These observations are consistent with the concept that the production of endothelin in endothelial cells may be regulated by intracellular Ca^{2+} and by protein kinase C. However, the mechanisms of regulation of endothelin-1 mRNA expression by other substances remain to be elucidated. Although no SSRE (shear stress responsive element) has been detected in the endothelin-1 gene, mRNA expression and endothelin-1 production in endothelial cells are regulated by fluid shear stress in a complicated manner: high shear stress (25 dynes/cm^2) sharply decreases mRNA level [37, 38], whereas low shear stress (5 dynes/cm^2) increases mRNA expression [39]. Pulsatile stretch also causes enhanced production of endothelin-1 in endothelial cells [40]. It has been shown that shear stress also increases intracellular free Ca^{2+} concentration in cultured endothelial cells by both stimulating the influx of extracellular Ca^{2+} and mobilizing intracellular Ca^{2+}, but it does not appear to be the sole mechanism in the regulation of endothelin-1 production by shear stress [41].

The human endothelin-1 gene contains five exons, four introns, and 5'- and 3'-flanking regions, and spans approximately 6.8 kb of DNA [22, 42]. Each of the five exons encodes a portion of preproendothelin-1. The primary transcription start site in endothelial cells was mapped by S1 nuclease protection to a position 98 base pairs downstream from the CAAT box and 31 base pairs from the TATAA box. The 5′-promoter region of human endothelin-1 gene contains several elements responsive to 12-O-tetradecanoylphorbol 13-acetate (AP-1/jun-binding elements) that are found in other genes, the expression of which can be induced by phorbol esters [43]. It also contains the nuclear factor-1 (NF-1)-binding elements that have recently been recognized to be involved in gene regulation in response to TGF-β [44]. Acute phase reaction regulatory element, another characteristic regulatory element, is also found within the endothelin-1 gene. The fluid shear stress appears to regulate endothelin-1 gene transcription via, in addition to Ca^{2+} signaling, an upstream cis element by a distinct mechanism not dependent on protein kinase C or cAMP pathways [41]. Further promoter mapping studies will help in clarifying the potential regulatory DNA elements in the endothelin-1 gene involved in the regulation of endothelin-1 production.

In contrast, the expression of endothelin-1 mRNA has been shown to be inhibited by EDRF (nitric oxide), prostacyclin and atrial natriuretic peptide, presumably via cGMP-mediated inhibition of phosphatidyl inositide metabolism [45]. Heparin also decreases endothelin-1 mRNA expression via inhibition of protein kinase C [46].

The level of endothelin-1 mRNA in endothelial cells may be controlled not only by transcriptional regulation but also by post-transcriptional regulation of mRNA degradation. Half-life studies using the transcription inhibitor actinomycin D have revealed that endothelin-1 mRNA is extremely labile, having an intracellular half-life of about 15 min [22, 42]. An important feature of endothelin-1 mRNA is the presence of two AUUUA sequences in the 3′-untranslated region, which is conserved among different species. These AU motifs are known to mediate selective mRNA degradation and are found in mRNAs encoding certain transiently expressed cytokines, growth factors and nuclear proto-oncogene products (usually those involved in programming of cellular growth and differentiation).

5.2. Endothelin Converting Enzyme

Sequence analysis of the endothelin-1 cDNA revealed the existence of a single copy gene that encoded for the precursor, preproendothelin-1. Human preproendothelin-1 is a 212-amino acid protein that is proteolytically cleaved, as in the case of many other bioactive peptides, at paired basic amino acids, specifically at adjacent positions of Lys^{51}-Arg^{52} and Arg^{92}-Arg^{93} by a dibasic-pair specific endopeptidase, thereby producing a 38-amino acid residue intermediate peptide, termed "big endothelin-1." As an endopeptidase professing preproendothelin-1, furin has been predicted as a definite enzyme [47]. Big endothelin-1 is subsequently cleaved at Trp^{21}-Val^{22} by a novel endopeptidase, which appears to be specific for endothelin and was putatively termed "endothelin converting enzyme (ECE)."

By sequencing the ECE fragment purified from bovine adrenal cortex and using a cDNA product obtained by reverse transcription-polymerase chain reaction (RT-PCR) as a probe, the ECE gene was cloned from bovine adrenal cortex cDNA libraries [48]. The structure of the ECE gene reflects a 758-amino acid metalloprotease enzyme (approximately 130 kD in molecular mass), containing a single membrane-spanning sequence with only a 56-residue N-terminal cytoplasmic tail and an extracellular C-terminal of 681-amino acid residues that contains the catalytic domain. Amino acid residues 593–601 match the highly conserved consensus sequence of a zinc-binding motif, HEXXH, which is shared by many known metalloproteases. The presence of ten predicted sites for N-glycosylation in the extracellular domain suggests that ECE is a highly glycosylated protein. From the same ECE gene, another type of ECE, the intracellular domain of which is different from that of former ECE, was also produced via an alternative splicing, and termed ECE-1β. The former is now called ECE-1α and it is abundantly expressed mainly on the membrane of endothelial and vascular smooth muscle cells. Both ECEs are present as a homodimer through making numerous S-S bonds between the two molecules, and cleave big endothelin-1 more efficiently than the other family peptides [49, 50].

5.3. Release and Inactivation of Endothelin-1

The conversion from big endothelin-1 to endothelin-1 is essential for exertion of biological activity, because the potent pressor action of big endothelin-1 was almost completely inhibited by a relatively large dose of phosphoramidon [51, 52]. In the culture medium of endothelial cells, the C-terminal fragment of big endothelin-1 was detected to exist at a stoichiometoric level with endothelin-1, and a small amount of big endothelin-1 was also detected [53, 54]. Taken together, it can be speculated that big endothelin-1 is cleaved by ECEs when it crosses the endothelial cell membrane.

Endothelin-1 may be released in both luminal and abluminal directions from endothelial cells *in vivo*. Abluminally released endothelin-1 may be diluted by the bloodstream, and its circulating concentration is approximately 1 ~ 2 pg/ml, which is far below the threshold concentration producing vasoconstriction. Although the exact concentration in the abluminal space (vascular smooth muscle surface) is not known, endothelin-1 is more likely to be a locally acting rather than a circulating hormone. When endothelin-1 is intravenously administered as a bolus, it disappears quite rapidly from the bloodstream with a half-life of a few minutes [55, 56]. This rapid removal of endothelin-1 from the circulation is due to uptake into various tissues, including the lung, kidney, spleen, and liver. The lung appears to be one of the most important tissues for uptake, since approximately 60% of endothelin-1 is removed after a single passage through the pulmonary circulation [56]. Endothelin-1-receptor complex might be taken up into cells through an internalization process and degraded by lysozomal enzymes, e.g., aspartic proteases [57]. Neutral endopeptidase (encephalinase) [58, 59] and lysozomal cathepsin G [60] are also concerned with enzymatic degradation of endothelin-1. In spite of such a rapid disappearance from the circulation, the blood pressure-elevating effect of endothelin-1 continues for an extremely long period; the exact reason for this phenomenon is unknown.

6. Endothelin Receptors and Intracellular Signal Transduction

Endothelin-induced responses can be divided into two major groups, according to the pharmacological potency order of the three isopeptides. In the first group of responses, including vasoconstriction, bronchoconstriction, uterine smooth muscle contraction, and stimulation of aldosterone secretion, endothelin-1 and endothelin-2 act as far more potent agonists than endothelin-3. In the second group, which includes endothelium-dependent vasorelaxation, astrocyte proliferation, and inhibition of *ex vivo* platelet aggregation, the three isopeptides possess almost equal potency. These findings suggest that there are at least two distinct endothelin receptors mediating these two distinct groups of pharmacological responses [61]. The existence of multiple receptors was also supported by biochemical studies, e.g., cross-linking experiments [62], and radioligand binding affinity study [63]. Subsequently, two cDNAs encoding for endothelin receptors were cloned and their amino acid sequences were deduced [64, 65]. The order of affinity of the three endothelin isopeptides for one receptor type, designated ET_A, is endothelin-1 $\geq$ endothelin-2 $\gg$ endothelin-3. The other type of receptor, designated ET_B, shows equipotent affinity for all three endothelins. These results are consistent with the results of previous pharmacological and biochemical studies and, thus, definite molecular evidence for the existence of two distinct subtypes of endothelin receptors was finally provided.

The existence of more subtypes of endothelin receptor has been a matter of controversy. Much pharmacological evidence has accumulated to suggest that there may exist an ET_C receptor which is specific for endothelin-3, or further sub-subtypes of ET_A and ET_B receptors. However, there has been no report thus far on isolation of the cDNA clone encoding for such a subtype or sub-subtype of endothelin receptor from mammalian tissues. Further pharmacological, biochemical or molecular biological analysis may be necessary to solve such apparent discrepancies.

Upon activation of either ET_A or ET_B receptor, many cells show a rapid transient increase in cintracellular Ca^{2+} concentration ($[Ca^{2+}]_i$) and a subsequent sustained increase in ($[Ca^{2+}]_i$). It is recognized generally that the initial transient phase is the result of IP_3-induced mobilization of Ca^{2+} from intracellular stores and that the sustained increase is due to an influx of extracellular Ca^{2+} through either dihydropyridine-sensitive, voltage-dependent L-type Ca^{2+} channels or receptor-operated (nonselective) cation channels that are insensitive to dihydropyridine. In addition to the increase in ($[Ca^{2+}]_i$), endothelin receptor activation is also known to stimulate arachidonic acid metabolism and to activate phospholipase A_2 or D [66]. In addition to short-term actions, endothelin-1 exerts potent mitogenic actions on vascular smooth muscle cells, cardiac myocytes, glomerular mesangial cells and other cells. These effects might be provoked through sequential activation of an intracellular kinase cascade, including raf-1, mitogen-activated protein kinase (MAPK) kinase, MAPK and S6 kinase. Since both ET_A and ET_B receptors are capable of coupling to various GTP-binding proteins (G_q, G_s, G_i, G_o, etc.), the downstream signal transduction may differ, depending on the type of GTP-binding proteins coupled, thereby resulting in different responses in various tissues [67].

7. Measurement of Plasma and Tissue Endothelin-1

The method for measurement of endothelin levels in plasma or various tissues by means of radioimmunoassay (RIA) [68, 69] or enzyme immuno-

Table 1. Plasma levels of endothelin-1 immunoreactivity in humans

Control (pg/ml)	Abnormal States (pg/ml)		Reference
1.5			68
1.7			69
1.5			70
1.5	7.3	(Surgical operation)	71
0.3	3.7	(Cardiac shock)	72
	1.1	(Chronic dialysis)	
	1.5	(Pulmonary hypertension)	
1.5	3.5	(Acute myocardial infarction)	73
6.0	11.0	(Acute myocardial infarction)	74
1.2	8.2	(Cold pressor test)	75
< 7	10.7	(Haemodialysis)	76
< 3.5	12.3	(Uremia + Haemodialysis)	77
1.5	10.4	(Acute renal failure)	78
18.5	33.9	(Essential hypertension)	79
0.2	1.1	(Essential hypertension)	80
1.6	1.7	(Essential hypertension)	81
4.8	12.0	(Subarachnoid haemorrhage)	82
0.7	1.2	(Subarachnoid haemorrhage)	83
1.7	5.3	(Raynaud's phenomenon)	84
2.7	10.3	(Raynaud's phenomenon + Cold pressor)	
2.8	7.5	(Diabetes mellitus)	85
1.5	3.1	(Disseminated intravascular coagulation)	86
1.3	1.9	(Systemic sclerosis)	87
0.9	1.9	(Labour pain)	88
6.1	11.4	(Pre-eclampsis)	89

Control: Age-matched normal subjects.

assay was established soon after the discovery of endothelin (EIA) [70]. RIA employs only one kind of antibody against endothelin-1, while EIA uses two kinds of antibodies, each recognizing the N-terminal and C-terminal portions of endothelin-1, respectively. As can be seen in Table 1 [68–89], the plasma levels of endothelin-1 in humans were fairly divergent, ranging from 0.2 to 18.5 pg/kg in healthy subjects. In order to avoid unnecessary confusion, the difference in the values of plasma endothelin-1 level was examined thoroughly and, as a result, it was generally agreed that the divergence mostly originated from differences in specificity of antibodies used and in the recovery rate during extraction of endothelin-1 from plasma or tissues. It is accepted now that the plasma concentration of endothelin-1 in healthy subjects is 1.0 ~ 2.0 pg/ml, which is at least one order of magnitude lower than that of circulating human atrial natriuretic peptide and several times less than that of angiotentin II [90].

8. Endothelin Receptor Antagonists

The first important milestone in the development of endothelin receptor antagonists was the isolation of a cyclic pentapeptide, BE-18257B, from the fermentation products of *Streptomyces misakiensis* [91]. Further modification of BE-18257B has led to the identification of potent ET_A receptor-selective antagonists including cyclic pentapeptide BQ-123 [92], linear tripeptides FR139317 [93] and PD151242 [94]. Interestingly, *Streptomyces misakiensis* is found in the soil of Tsukuba, as does *Streptomyces tsukubaensis* which produces FK506. Other peptides exhibiting ET_B receptor-selective antagonistic action (IRL2500 [95], BQ-788 [96]) and ET_A/ET_B receptor-nonselective antagonists (TAK-044 [97], PD145065 [98]) have also been developed recently. These peptide antagonists have been used in a wide variety of pharmacological and biochemical studies to characterize the endothelin receptors and investigate their tissue distribution. Nevertheless, there are well-known limitations of the use of peptides *in vivo* for the chronic treatment of animals or for clinical application in human patients.

Much effort has been directed toward the development of nonpeptide and orally active endothelin receptor antagonists. The first compound was Ro46-2005 [99], which was produced by structural modification of sulfonamides first synthesized in the search for antidiabetic drugs. Now, many sulfonamide derivatives exhibiting potent endothelin receptor antagonistic activity have been developed, including Ro47-0203 (bosentan) [100], L754142 and T-0115 [101]. The second important lead compound was found in diphenylindane derivatives by means of three-dimensional structural analysis including ^1H NMR-based conformations. The first compound exhibiting extremely potent antagonistic action was SB209670 [102] and, subsequently, PD156707 and J-104121 were identified [101].

All these compounds have ET_A/ET_B receptor-nonselective antagonistic actions, although their affinities to each receptor subtype are quite divergent.

9. Pathophysiological Significance of Endogenous Endothelin-1

An extraordinary finding concerning an endothelin-secreting tumor was reported by Yokokawa et al. at the 2nd International Conference on Endothelin held in Tsukuba in 1990 [103]. There was a good correlation between the plasma levels of endothelin-1 and elevated blood pressure in two patients with hemangioendothelioma, a rare form of malignant tumor of endothelial cells. Furthermore, in these two patients, the elevated plasma levels of endothelin-1 and hypertension returned to normal following surgical removal of the tumor, and both parameters rose again along with the recurrence of tumor in one of the patients. Another impressive finding presented by McMahon and co-workers at the same meeting was that phosphoramidon, which exhibits an ECE-inhibiting action, lowers blood pressure in spontaneously hypertensive rats but not in normotensive rats [104]. Thus, it was somehow anticipated at an early stage after the discovery of endothelin that endothelin-1 might be involved in the pathogenesis of hypertension.

Subsequently, there appeared many reports regarding plasma levels of endothelin-1, and the results concerning the relationship between the plasma level and severity of hypertension were conflicting. Furthermore, in a hypertensive animal model (SHR), BQ-123 was not efficacious [105]. Although these divergent observations have gradually led to a decrease in the belief that endothelin-1 plays a significant pathophysiological role in the development of hypertension, recent studies have demonstrated that endothelin-1 may actually be involved as an etiologic or aggravating factor in a number of cardiovascular diseases, including not only hypertension but also pulmonary hypertension, acute renal failure, cerebral vasospasm after subarachnoid hemorrhage, vascular remodeling, cardiac hypertrophy, and chronic heart failure [101, 106]. In the future, the investigation of some orally active receptor antagonists and/or ECE inhibitors in clinical trials are expected to clarify the role of endothelin-1 in cardiovascular and non-cardiovascular disorders.

10. Production of Endothelin Gene-targeted Animals

The production of endothelin-1-deficient and ET_B receptor-deficient mice by means of gene targeting has shown unexpectedly that the endothelins play crucial roles in normal embryonic development. Endothelin-1-deficient mice (homozygotes) have craniofacial and cardiac abnormalities at birth and die of respiratory failure soon afterwards [107]. Although endothelin-1-deficient mice (heterozygotes), in which only half the normal level of endothelin-1 is expressed, exhibited slightly but significantly higher blood pressure compared with wild type mice, it would be premature at present to conclude that endothelin-1 plays a hypotensive role in normal physiological conditions. In fact, it was recently demonstrated that overexpression of human preproendothelin-1 only locally in the liver of rats caused systemic hypertension [108]. On the other hand, ET_B receptor-deficient mice exhibited aganglionic megacolon associated with coat color spotting, resembling a hereditary syndrome in humans, Hirschsprung's disease [109, 110]. Surprisingly, targeted disruption of endothelin-3 resulted in the development of an identical syndrome, suggesting that interaction of endothelin-3 and the ET_B receptor is essential for development of enteric neurons and epidermal melanocytes [111]. The ET_B receptor-deficient mice can be saved from the lethal syndrome of aganglionic megacolon by local re-expression of ET_B receptor in neurons (personal communication, M. Yanagisawa). These molecular biological studies open a new field of endothelin research and are expected to greatly contribute to the delineation of the physiological and/or pathophysiological roles of endogenous endothelins.

11. Conclusions and Perspectives

During the almost 10 years since endothelin was discovered, many of its features concerned with basic medical sciences have been clarified, including regulation of endothelin gene expression, its biosynthesis and secretion, identification of endothelin receptors and intracellular signal transduction following their activation, and discovery of various receptor agonists and antagonists. Significant pathophysiological roles of endogenous endothelin-1 in various kinds of cardiovascular and non-cardiovascular diseases have gradually been uncovered, suggesting that interference with the endothelin-1 system by either specific and/or selective receptor antagonists or ECE inhibitors may provide a new therapeutic intervention for several diseases. The development of nonpeptide endothelin receptor antagonists, as well as production of endothelin gene-manipulated animals, would also provide promising tools for further clarification of the physiological and/or pathophysiological roles of endogenous endothelins.

References

1 Kawasaki H, Takasak K, Saito A, Goto K (1988) Calcitonin gene-related peptide acts as a novel vasodilator neurotransmitter in mesenteric resistance vessels of the rat. *Nature* 335: 164–167
2 Toda N, Okamura T (1991) Role of nitric oxide in neurally induced cerebroarterial relaxation. *J Pharmacol Exp Ther* 258: 1027–1032

3 Kangawa K, Matsuo H (1984) Purification and complete amino acid sequence of α-human atrial natriuretic polypeptide. *Biochem Biophys Res Comm* 118: 131–139

4 Kitamura K, Kangawa K, Kawamoto M, Ichiki Y, Nakamura S, Matsuo H, Eto T (1993) Adrenomedullin: a novel hypotensive peptide isolated from human pheochromocytoma. *Biochem Biophys Res Comm* 192: 553–560

5 Furchgott RF, Zawadzki JV (1980) The obligatory role of endothelial cells in the relaxation of arterial smooth muscle by acetylcholine. *Nature* 288: 373–376

6 Moncada S, Gryglewski R, Bunting S, Vane JR (1976) An enzyme isolated from arteries transforms prostaglandin endoperoxides to an unstable substance that inhibits platelet aggregation. *Nature* 263: 663–665

7 Palmer RMJ, Ferrige AG, Moncada S (1987) Nitric oxide release accounts for the biological activity of endothelium-derived relaxing factor. *Nature* 327: 524–526

8 Palmer RMJ, Ashton DS, Moncada S (1988) Vascular endothelial cells synthesize nitric oxide from L-arginine. *Nature* 333: 664–666

9 DeMey JG, Vanhoutte PM (1982) Heterogenous behavior of the canine arterial and venous wall: importance of endothelium. *Circ Res* 51: 439–447

10 Rubanyi GM, Vanhoutte PM (1986) Hypoxia releases a vasoconstrictor substance from the canine vascular endothelium. *J Physiol* 364: 46–56

11 Harder DR (1987) Pressure-induced myogenic activation of cat cerebral arteries is dependent on intact endothelium. *Circ Res* 60: 102–107

12 Nakayama Y, Tanaka Y (1990) Implications of calcium ions and endothelium in the stretch-induced contractile activation of cerebral vascular tissues. *Eur J Pharmacol* 183: 174–175

13 Shirahase H, Usui H, Kurahashi K, Fujiwara M (1987) Possible role of endothelial thromboxane A2 in the resting and contractile responses to acetylcholine and arachidonic acid in canine cerebral arteries. *J Cardiovasc Pharmacol* 10: 517–522

14 Katusic ZS, Shepherd JT, Vanhoutte PM (1987) Endothelium-dependent contraction to stretch in canine basilar arteries. *Am J Physiol* 252: H671–H673

15 Hickey KA, Rubanyi GM, Paul RJ, Highsmith RF (1985) Characterization of a coronary vasoconstrictor produced by cultured endothelial cells. *Am J Physiol* 248: C550–C556

16 Gillespie MN, Owasoyo JD, McMurthy IF, O'Brien RF (1986) Sustained coronary vasoconstriction provoked by a peptidergic substance released from endothelial cells in culture. *J Pharmacol Exp Ther* 236: 339–343

17 O'Brien RF, Robbins RJ, McMurthy IF (1987) Endothelial cells in culture produce a vasoconstrictor substance. *J Cell Physiol* 132: 263–270

18 Yanagisawa M, Kurihara H, Kimura S, Tomobe Y, Kobayashi M, Mitsui Y, Yazaki Y, Goto K, Masaki T (1988) A novel potent vasoconstrictor peptide produced by vascular endothelial cells. *Nature* 332: 411–415

19 Inoue A, Yanagisawa M, Takuwa Y, Mitsui Y, Kobayashi M, Masaki T (1989) The human preproendothelin-1 gene: Complete nucleotide sequence and regulation of expression. *J Biol Chem* 264: 14954–14959

20 Kimura S, Sugita Y, Kanazawa I, Saito A, Goto K (1987) Isolation and amino acid sequence of calcitonin gene-related peptide from porcine spinal cord. *Neuropeptides* 9: 75–82

21 Amara SG, Jonas J, Rosenfeld MG, Ong ES, Evans RM (1982) Alternative RNA processing in calcitonin gene expression generates mRNAs encoding different polypeptide products. *Nature* 298: 240–244

22 Inoue A, Yanagisawa M, Kimura S, Kasuya Y, Miyauchi T, Goto K, Masaki M (1989) The human endothelin family: Three structurally and pharmacologically distinct isopeptides predicted by three separate genes. *Proc Natl Acad Sci USA* 86: 2863–2867

23 Yanagisawa M, Inoue A, Ishikawa T, Kasuya Y, Kimura S, Kumagaya S, Nakajima K, Watanabe TX., Sakakibara S, Goto K, Masaki T (1988) Primary structure, synthesis and biological activity of rat endothelin, an endothelium derived vasoconstrictor peptide. *Proc Natl Acad Sci USA* 85: 6964–6967

24 Takasaki C, Tamiya N, Bdalah A, Kllog Y, Sokolovsky M (1988) Sarafotoxin S6: several isotoxins from *Atractaspis engaddensis* (burrowing asp) venom that affect the heart. *Toxicon* 26: 543–548

25 Takasaki C, Yanagisawa M, Kimura S, Goto K, Masaki T (1988) Similarity of endothelin to snake venom toxin. *Nature* 335: 303

26 Kasuya Y, Kobayashi H, Uemura H (1991) Endothelin-like immunoreactivity in the nervous system of invertebrates and fish. *J Cardiovasc Pharmacol* 17 (Suppl 7): S463–S466

27 Uemura H, Naruse M, Naruse K (1991) Immunoreactive endothelin in plasma of non-mammalian vertebrates. *J Cardiovasc Pharmacol* 17 (Suppl 7): S414–S416

28 Emori T, Hirata Y, Imai T, Ohta K, Kanno K, Eguchi S, Marumo F (1992) Cellular mechanisms of thrombin on endothelin-1 biosynthesis and release in bovine endothelial cell. *Biochem Pharmacol* 44: 2409–2411

29 Kurihara H, Yoshizumi M, Sugiyama T, Takaku F, Yanagisawa M, Masaki T, Hamaoki M, Kato H, Yazaki Y (1989) Transforming growth factor beta stimulates the expression of endothelin mRNA from vascular endothelial cells. *Biochem Biophys Res Comm* 159: 1435–1440

30 Marsden PA, Brenner BM (1992) Transcriptional regulation of the endothelin gene by TNF alpha. *Am J Physiol* 262: C854–C861

31 Maemura K, Kurihara H, Morita T, Hayashi Y, Yazaki Y (1992) Production of endothelin-1 in vascular endothelial cells is regulated by factors associated with vascular injury. *Gerontology* 38 (Suppl 1): 29–35

32 Marsden PA, Dorfman DM, Collins T, Brenner BM, Orkin SH, Ballermann BJ (1991) Regulated expression of endothelin-1 in glomerular capillary endothelial cells. *Am J Physiol* 261: F117–F125

33 Imai T, Hirata Y, Emori T, Yanagisawa Y, Masaki T, Marumo F (1992) Induction of endothelin-1 gene by angiotensin and vasopressin in endothelial cells. *Hypertension* 19: 753–757

34 Boulanger CM, Tanner FC, Beau MI, Han AWA, Warner A, Luscher TF (1992) Oxidized low density lipoproteins induce mRNA expression and release of endothelin from human and porcine endothelium. *Circ Res* 70: 1191–1197

35 Emori T, Hirata Y, Ohta K, Schichiri M, Marumo F (1989) Secretory mechanism of immunoreactive endothelin in cultured bovine endothelial cells. *Biochem Biophys Res Comm* 160: 93–100

36 Yanagisawa M, Inoue A, Takuwa Y, Mitsui Y, Kobayashi M, Masaki T (1989) The human preproendothelin-1 gene: possible regulation by endothelial phosphoinositide turnover signaling. *J Cardiovasc Pharmacol* 13 (Suppl 5): S13–S17

37 Sharefki JB, Diamond SL, Eskin SG, McIntire LV, Fenbach CW (1991) Fluid flow decreases preproendothelin mRNA levels and suppresses endothelin-1 peptide release in cultured human endothelial cells. *J Vasc Surg* 14: 1–9

38 Malek A, Izumo S (1992) Physiological fluid shear stress causes downregulation of endothelin-1 mRNA in bovine aortic endothelium. *Am J Physiol* 263: C389–C396

39 Yoshizumi M, Kurihara H, Sugiyama T, Takaku F, Yanagisawa M, Masaki T, Yazaki Y (1989) Hemodynamic shear stress stimulates endothelin production by cultured endothelial cells. *Biochem Biophys Res Comm* 161: 859–864

40 Sumpio BE, Widmann MD (1990) Enhanced production of endothelium-derived contracting factor by endothelial cells subjected to pulsatile stretch. *Surgery* 108: 277–281

41 Malek AM, Greene AL, Izumo S (1993) Regulation of endothelin-1 gene by fluid shear stress is transcriptionally mediated and independent of protein kinase C and cAMP. *Proc Natl Acad Sci USA* 90: 5999–6003

42 Fabbrini MS, Valsasina B, Nitti G, Benatti L, Vitale A (1991) The signal peptide of human preproendothelin-1. *FEBS Lett* 286: 91–94

43 Bloch KD, Friedrich SP, Lee ME, Eddy RL, Shows TB, Quertermous T (1989) Structural organization and chromosomal assignment of the gene encoding endothelin. *J Biol Chem* 264: 10851–10857

44 Rossi P, Karsenty G, Roberts AB, Roche NS, Sporn MB, Crombrugghe B (1988) A nuclear factor 1 binding site mediates the transcriptional activation of type I collagen promoter by transforming growth factor-β. *Cell* 52: 405–414

45 Emori T, Hirata Y, Imai T, Eguchi S, Kanno K, Marumo F (1993) Cellular mechanism of natriuretic peptides-induced inhibition of endothelin-1 biosynthesis in rat endothelial cells. *Endocrinology* 133: 2474–2480

46 Imai T, Hirata Y, Emori T, Marumo F (1993) Heparin has an inhibitory effect on endothelin-1 synthesis and release by endothelial cells. *Hypertension* 21: 353–358

47 Denault J-B, Claing A, D'Orlean-Juste P, Sawamura T, Kido T, Masaki T, Leduc R (1995) Processing of proendothelin-1 by human furin convertase. *FEBS Lett* 362: 276–280

48 Xu D, Emoto N, Giaid A, Slaughter C, Kaw S, deWit D, Yanagisawa M (1994) ECE-1: a membrane-bound metalloprotease that catalyzes the proteolytic activation of big endothelin-1. *Cell* 78: 473–485

49 Shimada K, Takahashi M, Ikeda M, Tanzawa K (1995) Identification and characterization of two isoforms of an endothelin-converting enzyme-1. *FEBS Lett* 371: 140–144

50 Valdenaire O, Rohrbacher E, Mattei M-G (1995) Organization of the gene encoding the human endothelin-converting enzyme (ECE-1). *J Biol Chem* 270: 29794–29798

51 Fukuroda T, Noguchi K, Tsuchida S, Nishikibe M, Ikemoto F, Okada K, Yano M (1990) Inhibition of biological actions of big endothelin-1 by phosphoramidon. *Biochem Biophys Res Comm* 172: 669–675

52 Matsumura Y, Hisaki K, Takaoka M, Morimoto S (1990) Phosphoramidon, a metalloprotease inhibitor, suppresses the hypertensive effect of big endothelin-1. *Eur J Pharmacol* 185: 103–106

53 Sawamura T, Kimura S, Shinmi O, Sugita Y, Yanagisawa M, Masaki T (1989) Analysis of endothelin related peptides in culture supernatant of porcine aortic endothelial cells: Evidence for biosynthesis pathway of endothelin-1. *Biochem Biophys Res Comm* 162: 1287–1294

54 Ikegawa R, Matsumura Y, Tsukahara Y, Takaoka M, Morimoto S (1990) Phosphoramidon, a metalloprotease inhibitor, suppesses the secretion of endothelin from cultured endothelial cells by inhibiting a big endothelin-1 converting enzyme. *Biochem Biophys Res Comm* 171: 669–675

55 Shiba R, Yanagisawa M, Miyauchi T, Ishii Y, Kimura S, Uchiyama Y, Masaki T, Goto K (1989) Elimination of intravenously injected endothelin-1 from the circulation of the rat. *J Cardiovasc Pharmacol* 13 (Suppl 5): S98–S101

56 Anggard E, Galton S, Rae G, Thomas R, McLoughlin L, DeNucci G, Vane JR (1989) The fate of radioiodinated endothelin-1 and endothelin-3 in the rat. *J Cardiovasc Pharmacol* 13 (Suppl 5): S46–S49

57 Sawamura T, Kimura S, Shinmi O, Sugita Y, Yanagisawa M, Goto K, Masaki T (1990) Purification and characterization of putative endothelin converting enzyme in bovine adrenal medulla: evidence for a cathepsin D like enzyme. *Biochem Biophys Res Comm* 168: 1230–1236

58 Sokolovsky M, Galron R, Kloog Y, Bdolah A, Indig FE, Blumberg S, Fleming G (1990) Endothelins are more sensitive than sarafotoxins to neutral endopeptidase: possible physiological significance. *Proc Natl Acad Sci USA* 87: 4702–4706

59 Vijaraghavan J, Schichi AG, Careterio OA, Slaughter C, Moomaw C, Hersch LB (1990) The hydrolysis of endothelins by neutral endopeptidase 24.11. (enkephalinase). *J Biol Chem* 265: 14150–14155

60 Fagny C, Michel A, Nortier J, Deschodt Lanckman M (1992) Enzymatic degradation of endothelin-1 by activated human polymorphonuclear neutrophils. *Regul Pept* 42: 350–354

61 Sakurai T, Yanagisawa M, Masaki T (1992) Molecular characterization of endothelin receptors. *Trends Pharmacol Sci* 13: 103–108

62 Watanabe H, Miyazaki H, Kondoh M, Masuda Y, Kimura S, Yanagisawa M, Masaki T, Murakami K (1989) Two distinct types of endothelin receptors are present on chick cardiac membrane. *Biochem Biophys Res Comm* 161: 1252–1259

63 Sugiura M, Snajdar RM, Schwartberg M, Badr KF, Inagami T (1989) Identification of two types of specific endothelin receptors in rat mesangial cell. *Biochem Biophys Res Comm* 162: 1396–1401

64 Arai H, Hori S, Aramori I, Ohkubo H, Nakanishi S (1990) Cloning and expression of a cDNA encoding an endothelin receptor. *Nature* 348: 730–732

65 Sakurai T, Yanagisawa M, Takuwa Y, Miyazaki H, Kimura S, Goto K, Masaki T (1990) Cloning of cDNA encoding a non-isopeptide selective subtype of the endothelin receptor. *Nature* 348: 732–735

66 Simonson MS, Dunn MJ (1990) Cellular signaling by peptides of the endothelin gene family. *FASEB J* 4: 2989–3000

67 Takigawa M, Sakurai T, Kasuya Y, Abe Y, Masaki T, Goto K (1995) Molecular identification of guanine-nucleotide-binding regulatory proteins which couple to endothelin receptors. *Eur J Biochem* 228: 102–108

68 H tter N, Woloszczuk W (1989) Radioimmunoassay of endothelin. *Lancet* I: 909

69 Ando K, Hirata Y, Schichiri M, Emori T, Marumo F (1989) Presence of immunoreactive endothelin in human plasma. *FEBS Lett* 245: 164–166

70 Suzuki N, Matsumoto H, Kitada C, Masaki T, Fujino M (1989) A sensitive sandwich-enzyme immunoassay to detect immunoreactive big endothelin in plasma. *J Immunol Method* 118: 245–250

71 Hirata Y, Itoh K, Ando K, Endo M, Marumo F (1989) Plasma endothelin levels during surgery. *N Engl J Med* 321: 1686
72 Cernacek P, Stuwart J (1989) Immunoreactive endothelin in human plasma: marked elevation in patients in cardiogenic shock. *Biochem Biophys Res Comm* 161: 562–567
73 Miyauchi T, Yanagisawa M, Tomizawa T, Sugishita Y, Suzuki N, Fujino M, Goto K, Masaki T (1989) Increased plasma concentrations of endothelin-1 and big endothelin-1 in acute myocardial infarction. *Lancet* II: 53–54
74 Salminen K, Tikkanen I, Saijonmaa O, Neiminen M, Fyhrquist F, Frick M (1989) Modulation of coronary tone in acute myocardial infarction by endothelin. *Lancet* II: 747
75 Fyhrquist F, Saijonamaa O, Mesarinen K, Tikkanen I, Rosenlof K, Tikkanen T (1990) Raised plasma endothelin-1 concentration following cold pressor test. *Biochem Biophys Res Comm* 169: 217–221
76 Totsune K, Mouri T, Takahashi K, Ohneda M, Sone M, Saito T, Yoshinaga K (1989) Detection of immunoreactive endothelin in plasma of hemodialysis patients. *FEBS Lett* 249: 239–242
77 Koyama H, Tabata T, Nishizawa Y, Inoue K, Morii H, Yamaji T (1989) Plasma endothelin levels in patients with uraemia. *Lancet* I: 991–992
78 Tomita K, Ujiie K, Nakanishi T, Tomura S, Matsuda O, Ando K, Schichiri M, Hirata Y, Marumo F (1989) Plasma endothelin level in patients with acute renal failure. *N Engl J Med* 321: 1127
79 Saito Y, Nakao K, Mukoyama M, Imura H (1990) Increased plasma endothelin level in patients with essential hypertension. *N Engl J Med* 322: 205
80 Kohno M, Yasunari K, Murasawa K, Yokokawa K, Horio T, Fukui T, Takeda T (1990) Plasma immunoreactive endothelin in essential hypertension. *Am J Med* 88: 614–618
81 Miyauchi T, Yanagisawa M, Iida K, Ajisaka R, Suzuki N, Fujino M, Masaki T, Sugishita Y (1992) Age- and sex-related variation of plasma endothelin-1 concentration in normal and hypertensive subjects. *Am Heart J* 123: 1092–1093
82 Masaoka H, Suzuki R, Hirata Y, Emori T, Marumo F, Hirakawa K (1989) Raised plasma endothelin in aneurysmal subarachnoid haemorrhage. *Lancet* II: 1402
83 Fujimori A, Yanagisawa M, Saito A, Goto K, Masaki T, Mima T, Shigeno T, Takakura K (1990) Endothelin in plasma and cerebrospinal fluid of patients with subarachnoid haemorrhage. *Lancet* II: 633
84 Zamora M, O'Brien RF, Rutherford RB, Weil JV (1990) Serum endothelin-1 concentration and cold provocation in primary Raynaud's phenomenon. *Lancet* II: 1144–1147
85 Takahashi K, Ghatee MA, Lam HC, O'Halloran DJ, Bloom SK (1990) Elevated plasma endothelin in patients with diabetes mellitus *Diabetologia* 33: 306–310
86 Ishibashi K, Saito K, Watanabe K, Uesugi S, Furue H, Yamaji Y (1990) Plasma endothelin-1 levels in patients with disseminated intravascular coagulation. *N Engl J Med* 324: 1516–1517
87 Yamane K, Kashiwagi H, Suzuki N, Miyauchi T, Yanagisawa M, Goto K, Masaki T (1991) Elevated plasma levels of endothelin-1 in systemic sclerosis. *Arthr Rheum* 34: 243–244
88 Usuki S, Saitoh T, Sawamura T, Suzuki N, Shigemitsu S, Yanagisawa M, Goto K, Onda H, Fujino M, Masaki T (1990) Increased maternal plasma concentration of endothelin-1 during labor pain or on delivery and the existence of a large amount of endothelin-1 in amniotic fluid. *Gynecol Endocrinol* 4: 85–97
89 Taylor RN, Verma M, Teng NNH, Roberts JM (1990) Women with preeclampsia have higher plasma endothelin levels than women with normal pregnancies. *J Clin Endocrinol Metab* 71: 1675–1677
90 Battistini B, D'Orleans-Juste P, Sirois P (1993) Endothelins: circulating plasma levels and presence in other biologic fluids. *Lab Invest* 68: 600–628
91 Ihara M, Fukuroda T, Saeki T, Nishikibe M, Kojiri K, Suda H, Yano M (1997) An endothelin receptor (ET$_A$) antagonist isolated from Streptomyces misakiensis. *Biochem Biophys Res Comm* 178: 132–137
92 Ihara M, Noguchi K, Saeki T, Fukuroda M, Tsuchida S, Kimura S, Fukami T, Ishikawa K, Nishikibe M, Yano M (1992) Biological profiles of highly potent novel endothelin antagonist selective for the ET$_A$ receptor. *Life Sci* 50: 247–255

93 Aramori I, Nirei H, Shoubo M, Sogabe K, Nakamura K, Kojo H, Notsu Y, Ono T, Naka-nishi S (1993) Subtype selectivity of a novel endothelin antagonist, FR 139317, for the two endothelin receptors in transfected Chinese hamster ovary cells. *Mol Pharmacol* 43: 127–131

94 Davenport AP, Kuc RE, Fitzgerald F, Maguire JJ, Berryman K, Doherty AM (1994) [^{125}I]-PD151241: a selective radioligand for human ET_A receptors. *Br J Pharmacol* 111: 4–6

95 Webb RL, Navarrete AE, Ksander GM (1995) Effects of the ET_B-selective antagonist IRL 2500 in conscious spontaneously hypertensive and Wistar-Kyoto rats. *J Cardiovasc Pharmacol* 26 (Suppl 3): S389–S392

96 Ishikawa K, Ihara M, Noguchi K, Mase T, Mino N, Saeki T, Fukuroda T, Fukami T, Ozaki S, Nagase T, Nishikibe M, Yano M (1994) Biochemical and pharmacological profiles of a potent and selective endothelin B-receptor antagonist, BQ-788. *Proc Natl Acad Sci USA* 91: 4892–4896

97 Ikeda S, Awane Y, Kusumoto K, Wakimasu M, Watanabe T, Fujino M (1994) A new endo-thelin receptor antagonist, TAK-044, shows long-lasting inhibition of both ET_A and ET_B mediated blood pressure responses in rats. *J Pharmacol Exp Ther* 270: 728–733

98 Doherty AM, Cody WL, He JX, DePue PL, Leonard DM, Dunber JB jr, Hill RE, Flynn MA, Reynolds EE (1993) Design of C-terminal peptide antagonists of endothelin: struc-ture-activity relationships of ET-1[16-21, D-His 16]. *Bioorg Med Chem Lett* 3: 497–502

99 Clozel M, Brew V, Burri K, Cassal JM, Fischli W, Gray GA, Hirth G, Loffler BM, Muller M, Neidhart W, Ramuz H (1993) Pathophysiological role of endothelin revealed by the first orally active endothelin receptor antagonist. *Nature* 365: 759–761

100 Clozel M, Breu V, Gray GA, Kalina B, Loffler BM, Burri K, Cassal JM, Hirth G, Muller M, Neidhard W, Ramuz H (1994) Pharmacological characterization of bosentan, a new potent orally active nonpeptide endothelin antagonist. *J Pharmacol Exp Ther* 270: 228–235

101 Goto K, Hama H, Kasuya Y (1996) Molecular pharmacology and pathophysiological sig-nificance of endothelin. *Jap J Pharmacol* 72: 261–290

102 Ohlstein EH, Nambi P, Douglas SA, Edwards RM, Gellai M, Lago A (1994) SB 209670, a rationally designed potent nonpeptide endothelin receptor antagonist. *Proc Natl Acad Sci USA* 91: 8052–8056

103 Yokokawa K, Tahara H, Kohno M, Murakawa K, Yasunari K, Nakagawa K, Hamada T, Otani S, Yanagisawa M, Takeda T (1991) Endothelin secreting tumor. *J Cardiovasc Phar-macol* 17 (Suppl 7): S398–S401

104 McMahon EG, Palomo MA, Moore WM (1991) Phosphoramidon blocks the pressor activity. *J Cardiovasc Pharmacol* 17 (Suppl 7): S29–S33)

105 Nishikibe M, Tsuchida S, Okada M, Fukuroda T, Shimamoto K, Yano M, Ishikawa K, Ikemoto F (1993) Antihypertensive effect of a newly synthesized endothelin antagonist, BQ-123, in a genetic hypertensive model. *Life Sci* 52: 717–724

106 Sakai S, Miyauchi T, Kobayashi M, Yamaguchi I, Goto K, Sugishita Y (1996) Inhibition of myocardial endothelin pathway improves long-term survival in heart failure. *Nature* 384: 353–355

107 Kurihara Y, Kurihara H, Suzuki H, Kodama T, Maemura K, Nagai R, Oda H, Kuwaki T, Cao WH, Kamada N, Jishage K, Ouchi Y, Azuma S, Toyoda Y, Ishikawa T, Kumada M, Yazaki Y (1994) Elevated blood pressure and craniofacial abnormalities in mice deficient in endothelin-1. *Nature* 368: 703–710

108 Niranjan V, Telemaque S, deWit D, Gerard RD, Yanagisawa M (1996) Systemic hyperten-sion induced by hepatic overexpression of human preproendothelin-1 in rats. *J Clin Invest* 98: 2364–2372

109 Hosoda K, Hammer RE, Richardson JA, Greenstein Baynash A, Cheung JC, Giaid A, Yanagisawa M (1994) Targeted and natural (piebald-lethal) mutations of endothelin-B receptor gene produce megacolon associated with spotted coat color in mice. *Cell* 79: 1267–1276

110 Puffenberger EG, Hosoda K, Washington SS, Nakao K, deWit D, Yanagisawa M, Chakra-varti A (1994) A missense mutation of the endothelin-B receptor gene in multigenetic Hirschsprung disease. *Cell* 79: 1257–1266

111 Greenstein Baynash A, Hosoda K, Giaid A, Richardson JA, Emoto N, Hammer RE, Yanagisawa M (1994) Interaction of endothelin-3 with endothelin-B receptor is essen-tial for development of epidermal melanocytes and enteric neurons. *Cell* 79: 1277–1285

Pulmonary Actions of the Endothelins
ed. by R. G. Goudie and D. W. P. Hay
© 1999 Birkhäuser Verlag Basel/Switzerland

CHAPTER 2
Endothelin Receptors and Ligands

Timothy D. Warner

The William Harvey Research Institute, St. Bartholomew's and the Royal London School of Medicine and Dentistry, Charterhouse Square, London ECIM 6BQ, UK

1   Introduction
2   Endothelin Receptors
2.1   Endothelin Receptor Subtypes
2.2   Chimeric Receptors
2.3   Additional Endothelin Receptors
2.4   Endothelin Receptors in the Lung
2.4.1 Receptor Localization by Immunostaining and Autoradiography
2.4.2 Purification of ET receptors from the lung
2.4.3 Cloning of ET receptors in lungs
3   Endothelin Receptor Agonists
4   Endothelin Receptor Antagonists
4.1 ET_A Receptor-selective Antagonists
4.2 ET_B Receptor-selective Antagonists
4.3   Mixed $ET_{A/B}$ Receptor Antagonists
5   Effects of Endothelin Receptor Antagonists
6   Endothelin Receptors Involved in Pulmonary Responses
7   Endothelin in Disease Pathophysiology: Effects of Receptor Antagonists
7.1   Lung Diseases
7.2   Myocardial Infarction and Heart Disease
7.3   Renal Disease
7.4   Hypertension
7.5   Vascular Restenosis
8   Summary
      References

1. Introduction

The discovery of endothelin led to a huge body of research, much of it aimed at elucidating the receptors through which the endothelins act. Clearly, it was reasoned, the ability to antagonise the effects of such a powerful vasoconstrictor could be of potential benefit in a number of disease states. However, before antagonists could be produced and targetted the endothelin receptors needed to be characterised, their distributions determined, and the responses they mediated understood. Thus, much of the early research in the endothelin area was associated with isolating and characterising endothelin receptors, and examining the effects of selective and non-selective endothelin receptor agonists. More latterly experiments have focused increasingly on the effectiveness of endothelin receptor antagonists in a range of animal disease models, and most recently human disease states.

2. Endothelin Receptors

2.1. Endothelin Receptor Subtypes

Soon after the characterisation of the endothelin family of peptides it became apparent that there was more than one endothelin receptor. For
instance, although *in vivo*, and in many vascular beds and isolated vascular
tissues, ET-3 is less potent than ET-1 as a vasoconstrictor, these two peptides are equipotent as vasodilatators and with respect to inducing the
release of nitric oxide [1–3]. From this it appeared that there could be two
different endothelin receptors, an isopeptide selective receptor (ET-1 > ET-3)
that mediated the pressor effects of the endothelins and a non-selective
receptor (ET-1 = ET-3) present on the endothelium. Similarly, receptor
binding assays indicated the presence of heterogeneous populations of
endothelin receptors in a variety of tissues (see [4]). Molecular biology
soon confirmed these suggestions with the cloning and expression of two
endothelin receptors, the ET_A receptor, that is selective for ET-1, ET-2 or
sarafotoxin (Stx) 6b over ET-3 or Stx6c [5], and the ET_B receptor, that does
not discriminate between the endothelin/sarafotoxin peptides [6]. These
receptors have very similar molecular weights of approximately 47 kDa
and contain seven transmembrane domains of 20–27 hydrophobic amino
acid residues, typical of the rhodopsin-type superfamily of G-protein-
coupled receptors. As may be expected the genes encoding these receptors
have been suggested to be expressed in a wide variety of tissues and species
(see [7]).

2.2. Chimeric Receptors

Isolation of the genes for the ET_A and ET_B receptor permitted the production of chimeric receptors which have provided important insights into the
functions of different receptor regions. For example, early studies indicated that the first extracellular loop domain of the ET_A receptor is important
for ligand binding [8], while a part of the N-terminal domain in close proximity to transmembrane domain I is required for the ligand binding activity of the ET_A receptor; the C-terminal domain appears involved as an
anchor for the maintenance of the binding site [9]. Further experiments
indicated that in both ET_A and ET_B receptors the transmembrane domains
IV–VI, together with adjoining loops of the receptors, determine the iso-
peptide/subtype selectivity. Transmembrane domain I but also transmembrane domains I–III and VII, plus adjacent loops of the receptors together
with the C-terminal linear portion of the isopeptides, are probably involved
in ligand-receptor binding [10]. When used to determine the underlying
reasons for differences in the binding to ET_A and ET_B receptors these
chimeric receptors indicated that binding determinants for the ET_B recep-

tor-selective agonists ET-3, BQ 3020, and IRL 1620, reside within the region spanning the transmembrane helices IV–VI and the adjacent loop regions, whereas the transmembrane helices I, II, III, and VII, plus the intervening loop regions, specify the selectivity for the ET_A receptor antagonist BQ-123 [10]. Reduction to the level of the individual amino acids has shown that differences between ET_A and ET_B receptor binding can be determined by single substitutions. For example, mutation of a single amino acid, Tyr^{129} to Ala, in the second transmembrane region of the ET_A receptor changes the binding characteristics of the receptor towards that of the ET_B receptor [11, 12]. In addition, mutation of Asp^{126} to Ala in the ET_B receptor increases ET-3, but not ET-1 binding, and interestingly abolishes ET-1 activation of phospholipase C, as does mutation of Asp^{147} to Ala [13]. Similarly, on the subject of the functions of key amino acid residues, Cys^{402} is important for intracellular calcium signalling associated with ET_B receptor activation [14]. Finally, post-translational modifications affect receptor function. For instance, palmitoylation of residues in the cytoplasmic tail of the ET_B receptor also appears important for activation of phospholipase C [15].

2.3. Additional Endothelin Receptors

Despite the cloning of only two mammalian endothelin receptors functional evidence suggests the presence of additional subtypes. The most widely proposed of these is the ET_C receptor [16], classified as being selective for ET-3 over ET-1. This may be present on some bovine endothelial cells and its activation appears to lead to the release of nitric oxide [17, 18]. It may also be expressed elsewhere, as receptor binding assays have identified high affinity receptors selective for ET-3 in rat brain and atria [19, 20] and similar receptors have been detected in anterior pituitary cells [21]. However, to date such a receptor has only been cloned from non-mammalian species [22]. There is also accumulating data suggesting that the ET_B receptor present on the endothelium, that mediates the release of nitric oxide in response to the endothelins, is functionally different from the ET_B receptor which mediates the vasoconstrictor effects of the endothelins [23–25]. Thus, these have tentatively been classified as ET_{B1} (present on the endothelium) and ET_{B2} (present on vascular and non-vascular smooth muscle) [24]. Similarly, it has also been suggested that there may exist ET_{A1} and ET_{A2} receptors, where the former are BQ-123-sensitive and the latter BQ-123-insensitive [26]. Other functional studies have suggested that non ET_A/ET_B receptors mediate responses to the endothelins in smooth muscle preparations from the rat, guinea pig and rabbit (see [7]). It must, however, also be noted that experiments with antagonists of the ET_A and ET_B receptors (respectively BQ-123 and BQ-788, see below) have led to the suggestion (see [27]) that in fact the two cloned mammalian receptors may account for all the responses reported and that variations in the relative populations of these receptors [28] and in the kinetics of binding to each receptor subtype [29] in different tissues may confuse receptor characterisation. One other cause for the apparent presence of additional ET receptors may well be that ET-1, and other endothelins, bind very tightly, in fact almost irreversibly, to their receptors [30, 31]. Indeed, in intact cells ET-1 only becomes dissociated from its receptors following receptor internalisation, with about 40% of receptors being recycled after 1 h [32]. These binding characteristics mean that determinations of pA_2 values, for example, can be flawed, since the assumptions underlying these calculations are not met. In addition, the prolonged binding of the endothelins to their receptors may make responses in certain tissues particularly resistant to antagonism, since antagonists can only act to prevent new binding of ET-1 following re-externalization of the receptors [32]. This resistance may give the impression of the tissue containing additional, antagonist-resistant, receptors.

Better answers to the questions of whether or not there are additional endothelin receptors may well be supplied from experiments employing tissues derived from endothelin receptor deficient animals [33]. For example, analysis of responses in tissues from normal and in ET_B receptor gene knockout mice indicate that although pharmacological distinctions may be made between ET_B receptors present on smooth muscle and endothelium, both receptor types were lost when the single ET_B receptor gene was disrupted [34]. Clearly this indicates that the putative ET_{B1} and ET_{B2} receptors are in fact derived from the same ET_B receptor gene. Physical differences, as opposed to functional differences, between these receptors would, therefore, depend upon selective post-transcriptional modifications.

2.4. Endothelin Receptors in the Lung

2.4.1. Receptor localization by immunostaining and autoradiography: ET receptors have been localized and characterized throughout the airways and pulmonary vasculature (Table 1). Initial immunohistochemical studies showed the presence of ET receptors in bovine pulmonary artery endothelial cells [35] and on mucous, serous, and Clara cells, as well as on a few type II alveolar pneumocytes. Receptors were not generally found on most basal and ciliated cells in rat and mouse bronchoalveolar epithelium [36]. Similarly, autoradiography has demonstrated endothelin binding sites throughout pulmonary tissues, with particularly high density in human [37] and guinea-pig [38] pulmonary artery endothelium, in human airway [39–43] and rat trachea [39] smooth muscle, in human pulmonary nerve trunks, and in human alveoloar septa [39, 44]. Autoradiographical studies have also shown high affinity binding sites for ET-1 in the smooth muscle, submucosa, and epithelium of guinea-pig trachea [45]. Binding studies also confirm the presence of high affinity ET receptors within the lung (see [46]).

Table 1. Localization of high affinity ET-1 binding sites in the airways of various species

Sites/Species	Mouse	Rat	Guinea-pig	Bovine	Human
Airway SM	++	++	++++		++
Alveolar fibroblasts					
Alveolar wall/septa		++	++		++++
Basal cells	−				
Cartilage		−			±
Ciliated cells	−				
Clara cells	+				
Connective tissue		−			±
Endothelium				++++	++++
Epithelium	++	++	++		±
Intrapulmonary blood vessels					++++
Mucous cells	++	++			
Nerve trunks		++	−		++
Serous cells	++	++			
Submucosal layer			++		±
Type I/II pneumocytes	++				
Vascular sm					++++

2.4.2. Purification of ET receptors from the lung: As for the characterisations of ET receptors in other tissues, discussed above, early experiments in lung tissue provided confusing results. For example, cross-linking experiments suggested proteins of different molecular weights to be ET receptors in the lung. In the rat lung, a 44-kDa protein was reported to have a higher affinity for ET-1 than ET-3, whereas another 32-kDa protein preferentially interacted with ET-3 [47]. In bovine lung two proteins of 34 and 52-kDa [48, 49], in rat and bovine lungs a 34 kDa protein [50], and in human bronchial smooth muscle cells, a 70 kDa protein [51] were also suggested to be ET receptors. The exact nature of these proteins, however, is now clear, since both ET_A and ET_B receptors are now well characterised as having molecular weights of around 50-kDa.

2.4.3. Cloning of ET receptors in lungs: It is worth noting that the initial clonings of both ET_A- (427 residues) and ET_B- (441 residues) receptor subtypes were achieved using bovine and rat lungs, respectively [5, 6]. Since then the ET_A receptor subtype has also been cloned from rat (426 residues; [52]) and human (427 residues; [53]) lungs, while the ET_B receptor subtype has been cloned again from rat (442 residues; [52]), bovine (441 residues; [54]), and human (442 residues; [53]) lungs.

3. Endothelin Receptor Agonists

Clearly it would appear that the natural agonists for endothelin receptors are the endothelins, and as described above the different potencies of ET-1,

ET-2 and ET-3 on a number of preparations supplied some of the first indications that the endothelins acted on more than one receptor. The discovery
of the marked similarity between the sarafotoxins (Stx) and the endothelins
also supplied further agonists, particularly Stx6c, which were of much
use in making these comparisons. In particular, Stx6c is widely reported
as being as potent an agonist as ET-1 or ET-3 at ET_B receptors, but to be
almost entirely inactive as an agonist at ET_A receptors [23, 24, 55]. In
searching for synthetic ligands which are selective for the endothelin receptors efforts have mainly produced agents which act on the ET_B receptor.
The first of these were direct modifications of ET-1, such as [Ala$^{1, 3, 11, 15}$]-
ET-1 in which the replacement of the four cysteine residues produced
a compound without the characteristic intra-chain disulphide bridges
[56, 57]. Experiments with this and related but truncated peptides, e.g.,
[Ala^{1,3,11,15}]-ET-1$_{(6-21)}$, [Ala^{1,3,11,15}]-ET-1$_{(8-21)}$, and N-acetyl-[Ala^{1,3,11,15}]-
ET-1$_{(10-21)}$, showed that the parent compound, [Ala$^{1, 3, 11, 15}$]-ET-1 bound to
ET_B receptors with an affinity 1700 times higher than at ET_A receptors, and
that in general terms ET-1 requires the Glu^{10}-Trp^{21} sequence for ET_B binding, but not the disulphide bridges [56]. Similarly, the ET-1 derivative Suc-
[Glu^9,Ala^{11,15}]-ET-1$_{(8-21)}$, IRL 1620, has been shown to be highly selective
for the ET_B receptor (with K_iET_A/K_iET_B ratio of 120000) [58]. This compound is, for example, a potent constrictor of the guinea-pig trachea,
consistent with an action on ET_B receptors [58]. Binding assays have
shown that BQ 3020, (N-acetyl-LeuMetAspLysGluAlaValTyrPheAlaHis-
Leu-AspIleIleTrp), is 4500-fold more selective for ET_B than for ET_A receptors [59]. This compound contracts preparations such as the rabbit pulmonary artery [59] and also promotes vasorelaxation *via* stimulation of
ET_B receptors present on the vascular endothelium [60].

4. Endothelin Receptor Antagonists

As outlined above there are strong indications that an increase in the production of endothelin is a promoting factor in a number of disease states.
Taken together with the characterisation of endothelin receptors, there is
therefore a clear rationale and a good scientific background for the development of endothelin receptor antagonists. This review will concentrate on
those that have been used most widely, and those, particularly non-peptidic
antagonists, which will probably be of the greatest future interest.

4.1. ET_A Receptor-selective Antagonists

The most well known of the ET_A receptor-selective antagonists, in particular BQ-123 and BQ-153, were developed from peptides discovered in the
broth from the mycelium of *Streptomyces misakiensis* [61–64]. These are

cyclic pentapeptides that have a high affinity for ET_A receptors in porcine and rat vascular smooth muscle [62, 63, 65], whereas they are ineffective at reducing ET-1 binding to ET_B receptors, for example, in cerebellar membranes. Related to these are two linear tripeptidic, ET_A receptor-selective antagonists, FR 139317 [66−69] and BQ-610 [70] that were developed from the same natural product. A number of other ET_A-selective antagonists have also been reported including TTA-101 (*t*-Bu OCO-Leu-Trp-Ala), TTA-386 [71], myriceron caffeoyl ester (50-235), a non-peptide isolated from the bayberry *Myrica cerifera* [72], and asterric acid [73]. However, in contrast to BQ-123 and FR 139317, which are the compounds that have been mainly used, very few studies have been published using these latter antagonists and thus their selectivity or possible utility cannot be assessed here.

In addition to the peptide antagonists, there are now non-peptide antagonists selective for the ET_A receptor. The first of these was a sulphonamide, BMS-182874 (5-(dimethylamino)-N-(3,4-dimethyl-5-isoxazolyl)-1-napthalenesulfonamide) (Fig. 1), that is also orally active [74]. This agent inhibits the binding of ET-1 to ET_A receptors with an IC_{50} value of 55 nM, whereas it has little activity at ET_B receptors, such that overall it has a 3600-fold selectivity for the ET_A site. More recently PD 155080, PD 155719 and

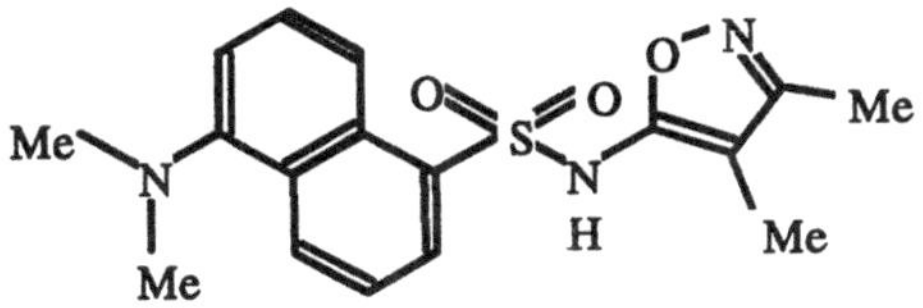

BMS-182874, selective ET_A receptor antagonist [74]

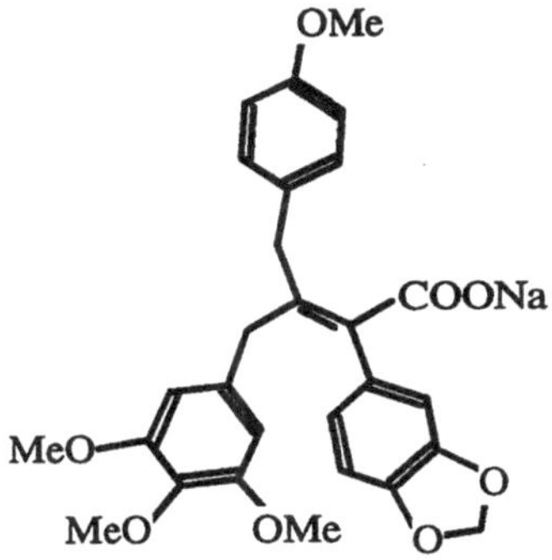

PD 156707, selective ET_A receptor antagonist [75]

Figure 1. Structures of representative non-peptide ET receptor antagonists.

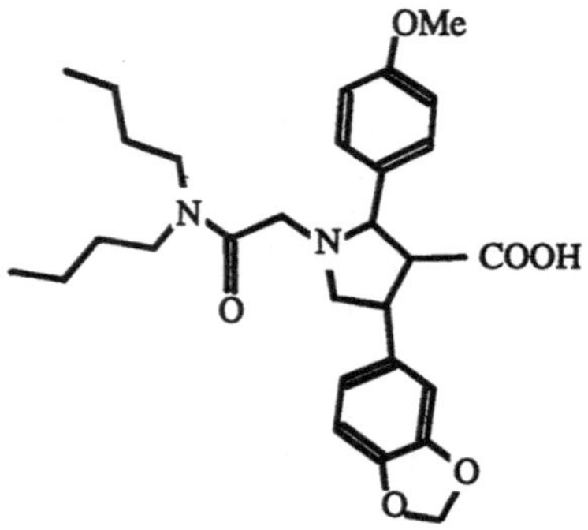

A-127722, selective ET$_A$ receptor antagonist [76, 77]

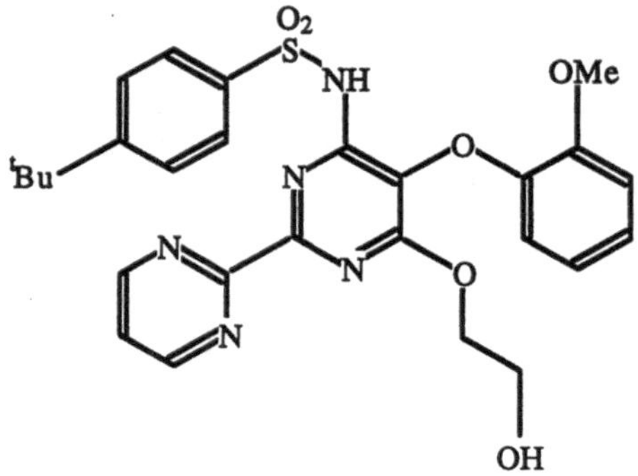

bosentan (Ro 47-0203), non-selective ET receptor antagonist [99]

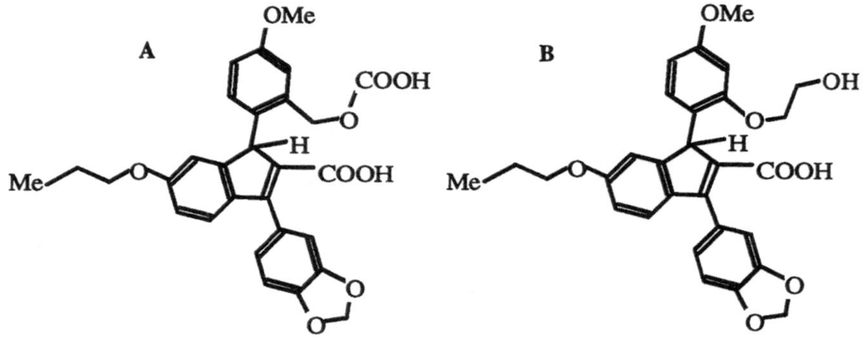

(A) SB 209670 [102] and (B) SB 217242 [103], non-selective ET receptor antagonists

Figure 1 (continued)

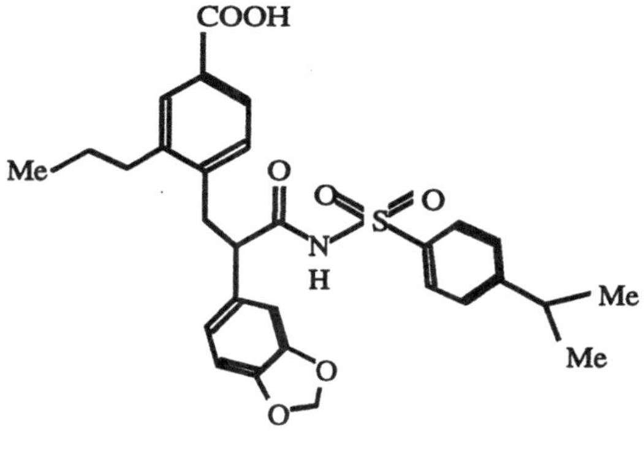

L751281, non-selective ET receptor antagonist [188]

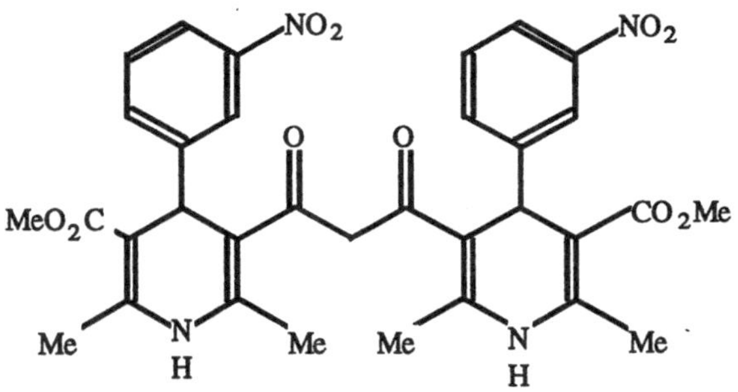

CGS 27830, non-selective ET receptor antagonist [106]

Figure 1 (continued).

PD 156707 (Fig. 1) have also been reported to be potent non-peptide ET_A receptor antagonists [75, 76]. PD 156707, for example, inhibits the binding of ET-1 to cloned human ET_A and ET_B receptors, with K_i values of 0.17 and 133.8 nM, and the vasoconstrictor response induced by ET-1 in rabbit blood vessels, with pA_2 values of 7.5 and 4.7, respectively [75]. A-127722 is a more recently described non-peptide ET_A receptor antagonist [77, 78] (Fig. 1). This is an extremely potent compound that competitively inhibits [^{125}I]ET-1 binding to cloned human ET_A and ET_B receptors with K_i values of 69 pM and 139 nM, respectively. In functional assays A-127722 inhibits ET-1-induced vasoconstriction in rat isolated aorta with a pA_2 value of 9.2. Furthermore, *in vivo*, A-127722 dose-dependently blocks the pressor response to ET-1 (0.3 nmol/kg, i.v.) in conscious rats at doses greater than 0.1 mg/kg p.o.: maximal inhibition was achieved at 10 mg/kg, at which dose the inhibition remains constant for at least 8 h. These characteristics make A-127722 the most potent ET_A receptor antagonist reported to date.

4.2. ET_B Receptor-selective Antagonists

There has been less effort applied to the production of ET_B receptor-selective antagonists than other classes of ET receptor antagonists and many of these have peptidic structures. IRL 1038, ([Cys^{11}-Cys^{15}]-ET-1$_{(11-21)}$), was one of the first reported of these peptide antagonists, and was stated to have a much higher affinity for ET_B receptors ($K_i = 6-11$ nM) than for ET_A receptors ($K_i = 400-700$ nM) and to antagonise functional responses mediated by ET_B receptors [79]. However, it has subsequently appeared that conclusions drawn from experiments using this compound should be regarded very cautiously [80]. BQ-788 has also been disclosed as a selective ET_B receptor antagonist. This is a linear peptide which inhibits binding to ET_B receptors with an IC_{50} value of 1.2 nM and to ET_A receptors with an IC_{50} value of 280 nM [81]. In isolated rabbit pulmonary arteries BQ-788 competitively antagonises the vasoconstrictions induced by an ET_B receptor-selective agonist BQ-3020 with a pA_2 of 8.4 [82]. Interestingly, when administered *in vivo* it abolished the initial transient depressor response induced by bolus injection of ET-1 and revealed an initial pressor effect. When given to the anaesthetised rat as an infusion together with ET-1, BQ-788 also increased the regional vasoconstrictor effects of ET-1, reinforcing the notion that ET_B receptors are particularly important in limiting the vasoconstrictor and pressor effects of endothelin [83]. The cyclic peptide antagonist RES-701-1, isolated from *Streptomyces sp.*, has also been reported to be a selective ET_B receptor antagonist [26, 84] and to inhibit the binding of ET-1 to ET_B receptors with an IC_{50} value of 10 nM. This compound did not affect binding of ET-1 to ET_A receptors even when used at concentrations up to 5 µM. Like BQ-788, RES-701-1 also attenuated the depressor and enhances the pressor effects of intravenously applied ET-1 in anaesthetized rats. More recently the first non-peptide, selective ET_B receptor antagonist has been reported, Ro 46-8443 [85]. This displayed up to 2000-fold selectivity for ET_B receptors, and should supply a valuable tool for clarifying the role(s) of ET_B receptors in pathology.

4.3. Mixed $ET_{A/B}$ Receptor Antagonists

As both ET_A and ET_B receptors mediate the contractile effects of the endothelins in the vasculature [23, 24, 86−92], both have been implicated as mediating the deleterious effects of the endothelins. Mixed $ET_{A/B}$ receptor antagonists may well, therefore, be the most effective in a range of disease states. The first well known of these non-selective antagonists were those produced by Parke-Davis. These peptide antagonists were developed by a rational approach starting with $ET_{(16-21)}$, which is known to interact with ET receptors [93−96]. Modification of the C-terminal hexapeptide portion of ET-1 by substituting His^{16} with (β-Phenyl)-D-Phe produced PD 142893, and incorporation of D-Bhg (5H-dibenzyl[a,d]cycloheptene-10,11-dihydroglycine) in position 16, produced PD 145065 (see Table 2). This latter compound had a further increased binding affinity to both ET_A and ET_B receptors than did PD 142893. Other much larger ET-1-derivatives, such as [Thr[18], γ methyl Leu[19]]ET-1, have also been shown to bind with high affinities to both ET_A and ET_B receptors. However, there is little information

Table 2. Notable endothelin receptor antagonists

Compound	Type	Selectivity	In vivo activity	Ref.
[D-Arg[1], D-Phe[5],D-Trp[7,9],Leu[11]]-SP	peptidic	ET_A?		183
[Dpr[1],Asp[15]]-ET-1	peptidic	ET_A?		182
BE-18257B	peptidic	ET_A		61
BQ-123	peptidic	ET_A	i.v.	62
BQ-153	peptidic	ET_A	i.v.	62
BQ-485	peptidic	ET_A	i.v.	94
FR 139317	peptidic	ET_A	i.v.	69
50-235	non-peptidic (myricerone)	ET_A		72
TTA-386	peptidic	ET_A		71
Asterric acid	non-peptidic	ET_A?		73
BMS-182874	non-peptidic (sulphonamide)	ET_A	i.v., p.o.	74, 185
PD 155080	non-peptidic (butenolide)	ET_A	i.v., p.o.	76
PD 155719	non-peptidic (butenolide)	ET_A		76
PD 156707	non-peptidic (butenolide)	ET_A	i.v., p.o.	75, 76
PD 159433	non-peptidic (indole)	ET_A		186
A127722	non-peptidic	ET_A	i.v., p.o.	77, 78
IRL 1038	peptidic	ET_B	i.v.	79
IRL 2500	peptidic	ET_B	i.v.	187
IRL 2659	peptidic	ET_B		see 188
IRL 2796	peptidic	ET_B		see 188
RES-701-1	peptidic	ET_B	i.v.	84
BQ-788	peptidic	ET_B	i.v.	82
Ro 46-8443	non-peptidic (sulphonamide)	ET_B		85
TAK-044	peptidic	ET_A/ET_B	i.v.	98
PD 142893	peptidic	ET_A/ET_B	i.v.	93
PD 145065	peptidic	ET_A/ET_B	i.v.	94
Thr[(18)], γ-MeLeu[(19)]--endothelin-1	peptidic	ET_A/ET_B		119
cochinmicins	cyclodepsipeptides	ET_A/ET_B		107
pheophorbide A		ET_A/ET_B		189
Ro 46-2005	non-peptidic (sulphonamide)	ET_A/ET_B	i.v., p.o.	100
Ro 47-0203	non-peptidic (sulphonamide)	ET_A/ET_B	i.v., p.o.	101
SB 209670	non-peptidic	ET_A/ET_B	i.v., i.d.*	102
SB 217242	non-peptidic	ET_A/ET_B	i.v., i.d.*, p.o.	103
L745453	non-peptidic (α-phenoxyphenlyacetic acid derivative)	ET_A/ET_B	i.v., p.o.	105
L751281	non-peptidic (sulphonamide)	ET_A/ET_B		see 188
L754142	non-peptidic	ET_A/ET_B	i.v., p.o.	104
CGS 27830	non-peptidic (dihydropyridine)	ET_A/ET_B	i.v.	106

* i.d. = intraduodenally.

as to the usefulness of this latter compound as a functional antagonist of endothelin responses [97].

Compounds related to BQ-123 have also been developed as non-selective endothelin receptor antagonists [98]. The most active of these is TAK-044 (cyclo(-D-Asp-AlaPhp-Asp-D-Thg-Leu-D-Trp-) where Php = 3-[(4-phenylpiperazin-1-yl)carbonyl] and Thg = 2-(2-thienyl)glycine), which has an IC_{50} to inhibit binding of ET-1 to ET_A receptors of 0.08 nM and to ET_B receptors of 120 nM. TAK-044 also antagonises ET_A receptor-mediated constriction of porcine isolated coronary arteries and ET_B receptor-mediated constriction of porcine isolated coronary vein. As may be expected, when tested *in vivo* this compound blocks both the pressor and the depressor responses to bolus injection of ET-1 [99].

Probably more interestingly, product library screening has led to the discovery of the orally active non-peptide endothelin receptor antagonist, Ro 46-2005 [100]. Ro 46-2005 inhibits binding to both ET_A and ET_B receptors, although with considerably less affinity than the non-selective peptidic antagonist, PD 145065. On the other hand Ro 46-2005 has the advantage of being orally active (30% bioavailability), which peptide antagonists rarely are. Ro 46-2005 has been superceded by Ro 47-0203 (bosentan) (Fig. 1) which is a more potent receptor non-selective antagonist [101]. Bosentan inhibits binding of ET-1 to ET_A receptors with a K_i of 4.7 nM and to ET_B receptors with a K_i of 95 nM. As may be predicted, it inhibits the pressor effects of big ET-1, when both antagonist and peptide are administered intravenously to pithed rats, and it is also active when administered orally.

SB 209670 is another recently disclosed non-peptide non-selective endothelin receptor antagonist which was rationally designed using conformational models of ET-1 (Fig. 1; [102]). This agent inhibits binding to ET_A and ET_B receptors and has K_d values for binding to human ET_A and ET_B receptors of 1 and 10 nM [102]. *In vitro*, it inhibits contractions of the rat thoracic aorta induced by ET-1, an ET_A receptor-mediated response, with a K_B value of 0.4 nM and ET_B receptor-mediated contractions of rabbit pulmonary artery with a K_B value of 88 nM [102]. SB 209670 also inhibits both the pressor and depressor effects of intravenously applied ET-1 [102, 103]. Similarly, SB 217242 has been reported to be an extremely potent nonselective antagonist, displacing ET-1 binding to ET_A and ET_B receptors with K_i values of 1.1 and 111 nM, respectively [102, 103]. In rats, 30 mg/kg SB 217242 inhibits responses to intravenously applied ET-1 for at least 5.5 h. Finally, within the sulphonamide derivatives, L751,281 (Fig. 1) has also been reported as a non-selective antagonist. This is related to L745,142, which, like SB 217242, potently inhibits binding of ET-1 to cloned human receptors (K_is: ET_A = 0.062 nM, ET_B = 2.25 nM [104]. L745,142 is also active *in vivo*; for example, protecting against ET-1-induced lethality in mice (AD_{50}, 0.26 mg/kg). From the same company is L744,453, a compound from a new structural class, the dipropyl-α-phenoxyphenlacetic acid derivatives [105]. This, for example, protects against ET-1-induced lethality in mice with an AD_{50} of 13 mg/kg.

CGS 27830 (Fig. 1), which is derived from the quite structurally different dihydropyridine class of calcium ion modulators, potently inhibits binding to ET_A receptors (IC_{50} value 16 nM), but has less activity at ET_B receptors (IC_{50} value 295 nM) [106]. This agent also inhibits contractions of the rabbit aorta induced by ET-1 and the increases in inositol phosphate turnover stimulated by ET_A receptor activation. *In vivo*, CGS 27830 abolishes the pressor effect and reduces the depressor effect of a bolus injection of ET-1.

In a fashion analogous to the discovery of BQ-123 and related compounds the cochinmicins I, II and III, which are products of *Microspora sp.* ATCC55140, have also been reported to be non-selective $ET_{A/B}$ antagonists [107, 108].

5. Effects of Endothelin Receptor Antagonists

The ET_A receptor-selective antagonists are the compounds that have been most widely studied. As might be expected of it, BQ-123 antagonises constrictions of the isolated porcine coronary artery induced by ET-1, inhibits binding to ET_A receptors on vascular smooth muscle cells, and blunts, but does not ablate, the pressor effects of ET-1 in anaesthetized rats [62]. This incomplete suppression of the pressor effects of ET-1 is due to the presence of non-ET_A constrictor receptors within the rat circulation, most notably within the mesenteric and renal beds [109–112]. Thus, PD 145065, or TAK-044 block the renal constrictor effects of ET-1 in the anaesthetised rat illustrating the importance of ET_B vasoconstrictor receptors [99]. However, the other side of their potency on ET_B receptors is that PD 142893, PD 145065 or TAK-044 block the depressor effects of intravenously administered ET-1 [93, 99, 103, 112], as does the selective ET_B receptor antagonist, BQ-788 [82]. This is in contrast to BQ-123 or FR 139317 which either do not affect or tend to potentiate this portion of the ET-1 response [69, 113]. From these experiments it therefore appears to be true that in the rat, both ET_A and ET_B receptors mediate the vasoconstrictor, or pressor effects of the endothelins and that ET_B receptors, most probably present on the endothelium [2, 3, 23, 24], mediate the transient depressor response to the endothelins. In addition, these non-selective antagonists have been suggested to increase the circulating levels of ET-1 by displacing it from ET_B receptors [114].

In all other species so far tested it also appears that while both ET_A and ET_B receptors mediate vasoconstriction only ET_B receptors mediate vasodilatation. However, the relative importance of different endothelin receptors in various tissues differs among species. For instance, in the rat, the renal vasoconstrictor effects of the endothelins were mediated by a mixed population of ET_A and ET_B receptors. However, in the pig these responses were more dependent on ET_A receptors [115] and in the rabbit they were entirely blocked by BQ-123 [116]. Similarly, contractions of the pulmonary artery from the rabbit were mediated mainly by ET_B receptors and not sensitive to BQ-123 [22, 24, 28, 55, 117], whereas in porcine, guinea-pig and human vessels ET_A receptors predominated [25, 88, 118]. Furthermore, there were variations in the endothelin receptors mediating the effects of the endothelins in non-vascular tissue. For instance, BQ-123 or FR 139317 were not effective against contractions of the trachea or upper bronchi from the guinea-pig [119] or human bronchi [89] induced by ET-1, whereas they may reduce contractions induced by ET-1 in the rat trachea [120].

In most species, e.g., dog [121, 122], application of endothelin receptor antagonists does not generally affect the basal blood pressure. However, there are exceptions to this rule. For instance, in either anaesthetised or conscious guinea-pigs both bosentan and BQ-123 lowered blood pressure [123]. Also, in the rat removal of nitric oxide vasodilator tone, by application of inhibitors of nitric oxide synthase, caused an increase in blood pressure which was reduced by either BQ-123 or bosentan [124]. Most importantly, preliminary observations suggest that ET-1 maintains vascular tone in humans, as infusion of phosphoramidon or BQ-123 into the brachial artery increased forearm blood flow [125]. Furthermore, in healthy humans intravenous application of TAK-044 ($10-100$ mg, i.v.) has been reported to lower blood pressure by approximately 10%, an effect that was sustained for at least 4 h [126].

6. Endothelin Receptors Involved in Pulmonary Responses

As in the more general field of endothelin research, the initial characterization of the ET receptors involved in pulmonary responses was based on the efficacy of ET agonists and *in situ* hybridization. The use of selective ET_A- or ET_B-receptor antagonists and/or nonselective ET_A/ET_B-receptor antagonists has permitted the further characterization of ET receptors involved in various responses in the pulmonary system (see Table 3).

In guinea-pig isolated airways, contractions induced by ET/Stx peptides in the trachea are mediated by both ET_A- and ET_B-receptors, whereas mostly ET_B-receptors are involved in the bronchus and the lung parenchyma. In rat isolated airways, both receptors are involved. In human isolated bronchus, the consensus is that ET_B receptors predominate [43, 127]. In the pulmonary artery of the rabbit, nonselective $ET_{A/B}$-receptor agonists (e.g., ET-1, ET-3) and selective ET_B-receptor agonists (e.g., Stx6c) are equipotent [128]. Similarly, both ET-1 and ET-3 contract isolated pulmonary artery and veins [129, 130]. These data suggest mediation of these responses by ET_B-receptors. In the pulmonary artery of guinea-pigs or rabbits, desensitization experiments also suggest the presence of functional ET_A-receptors

Table 3. Populations of ET-receptor subtypes in the pulmonary system of various species

Tissue	Species	Responses	ETA	ETB	Ref.
bronchus	guinea-pig	contraction	++++	−	159
		contraction	−	++++	160, 161
		contraction	−	++++	89
		contraction	−	++++	162
	human	contraction	−	++++	43, 89, 163
		contraction	−	++++	127, 164
	lamb	binding	++++	−	165
	rat	contraction	+++	−	166
		in situ hybridization	+++	−	52
parenchyma	rat	binding	65	35	62
	guinea-pig	binding	−	++++	167
		binding	15	85	62
	human	binding	+++	+	44
	lamb	contraction	++++	−	165
	monkey	binding	45	55	62
	mouse	binding	60	40	62
	pig	binding	++	++	168
		binding	35	65	62
	rat	binding	55	45	62
		binding	60	40	169
		contraction	+	+++	166
		in situ hybridation	++	++	52
pulmonary artery	bovine	binding; PI hydrolysis	++	++	170
	guinea-pig	contraction	++	−	131
		contraction	+++	−	89
		contraction	+++	−	171
		contraction	++++	−	159
		contraction, cross-des	++	−	38
	human	contraction	++	−	172
		contraction	+++	−	163
		contraction	++	−	173
		SMC proliferation	++	−	174
	pig	contraction	+++	−	25
	rabbit	contraction; binding	40	60	128
	rat	contraction	++	−	175
		contraction	+++	+	166
		in situ hybridization	+++	+	52
pulmonary vein	pig	contraction	−	++	25
	human	contraction			174
trachea	dog	contraction, PI hydrolysis	++++	−	178
	guinea-pig	binding	++	++	177
		contraction	++	++	89
		contraction	++	+++	160, 161
		contraction	−	++++	126
		contraction	−	++++	171
		contraction, binding	++	++	45
		contraction, cross-desensitization	±	+++	38
trachea	lamb	contraction, binding	++++	−	167
	rat	contraction	++	++	178

Table 3 (continued)

Tissue	Species	Responses	ETA	ETB	Ref.
in vivo	guinea-pig	↑ PIP			131
		↑ PPP, ↑PIP			167
		↑ PIP			179
		↑ microvascular permeability			180
		↓ PPP			181
		↑ PIP			183
		↑ PGI$_2$			169
		↑ PPP			175

[128, 130, 131]. It can be concluded, therefore, that both ET_A- and ET_B-receptors mediate contractions induced by ET/Stx peptides in the pulmonary artery of rabbits and, most probably, in other species. In perfused lungs or *in vivo*, pulmonary responses in the rat, cat, or lamb are mainly mediated by ET_A-receptors, whereas in the guinea-pig, ET_B-receptors are involved.

7. Endothelin in Disease Pathophysiology: Effects of Receptor Antagonists

Endothelin receptor antagonists have been examined extensively in various models of different diseases. Such studies are of general interest as they provide information about the roles of the endothelins in these pathologies, and also the efficacies of the antagonists against endogenously produced endothelin.

7.1. Lung Diseases

There is no information about the use of endothelin receptor antagonists in lung diseases in humans. However, there have been a number of studies investigating the efficacies of endothelin receptor antagonists in animal models of lung diseases, in particular asthma and pulmonary hypertension. These studies are described in detail in chapters 13 and 14, but some specific examples of the effects of receptor antagonists in models of pulmonary hypertension are described. These follow from studies indicating that endothelin production is elevated in human lung in pulmonary hypertension [132, 133]. For example, in anaesthetised rats acute hypoxia-induced pulmonary hypertension was markedly attenuated by BQ-123 [134], and in conscious rats exposed to hypoxia for 15 days there was an increase in pulmonary artery pressure and a right ventricular hypertrophy that was reduced by BQ-123 applied i.p. by minipump [135] or by bosentan given orally [136]. The non-peptide, selective ET_A receptor antagonist,

TBC11251, prevented and reversed acute hypoxia-induced pulmonary hypertension in the rat [137]. Similarly, in the conscious lamb, bolus administration of BQ-123 into the pulmonary artery caused a reduction in the rise in pulmonary vascular resistance that resulted from alveolar hypoxia [138]. ET-1 may contribute to the high pulmonary vascular resistance in the normal foetus, for intrapulmonary infusions of BQ-123 increased left pulmonary artery flow in late gestation foetal lambs [139]. Finally, in the dog, pulmonary hypertension caused by injection of dehydromonocrotaline was reduced by FR 139317, but increased by RES-701-1, suggesting that ET-1 production was elevated in this model, but also partly mitigates the pulmonary hypertension by acting on ET_B receptors [140, 141]. PD 156707 has also been shown to reduce ischaemia and reperfusion injury in rat lungs [142].

7.2. Myocardial Infarction and Heart Disease

In the dog BQ-123 decreased the size of the infarction following 90 min of ischaemia and 5 h of reperfusion [143], and TAK-044 has been reported to decrease infarct size in a rat model of coronary ischaemia and reperfusion [144]. Such data clearly imply a role for endothelin in ischaemia/reperfusion injury. However, this conclusion must be weighted by other observations suggesting that endothelin only plays a minor role under these conditions. For instance, PD 156707 has been reported to have no effect on infarct size after coronary artery occlusion/reperfusion in pigs [145].

There is some evidence supportive of a role for endothelin in chronic heart failure. For example, in a conscious rat model chronic administration of bosentan significantly reduced arterial blood pressure even in the presence of cilazapril, an angiotensin converting enzyme inhibitor [146]. More significantly, in human patients with chronic heart failure bosentan reduced mean arterial pressure, pulmonary artery pressure, right arterial pressure and pulmonary wedge pressure, as well as lowering systemic and pulmonary vascular resistance [147].

7.3. Renal Disease

ET-1 is expressed throughout the kidney, in the glomeruli, small cortical vessels, early proximal tubules, collecting ducts and vasa recta, and ET binding sites are similarly found in the glomeruli, early proximal tubules, inner medullary collecting ducts and vasa recta. These binding sites represent both ET_A receptors, particularly in the vascular portion of the kidney, and ET_B receptors, particularly in the tubular portion.

Positive data have been shown in a model of severe post-ischaemic acute renal failure in which treatment with BQ-123 for 3 h 1 day after ischaemia, resulted in an improved survival rate (75%) and an improved reabsorption of sodium [148]. Similarly, the non-selective antagonist Ro 46-2005 partially restored the initial (20–30 min) fall in renal blood flow that followed ischaemia and reperfusion [100]. In one other model of renal dysfunction, SB 209670 was also found to prevent the renal vasoconstriction caused by the intrarenal infusion of the radiocontrast medium Hypaque to indomethacin-treated anaesthetized dogs [149].

7.4. Hypertension

BQ-123 has been shown to lower blood pressure in stroke prone spontaneously hypertensive rats [150], and blood pressure and peripheral resistance in spontaneously hypertensive rats (SHRs, [151]), and BMS-182874 to reduce the blood pressure of SHRs, but not sodium-depleted SHRs. Bosentan also reduced the blood pressure of DOCA-salt SHRs, a model of malignant hypertension [152]. In a canine model of renal hypertension, bosentan reduced blood pressure in hypertensive animals to a much greater extent than in their matched controls [121].

7.5. Vascular Restenosis

In humans the effectiveness of percutaneous transluminal coronary angioplasty (PTCA) is limited by the high incidence of neointimal formation and subsequent vascular restenosis. The changes after PTCA were characterised by an initial phase of platelet deposition and cellular infiltration followed by medial proliferation and cellular migration and matrix formation. *In vitro*, ET-1 will produce many of these effects. *In vivo* studies show that i.p. administration of SB 209670 (2.5 mg/kg, i.p., twice a day) for 3 days before and 14 days after balloon angioplasty of the rat carotid artery reduced neointima formation by approximately 50% [153]. Similarly, TAK-044 gave some protection against neointimal lesion in the same model [154] as does BMS-182874 [155].

8. Summary

As the endothelin system has become defined, often by the use of the ligands described above and in particular the newer orally active endothelin receptor antagonists, it has become increasingly clear that endothelin does have roles to play in some forms of hypertension, in the events following ischaemia and reperfusion, and possibly in pathologies associated with inappropriate vascular cell proliferation. Within the pulmonary system there is evidence that endothelin may have a contributory role to play in asthma. More interestingly it appears likely that endothelin could be a principal mediator of the events underlying the pathological changes of pulmonary hypertension. Studies with endothelin antagonists suggest that these compounds may be able to reverse these effects, although we should bear in mind that established responses to the endothelins are only very slowly overcome due to the almost irreversible binding of ET-1 to its receptors [5, 156–158]. This caveat aside we may still hope that the development of drugs that modulate the activities of endothelin by actions at the level of the endothelin receptors will yield beneficial new therapies.

Acknowledgements

The author holds a British Heart Foundation Lectureship (BS/95003).

References

1 Inoue A, Yanagisawa M, Kimura S, Kasuya Y, Miyauchi T, Goto K, Masaki T (1989) The human endothelin family: three structurally and pharmacologically distinct isopeptides predicted by three separate genes. *Proc Natl Acad Sci* 86: 2863–2867
2 Warner TD, de Nucci G, Vane JR (1989) Rat endothelin is a vasodilator in the isolated perfused mesentery of the rat. *Eur J Pharmacol* 159: 325–326
3 Warner TD, Mitchell JA, de Nucci G, Vane JR (1989) Endothelin-1 and endothelin-3 release EDRF from isolated perfused arterial vessels of the rat and rabbit. *J Cardiovasc Pharmacol* 13 (Suppl 5): S85–S88
4 Sokolovsky M (1992) Endothelins and sarafotoxins: physiological regulation; receptor subtypes and transmembrane signaling. *Pharmacol Ther* 54: 129–149
5 Arai H, Hori S, Aramori I, Ohkubo H, Nakanishi S (1990) Cloning and expression of a cDNA encoding an endothelin receptor. *Nature* 348: 730–732
6 Sakurai T, Yanagisawa M, Takuwa Y, Miyazaki H, Kimura S, Goto K, Masaki T (1990) Cloning of a cDNA encoding a non-isopeptide-selective subtype of the endothelin receptor. *Nature* 348: 732–735
7 Rubanyi GM, Polokoff MA (1994) Endothelins: Molecular biology, biochemistry, pharmacology, physiology, and pathophysiology. *Pharmacol Rev* 46: 325–415
8 Adachi M, Yang YZ, Trzeciak A, Furuichi Y, Miyamato C (1992) Identification of a domain of ET_A receptor required for ligand binding. *FEBS Lett* 311: 179–183
9 Hashido K, Gamou T, Adachi M, Tabuchi H, Watanabe T, Furuichi Y, Miyamato C (1992) Truncation of N-terminal extracellular or C-terminal intracellular domains of human ET_A receptor abrogated the binding activity to ET-1. *Biochem Biophys Res Commun* 187: 1241–1248
10 Sakamoto A, Yanagisawa M, Sakurai T, Nakao K, Toyo-oka T, Yano M, Masaki T (1993) The ligand-receptor interactions of the endothelin systems are mediated by distinct "message" and "address" domains. *J Cardiovasc Pharmacol* 22 (Suppl 8): S113–S116
11 Krystek SR jr, Patel PS, Rose PM, Fisher SM, Kienzle BK, Lach DA et al (1994) Mutation of peptide binding site in transmembrane region of a G protein-coupled receptor accounts for endothelin receptor subtype selectivity. *J Biol Chem* 269: 12383–12386
12 Lee JA, Elliott JD, Sutiphong JA, Friesen WJ, Ohlstein EH, Stadel JM et al (1994) Tyr-129 is important to the peptide ligand affinity and selectivity of human endothelin type A receptor. *Proc Natl Acad Sci USA* 91: 7164–7168
13 Rose PM, Krystek SR jr, Patel PS, Liu EC, Lynch JS, Lach DA et al (1996) Aspartate mutation distinguishes ET_A but not ET_B receptor subtype-selective ligand binding while abolishing phospholipase C activation in both receptors. *FEBS Lett* 361: 243–249
14 Koshimizu T, Tsujimoto G, Ono K, Masaki T, Sakamoto A (1995) Truncation of the receptor carboxyl terminus impairs membrane signalling but not ligand binding of human ET_B endothelin receptor. *Biochem Biophys Res Commun* 217: 354–362

15 Horstmeyer A, Cramer H, Sauer T, Muller-Esterl W, Schroeder C (1996) Palmitoylation of endothelin receptor A. Differential modulation of signal transduction activity by post-translational modification. *J Biol Chem* 271: 20811–20819

16 Sakurai T, Yanagisawa M, Masaki T (1992) Molecular characterization of endothelin receptors. *Trends Pharmacol Sci* 13: 103–108

17 Emori T, Hirata Y, Marumo F (1990) Specific receptors for endothelin-3 in cultured bovine endothelial cells and its cellular mechanism of action. *FEBS Lett* 263: 261–264

18 Warner TD, Schmidt HHHW, Murad F (1992) Interactions of endothelins and EDRF in bovine native endothelial cells: selective effects of endothelin-3. *Am J Physiol* 262: H1600–H1605

19 Nambi P, Pullen M, Feuerstein G (1990) Identification of endothelin receptors in various regions of rat brain. *Neuropeptides* 16: 195–199

20 Sokolovsky M, Ambar I, Galron R (1992) A novel subtype of endothelin receptors. *J Biol Chem* 267: 20551–20554

21 Samson WK, Skala D, Alexander BD, Huang FLS (1990) Pituitary site of action of endo-thelin: selective inhibition of prolactin release *in vitro*. *Biochem Biophys Res Commun* 169: 737–743

22 Karne S, Jayawickreme CK, Lerner MR (1993) Cloning and characterization of an endo-thelin-3 specific receptor (ET$_C$ receptor) from *Xenopus laevis* dermal melanophores. *J Biol Chem* 268: 19126–19133

23 Warner TD, Allcock GH, Corder R, Vane JR (1993) Use of the endothelin receptor antag-onists BQ-123 and PD 142893 to reveal three endothelin receptors mediating smooth muscle contraction and the release of EDRF. *Br J Pharmacol* 110: 777–782

24 Warner TD, Allcock GH, Mickley EJ, Corder R, Vane JR (1993) Comparative studies with the endothelin-receptor antagonists BQ-123 and PD 142893 indicate at least three endo-thelin-receptors. *J Cardiovasc Pharmacol* 22 (Suppl 8): S117–S120

25 Sudjarwo SA, Hori M, Takai M, Urade Y, Okada T, Karaki H (1993) A novel subtype of endothelin B receptor mediating contraction in swine pulmonary vein. *Life Sci* 53: 431–437

26 Sudjarwo SA, Hori M, Tanaka T, Matsuda Y, Okada T, Karaki H (1994) Subtypes of endo-thelin ET$_A$ and ET$_B$ receptors mediating venous smooth muscle contraction. *Biochem Bio-phys Res Commun* 200: 627–633

27 Warner, TD, Battistini B (1994) Advances in endothelin inhibitors. *Drug News Perspec-tives* 7: 429–436

28 Fukuroda T, Ozaki S, Nakajima A, Fujikawa T, Ihara M, Ishikawa K, Yano M, Nishikibe M (1994) Analysis of non-ET$_A$-mediated lung tissue responses using BQ-788, a novel selec-tive ET$_B$ receptor ligand. *Can J Pharmacol Physiol* 72 (Suppl I): 173

29 Devadason PSS, Henry PJ (1997) Comparison of the contractile effects and binding kine-tics of endothelin-1 and sarafotoxin S6c in rat isolated renal artery. *Br J Pharmacol* 121: 253–263

30 Hirata Y, Yoshimi J, Takata S, Watanabe TX, Kumagaye S, Nakajima K, Sakakibara S (1988) Cellular mechanism of action by a novel vasoconstrictor endothelin in cultured rat vascular smooth muscle cells. *Biochem Biophys Res Commun* 154: 868–875

31 Waggoner WG, Genova SL, Rash VA (1992) Kinetic analyses demonstrate that the equi-librium assumption does not apply to [^{125}I]endothelin-1 binding data. *Life Sci* 51: 1869–1876

32 Marsault R, Vigne P, Frelin C (1993) The irreversibility of endothelin action is a property of a late intracellular signalling event. *Biochem Biophys Res Commun* 179: 1408–1413

33 Baynash AG, Hosoda K, Giaid A, Richardson JA, Emoto N, Hammer RE, Yanagisawa M (1994) Interaction of endothelin-3 with endothelin-B receptor is essential for development of epidermal melanocytes and enteric neurons. *Cell* 79: 1277–1285

34 Mizuguchi T, Nishiyama M, Moroi K, Tanaka H, Saito T, Masuda Y, Masaki T, deWit D, Yanagisawa M, Kimura S (1997) Analysis of two pharmacologically predicted endothelin B receptor subtypes by using the endothelin receptor gene knockout mouse. *Br J Pharma-col* 120: 1427–1430

35 Naruse M, Narase K, Kurimoto F, Horiuchi J, Tsuchiya K, Kawana M, Kato Y, Zeng Z, Saku-rai H, Demura H, Schizume K (1989) Radioimmunoassay for endothelin and immunoreac-tive endothelin in culture medium of bovine endothelial cells. *Biochem Biophys Res Com-mun* 160: 662–668

36 Rozengurt N, Springall DR, Polak JM (1990) Localization of endothelin-like immunoreactivity in airway epithelium of rats and mice. *J Pathol* 160: 5–8

37 McKay KO, Black JL, Diment LM, Armour CL (1991) Functional and autoradiographic studies of endothelin-1 and endothelin-2 in human bronchi, pulmonary arteries, and airway parasympathetic ganglia. *J Cardiovasc Pharmacol* 17: S206–S209

38 Cardell LO, Uddman R, Edvinsson L (1991) Two functional endothelin receptors in guinea pig pulmonary arteries. *Neurochem Int* 18: 571–574

39 Power RF, Wharton J, Zhao Y, Bloom SR, Polak JM (1989) Autoradiographic localization of endothelin-1 binding sites in the cardiovascular and respiratory systems. *J Cardiovasc Pharmacol* 13: S50–S56

40 Neuser D, Steinke W, Theiss G, Stasch JP (1989) Autoradiographic localization of [^{125}I]endothelin-1 and [^{125}I]atrial natriuretic peptide in rat tissue: a comparative study. *J Cardiovasc Pharmacol* 13: S67–S73

41 Neuser D, Steinke W, Dellweg H, Kazda S, Stasch JP (1991) ^{125}I-endothelin and ^{125}I-big endothelin-1 in rat tissues: autoradiographic localization and receptor binding. *Histochemistry* 95: 621–628

42 Marciniak SJ, Plumpton C, Barker PJ, Huskisson NS, Davenport AP (1992) Localization of immunoreactive endothelin and proendothelin in the human lung. *Pulm Pharmacol* 5: 175–182

43 Goldie RG, Henry PJ, Knott PG, Self GJ, Luttmann MA, Hay DWP (1995) Endothelin receptor density, distribution and function in human isolated asthmatic airways. *Am J Respir Crit Care Med* 152: 1653–1658

44 Knott PG, D'Apile AC, Henry PJ, Hay DWP, Goldie RG (1995) Receptors for endothelin-1 in asthmatic human peripheral lung. *Br J Pharmacol* 114:1–3

45 Tschirhart EJ, Drijfhout JW, Pelton JT, Miller RC, Jones CR (1991) Endothelins: functional and autoradiographic studies in guinea pig trachea. *J Pharmacol Exp Ther* 258: 381–387

46 Battistini B, Warner TD (1997) Endothelin-induced responses in the pulmonary system. In: JP Huggins, JT Pelton (eds) *Endothelins in biology and medicine.* CRC Press, p 191–222

47 Masuda Y, Miyazaki H, Kohdoh M, Watanabe H, Yanagisawa M, Masaki T, Murakami K (1989) Two different forms of endothelin receptors in rat lung. *FEBS Lett* 257: 208–210

48 Hagiwara H, Kozuka M, Eguchi S, Shibabe S, Ito T, Hirose S (1990) Solubilization of endothelin receptors from bovine lung plasma membranes in a non-aggregated state and estimation of their minimal functional sizes. *Biochem Biophys Res Commun* 172: 576–581

49 Kozuka M, Ito T, Hirose S, Lodhi KM, Hagiwara H (1991) Purification and characterization of bovine lung endothelin receptor. *J Biol Chem* 266: 16892–16896

50 Kundu GC, Misono KS (1991) Affinity labeling of endothelin receptors in bovine and rat lung membranes by N$^{\epsilon 9}$-azidobenzoyl-^{125}I-endothelin-1. *Mol Cell Endocrinol* 79: 85–92

51 Mattoli S, Soloperto M, Marini M, Fasoli A (1991) Mechanisms of calcium mobilization and phosphoinositide hydrolysis in human bronchial smooth muscle cells by endothelin 1. *Am J Respir Cell Mol Biol* 5: 424–430

52 Hori S, Komatsu Y, Shigemoto R, Mizuno M, Nakanishi S (1992) Distinct tissue distribution and cellular localization of two messenger ribonucleic acids encoding different subtypes of rat endothelin receptors. *Endocrinology* 130: 1885–1895

53 Haendler B, Hechler U, Schleuning WD (1992) Molecular cloning of human endothelin (ET) receptors ET$_A$ and ET$_B$. *J Cardiovasc Pharmacol* 20: S1–S4

54 Saito Y, Mizuno T, Itakura M, Suzuki Y, Ito T, Hagisawa H et al (1991) Primary structure of bovine endothelin ET$_B$ receptor and identification of signal peptidase and metal proteinase cleavage sites. *J Biol Chem* 266: 23433–23437

55 Panek RL, Major TC, Hingorani GP, Doherty AM, Taylor DG, Rapundalo ST (1992) Endothelin and structurally related analogs distinguish between endothelin receptor subtypes. *Biochem Biophys Res Commun* 183: 566–571

56 Saeki T, Ihara M, Fukufoda T, Yamagiwa M, Yano M (1991) [Ala^{1,3,11,15}]endothelin-1 analogs with ET$_B$ agonistic activity. *Biochem Biophys Res Commun* 179(1): 286–92

57 Nakamichi K, Ihara M, Kobayashi M, Saeki T, Ishikawa K, Yano M (1992) Different distribution of endothelin receptor subtypes in pulmonary tissues revealed by the novel selective ligands BQ-123 and [Ala^{1,3,11,15}]ET-1. *Biochem Biophys Res Commun* 182: 144–150

58 Takai M, Umemura I, Yamasaki K, Watakabe T, Fujitani Y, Oda K et al (1992) A potent and specific agonist, Suc-[Glu^9,Ala$^{11,\ 15}$]-endothelin-1$_{(8-21)}$, IRL 1620, for the ET$_B$ receptor. *Biochem Biophys Res Commun* 184: 953–959

59 Ihara M, Saeki T, Fukuroda T, Kimura S, Ozaki S, Patel AC, Yano M (1992) A novel radioligand [^{125}I]BQ-3020 selective for endothelin (ET$_B$) receptors. *Life Sci* 51: PL47–PL52

60 Hirata Y, Emori T, Eguchi S, Kanno K, Imai T, Ohta K, Marumo F (1993) Endothelin receptor subtype B mediates synthesis of nitric oxide by cultured bovine endothelial cells. *J Clin Invest* 91: 1367–1373

61 Ihara M, Fukuroda T, Saeki T, Nishikibe M, Kojiri K, Suda H, Yano M (1991) An endothelin receptor (ET$_A$) antagonist isolated from *streptomyces misakiensis*. *Biochem Biophys Res Commun* 178: 132–137

62 Ihara M, Noguchi K, Saeki T, Fukuroda T, Tsuchida S, Kimura S et al (1992) Biological profiles of highly potent novel endothelin antagonists selective for the ET$_A$ receptor. *Life Sci* 50: 247–255

63 Ihara M, Ishikawa K, Fukuroda T, Saeki T, Funabashi K, Fukami T et al (1992) *In vitro* biological profile of a highly potent novel endothelin (ET) antagonist BQ-123 selective for the ET$_A$ receptor. *J Cardiovasc Pharmacol* 20 (Suppl 12): S11–S14

64 Moreland S (1994) BQ-123, a selective endothelin ETA receptor antagonist. *Cardiovasc Drug Rev* 12: 48–69

65 Ishikawa K, Fukami T, Nagase T, Fujita K, Hayama T, Niiyama K et al (1992) Cyclic pentapeptide endothelin antagonists with high ET$_A$ selectivity. Potency- and solubility-enhancing modifications. *J Med Chem* 35: 2139–2142

66 Aramori A, Nirei H, Shoubo M, Sogabe K, Nakamura K, Kojo H et al (1993) Subtype selectivity of a novel endothelin antagonist, FR 139317, for the two endothelin receptors in transfected chinese hamster ovary cells. *Mol Pharmacol* 43: 127–131

67 Hemmi K, Neya M, Fukami N, Hashimoto M, Tanaka H, Kayakiri N (1992) Peptides having endothelin antagonist activity, a process for preparation thereof and pharmaceutical compositions comprising the same. Eur Pat Appl No 0457195A2

68 Sogabe K, Nirei H, Shoubo M, Hamada K, Nomoto A, Henmi K et al (1992) A novel endothelial receptor antagonist: studies with FR 139317. *J Vasc Res* 29: 201

69 Sogabe K, Nirei H, Shoubo M, Nomoto A, Ao S, Notsu Y, Ono T (1993) Pharmacological profile of FR 139317, a novel, potent endothelin ET$_A$ receptor antagonist. *J Pharmacol Exp Ther* 264: 1040–1046

70 Ishikawa K, Fukami T, Nagase T, Mase T, Hayama T, Niiyama K et al (1993) Endothelin antagonistic peptide derivatives with high selectivity for ET$_A$ receptors. In: Schneider CH, Eberle AN (eds) *Peptides 1992 (XXII EPS) ESCOM*, Leiden, 685–686

71 Takeda M, Breyer MD, Noland TD, Homma T, Hoover RL, Inagami T, Kon V (1992) Endothelin-1 receptor antagonist: effects on endothelin- and cyclosporine-treated mesangial cells. *Kid Int* 42: 1713–1719

72 Fujimoto M, Mihara S-I, Nakajima S, Ueda M, Nakamura M, Sakurai K-S (1992) A novel non-peptide endothelin antagonist isolated from bayberry, *Myrica cerifera*. *FEBS Lett* 305: 41–44

73 Ohashi H, Akiyama H, Nishikori K, Mochizuki J-I (1992) Asterric acid, a new endothelin binding inhibitor. *J Antibiotics* 45: 1684–1685

74 Stein PD, Hunt JT, Floyd DM, Moreland S, Dickinson KEJ, Mitchell C et al (1994) The discovery of sulfonamide endothelin antagonists and the development of the orally active ET$_A$ antagonist 5-(dimethylamino)-N-(3,4-dimethyl-5-isoxazolyl)-1-napthalene-sulfonamide (BMS-182874). *J Med Chem* 37: 329–331

75 Reynolds EE, Keiser JA, Haleen SJ, Walker DM, Olszewski B, Schroeder RL et al (1995) Pharmacological characterization of PD-156707, an orally-active ET$_A$ receptor antagonist. *J Pharmacol Exp Ther* 273: 1410–1417

76 Doherty AM, Patt WC, Repine J, Edmunds JJ, Berryman KA, Reisdorph BR et al (1995) Structure-activity relationships of a novel series of orally active nonpeptide ET$_A$ and ETA/$_B$ endothelin receptor-selective antagonists. *J Cardiovasc Pharmacol* 26 (Suppl 3): S358–S361

77 Opgenorth TJ, Adler AL, Calzadilla SV, Chiou WJ, Dayton BD, Dixon DB et al (1996) Pharmacological characterization of A-127722 – an orally-active and highly potent ET$_A$-selective receptor antagonist. *J Pharmacol Exp Ther* 276: 473–481

78 Winn M, von Geldern TW, Opgenorth TJ, Jae HS, Tasker AS, Boyd SA et al (1996) 2,4-Diarylpyrrolidine-3-carboxylic acids-potent ET_A selective endothelin receptor antagonists. 1. Discovery of A-127722. *J Med Chem* 39: 1039–1048

79 Urade Y, Fujitani Y, Oda K, Watakabe T, Umemura I, Takai M et al (1992) An endothelin B receptor-selective antagonist: IRL 1038, [Cys[11]-Cys[15]]-endothelin-1 (11–21). *FEBS Lett* 311: 12–16

80 Urade Y, Fujitani Y, Oda K, Watakabe T, Umemura I, Takai M et al (1994) Retraction concerning an endothelin B receptor-selective antagonist: IRL 1038, [Cys[11]-Cys[15]]-endothelin-1 (11-21). *FEBS Lett* 342: 103

81 Mase T, Fukami T, Nagase T, Yamakawa T, Takahashi H, Naya A et al (1993) Structure activity relationships of endothelin B receptor-selective antagonists. 14th Symposium on Medicinal Chemistry, Pharmaceutical Society of Japan, P-28

82 Ishikawa K, Ihara M, Noguchi K, Mase T, Mino N, Saeki T et al (1994) Biochemical and pharmacological profile of a potent and selective endothelin B-receptor antagonist, BQ-788. *Proc Natl Acad Sci* 91: 4892–4896

83 Allcock GH, Warner TD, Vane JR (1995) Roles of endothelin A and B receptors in the regional and systemic vascular responses to ET-1 in the anaesthetised ganglion-blocked rat: use of the antagonists, BQ-123, BQ-788 and PD 145065. *Br J Pharmacol* 116: 2482–2486

84 Matsuda Y, Tanaka T, Morishita Y, Nozasa M, Ohno T, Yamada K (1993) Pharmacological profile of RES-701-1, a novel, potent endothelin subtype B receptor antagonist of microbial origin. *Circulation* 88: I-281

85 Breu V, Clozel M, Burri K, Hirth G, Neidhart W, Ramuz H (1996) Characterization of Ro-46-8443, the first nonpeptide antagonist selective for the endothelin ET(B) receptor. *FEBS Lett* 383: 37–41

86 Clozel M, Gray GA, Breu W, Löffler BM, Osterwalder R (1992) The endothelin ET_B receptor mediates both vasodilatation and vasoconstriction *in vivo*. *Biochem Biophys Res Commun* 186: 867–873

87 Harrison VJ, Randriantsoa A, Schoeffer P (1992) Heterogeneity of endothelin-sarafotoxin receptors mediating contraction of pig coronary artery. *Br J Pharmacol* 105: 511–513

88 Hay DWP (1992) Pharmacological evidence for distinct endothelin receptors in guinea-pig bronchus and aorta. *Br J Pharmacol* 106: 759–761

89 Hay DWP, Luttmann MA, Hubbard WC, Undem BJ (1993) Endothelin receptor subtypes in human and guinea-pig pulmonary tissues. *Br J Pharmacol* 110: 1175–1183

90 Moreland S, McMullen DM, Delaney CL, Lee VG, Hunt JT (1992) Venous smooth muscle contains vasoconstrictor ET_B-like receptors. *Biochem Biophys Res Commun* 184: 100–106

91 Sumner MJ, Cannon TR, Mundin JW, White DG, Watts IS (1992) Endothelin ET_A and ET_B receptors mediate vascular smooth muscle contraction. *Br J Pharmacol* 107: 858–860

92 Warner TD, Battistini B, Allcock GH, Vane JR (1993) Endothelin ET_A and ET_B receptors mediate vasoconstriction and prostanoid release in the isolated kidney of the rat. *Eur J Pharmacol* 250: 447–453

93 Cody WL, Doherty AM, He JX, DePue PL, Rapundalo ST, Hingorani GA et al (1992) Design of a functional hexapeptide antagonist of endothelin. *J Med Chem* 35: 3301–3303

94 Cody WL, Doherty AM, He JX, DePue PL, Waite LA, Topliss JG et al (1993) The rational design of a highly potent combined ET_A and ET_B receptor antagonist (PD 145065) and related analogues. *Med Chem Res* 3: 154–162

95 Doherty AM, Cody WL, He JX, Depue PL, Cheng XM, Welch KM et al (1993) *In vitro* and *in vivo* studies with a series of hexapeptide endothelin antagonists. *J Cardiovasc Pharmacol* 22: S98–102

96 Doherty AM, Cody WL, He X, DePue PL, Leonard DM, Dunbar JB jr et al (1993) Design of C-terminal peptide antagonists of endothelin: structure-activity relationships of ET-1[16-21, D-His[16]]. *Bioorg Med Chem Lett* 3: 497–502

97 Shimamoto N, Kubo K, Watanabe T, Suzuki N, Abe M, Kikuchi T et al (1993) Pharmacological profile of ET_A antagonist [Thr[18], γ Methyl Leu[19]]-endothelin-1. *J Cardiovasc Pharmacol* 22: S107–S110

98 Kikuchi T, Ohtaki T, Kawata A, Imada T, Asami T, Masuda Y et al (1994) Cyclic hexapeptide endothelin receptor antagonists highly potent for both receptor subtypes ET_A and ET_B. *Biochem Biophys Res Commun* 200: 1708–1712

99 Ikeda S, Awane Y, Kusumoto K, Wakimasu M, Watanabe T, Fujino M (1994) A new endothelin receptor antagonist, TAK-044, shows long-lasting inhibition of both ET_A- and ET_B-mediated blood pressure responses in rats. *J Pharmacol Exp Ther* 270: 728–733

100 Clozel M, Breu V, Burri K, Cassal JM, Fischli W, Gray GA et al (1993) Pathophysiological role of endothelin revealed by the first orally active endothelin receptor antagonist. *Nature* 365: 759–761

101 Clozel M, Breu V, Gray GA, Kalina B, Löffler B-M, Burri K et al (1994) Pharmacological characterization of bosentan, a new potent orally active non-peptide endothelin receptor antagonist. *J Pharmacol Exp Ther* 270: 228–235

102 Ohlstein EH, Beck GR jr, Douglas SA, Nambi P, Lago MA, Gleason JG et al (1994) Nonpeptide endothelin receptor antagonists. II. Pharmacological characterization of SB 209670. *J Pharmacol Exp Ther* 271: 762–768

103 Ohlstein EH, Nambi P, Lago A, Hay DW, Beck G, Fong KL et al (1996) Nonpeptide endothelin receptor antagonists. VI: Pharmacological characterization of SB 217242, a potent and highly bioavailable endothelin receptor antagonist. *J Pharmacol Exp Ther* 76: 609–615

104 Williams DL, Murphy KL, Nolan NA, O'Brien JA, Pettibone DJ, Kivlighn SD et al (1995) Pharmacology of L-745,142, a highly potent, orally active, nonpeptidyl endothelin antagonist. *J Pharmacol Exp Ther* 275: 1518–1526

105 Williams DL, Murphy KL, Nolan NA, O'Brien JA, Lis EV, Pettibone DJ et al (1996) Pharmacology of L-745,453, a novel nonpeptidly endothelin antagonist. *Life Sci* 58: 1149–1157

106 Mugrage B, Moliterni J, Robinson L, Webb RL, Shetty SS, Lipson KE et al (1993) CGS 27830, a potent nonpeptide endothelin receptor antagonist. *Bioorganic Med Chem Lett* 3: 2099–2104

107 Lam YKT, Williams DL, Sigmund JM, Sanchez M, Genilloud O, Kong YL et al (1992) Cochinmicins, novel and potent cyclodepsipeptide endothelin antagonists from a Microbispora sp. *J Antibiotics* 45: 1709–1716

108 Zink D, Hensens OD, Lam YKT, Reamer R, Liesch JM (1992) Cochinmicins, novel and potent cyclodepsipeptide endothelin antagonists from a microspora sp2. structure determination. *J Antibiotic* 45: 1717–1722

109 Bigaud M, Pelton JT (1992) Discrimination between ET_A- and ET_B-receptor-mediated effects of endothelin-1 and [Ala[1, 3, 11, 15]]endothelin-1 by BQ-123 in the anaesthetized rat. *Br J Pharmacol* 107: 912–918

110 Cristol JP, Warner TD, Thiemermann C, Vane JR (1993) Mediation via different receptors of the vasoconstrictor effects of endothelins and sarafotoxins in the systemic circulation and renal vasculature of the anaesthetized rat. *Br J Pharmacol* 108: 776–779

111 Pollock DM, Opgenorth TJ (1993) Evidence for endothelin-induced renal vasoconstriction independent of ET_A receptor activation. *Am J Physiol* 264: R222–R226

112 Wellings RP, Warner TD, Thiemermann C, Corder R, Vane JR (1993) Vasoconstriction in the rat kidney induced by endothelin-1 is blocked by PD 145065. *J Cardiovasc Pharmacol* 22 (Suppl 8): S103–S106s

113 McMurdo L, Corder R, Thiemermann C, Vane JR (1993) Incomplete inhibition of the pressor effects of endothelin-1 and related peptides in the anaesthetised rat with BQ-123 provides evidence for more than one vasoconstrictor receptor. *Br J Pharmacol* 108: 557–561

114 Löffler B-M, Breu, V, Clozel M (1993) Effect of different endothelin receptor antagonists and of the novel non-peptide antagonist Ro 46-2005 on endothelin levels in rat plasma. *FEBS Lett* 333: 108–110

115 Cirino M, Motz C, Maw J, Ford-Hutchinson AW, Yano M (1992) BQ-153, a novel endothelin (ET)A antagonist, attenuates the renal vascular effects of endothelin-1. *J Pharm Pharmacol* 44: 782–785

116 Télémaque S, Gratton J-P, Claing A, D'Orleans-Juste P (1993) Endothelin-1 induces vasoconstriction and prostacyclin release via the activation of endothelin ET_A receptors in the perfused rabbit kidney. *Eur J Pharmacol* 237: 275–281

117 Maggi CA, Giuliani S, Patacchini R, Santicioli P, Giachetti A, Meli A (1990) Further studies on the response of the guinea-pig isolated bronchus to endothelins and sarafotoxin 6b. *Eur J Pharmacol* 176: 1–9

118 Fukuroda T, Nishikibe M, Ohta Y, Ihara M, Yano M, Ishikawa K et al (1992) Analysis of responses to endothelins in isolated porcine blood vessels by using a novel endothelin antagonist, BQ-153. *Life Sci* 50: PL107–PL112

119 Battistini B, Warner TD, Fournier A, Vane JR (1994) Characterization of ET_B receptors mediating contractions induced by endothelin-1 or IRL 1620 in guinea-pig isolated airways: effects of BQ-123, FR 139317 or PD 145065. *Br J Pharmacol* 111: 1009–1016

120 Henry PJ (1993) Endothelin-1 (ET-1)-induced contraction in rat isolated trachea: involvement of ET_A and ET_B receptors and multiple signal transduction systems. *Br J Pharmacol* 110: 435–441

121 Donckier J, Stoleru L, Hayashida W, Vanmechelen H, Selvais P, Galanti L et al (1995) Role of endogenous endothelin-1 in experimental renal hypertension in dogs. *Circ* 92: 106–113

122 Teerlink JR, Carteaux JP, Sprecher U, Loffler BM, Clozel M, Clozel JP (1995) Role of endogenous endothelin in normal hemodynamic status of anesthetized dogs. *Am J Physiol* 268: H432–440

123 Veniant M, Clozel JP, Hess P, Clozel M (1994) Endothelin plays a role in the maintenance of blood pressure in normotensive guinea-pigs. *Life Sci* 55: 445–454

124 Richard V, Hogie M, Clozel M, Loffler BM, Thuillez C (1995) *In vivo* evidence of an endothelin-induced vasopressor tone after injection nitric oxide synthesis in rats. *Circulation* 91: 771–775

125 Webb DJ (1995) Endogenous endothelin generation maintains vascular tone in humans. *J Human Hypertens* 9: 459–463

126 Haynes WG, Ferro CJ, O'Kane KP, Somerville D, Lomax CC, Webb DJ (1996) Systemic endothelin receptor blockade decreases peripheral vascular resistance and blood pressure in humans. *Circulation* 93: 1860–1870

127 Fuduroda T, Ozaki S, Ihara M, Ishikawa K, Yano M, Miyauchi T, Ishikawa S, Onizuka M, Goto K, Nishikibe M (1996) Necessity of dual blockade of endothelin ET_A and ET_B receptor subtypes for antagonism of endothelin-1-induced contraction in human bronchi. *Br J Pharmacol* 117: 995–999

128 LaDouceur DM, Flynn MA, Keiser JA, Reynolds EE, Haleen SJ (1993) ET_A and ET_B receptors coexist on rabbit pulmonary artery vascular smooth muscle mediating contraction. *Biochem Biophys Res Commun* 196: 209–215

129 Rodman DM, McMurtry IF, Peach JL, O'Brien RF (1989) Comparative pharmacology of rat and porcine endothelin in rat aorta and pulmonary artery. *Eur J Pharmacol* 165: 297–300

130 Lippton HL, Ohlstein EH, Sumner WR, Hyman AL (1991) Analysis of responses to endothelins in the rabbit pulmonary and systemic vascular beds. *J Appl Physiol* 70: 331–341

131 Cardell LO, Uddman R, Edvinsson L (1992) Evidence for multiple endothelin receptors in the guinea pig pulmonary artery and trachea. *Br J Pharmacol* 105: 376–380

132 Stewart DJ, Levy RD, Cernacek P, Langleben D (1991) Increased plasma endothelin-1 in pulmonary hypertension: marker or mediator of disease? *Ann Intern Med* 114: 464–469

133 Giaid A, Yanagisawa M, Langleben D, Michel RP, Levy R, Shennib H, Kimura S, Masaki T, Duguid WP, Stewart DJ (1993) Expression of endothelin-1 in the lungs of patients with pulmonary hypertension. *N Engl J Med* 328: 1732–1739

134 Oparil S, Chen SJ, Meng QC, Elton TS, Yano M, Chen YF (1995) Endothelin-A receptor antagonist prevents acute hypoxia-induced pulmonary hypertension in the rat. *Am J Physiol* 268: L95–L100

135 Bonvallet ST, Zamora MR, Hasunuma K, Sato K, Hanasato N, Anderson D, Sato K, Stelzner TJ (1994) BQ123, an ET_A-receptor antagonist, attenuates hypoxic pulmonary hypertension in rats. *Am J Physiol* 266: H1327–H1331

136 Eddahibi S, Raffestin B, Clozel M, Levame M, Adnot S (1995) Protection from pulmonary hypertension with an orally active endothelin receptor antagonist in hypoxic rats. *Am J Physiol* 268: H828–H835

137 Chen SJ, Brock T, Stavros F, Okun I, Wu C, Chan F et al (1996) TBC11251, a highly selective endothelin-A receptor antagonist prevents and reverses acute hypoxia-induce pulmonary hypertension in the rat. *FASEB J* 10: 601

138 Wang Y, Coe Y, Toyoda O, Coceani F (1995) Involvement of endothelin-1 in hypoxic pulmonary vasoconstriction in the lamb. *J Physiol* 482: 421–434

139 Ivy DD, Kinsella JP, Abman SH (1994) Physiologic characterization of endothelin A and B receptor activity in the ovine fetal pulmonary circulation. *J Clin Invest* 93: 2141–2148

140 Okada M, Yamashita C, Okada M, Okada K (1995a) Role of endothelin-1 in beagles with dehydromonocrotaline-induced pulmonary hypertension. *Circ* 92: 114–119
141 Okada M, Yamashita C, Okada M, Okada K (1993) Endothelin receptor antagonists in a beagle model of pulmonary-hypertension – contribution to possible potential therapy. *J Amer Coll Cardiol* 25: 1213–1217
142 Khimenko PL, Moore TM, Taylor AE (1996) Blocked ET_A receptors prevent ischemia and reperfusion injury in rat lungs. *J Appl Physiol* 80: 203–207
143 Grover GJ, Dzwonczyk S, Parham CS (1993) The endothelin-1 receptor antagonist BQ-123 reduces infarct size in a canine model of coronary occlusion and reperfusion. *Cardiovasc Res* 27: 1613–1618
144 Watanabe T, Awane Y, Ikeda S, Fujiwara S, Kubo K, Kikuchi K et al (1995) Pharmacology of a non-selective ET_A and ET_B receptor antagonist, TAK-044 and the inhibition of myocardial infarct size in rats. *Br J Pharmacol* 114: 949–954
145 Mertz TE, McClanahan TB, Flynn MA, Juneau P, Reynolds EE, Hallak H et al (1996) Endothelin A receptor antagonism by PD 156707 does not reduce infarct size after coronary artery occlusion/reperfusion in pigs. *J Pharmacol Exp Ther* 278:42–49
146 Teerlink JR, Loffler BM, Hess P, Maire JP, Clozel M, Clozel JP (1994) Role of endothelin in the maintenance of blood pressure in conscious rats with chronic heart failure. Acute effects of the endothelin receptor antagonist Ro 47-0203 (bosentan). *Circ* 90: 2510–2518
147 Kiowski W, Sutsch G, Hunziker P, Muller P, Kim J, Oechslin E et al (1995) Evidence for endothelin-1-mediated vasoconstriction in severe chronic heart failure. *Lancet* 346: 732–736
148 Gellai M, Jugus M, Fletcher T, DeWolf R, Nambi P (1994) Reversal of postischemic acute renal failure with a selective endothelin A receptor antagonist in the rat. *J Clin Invest* 93: 900–906
149 Brooks DP, DePalma PD (1996) Blockade of radiocontrast-induced nephrotoxicity by the endothelin receptor antagonist, SB 209670. *Nephron* 72: 629–636
150 Nishikibe M, Tsuchida S, Okada M, Fukuroda T, Shimamoto K, Yano M et al (1993) Antihypertensive effect of a newly synthesized endothelin antagonist, BQ-123, in a genetic hypertensive model. *Life Sci* 52: 717–724
151 Ohlstein EH, Douglas SA, Ezekiel M, Gellai M (1993) Antihypertensive effects of the endothelin receptor antagonist BQ-123 in conscious spontaneously hypertensive rats. *J Cardiovasc Pharmacol* 22 (Suppl 8): S321–S324
152 Schiffrin EL, Sventek P, Li J-S, Turgeon A, Reudelhuber T (1995) Antihypertensive effect of an endothelin receptor antagonist in DOCA-salt spontaneously hypertensive rats. *Br J Pharmacol* 115: 1377–1381
153 Douglas SA, Louden C, Vickery-Clarke LM, Storer BL, Hart T, Feuerstein GZ et al (1994) A role for endogenous endothelin-1 in neointimal formation after rat carotid artery balloon angioplasty. Protective effects of the novel nonpeptide endothelin receptor antagonist SB 209670. *Circ Res* 75: 190–197
154 Tsujino M, Hirata Y, Eguchi S, Watanabe T, Chatani F, Marumo F (1995) Nonselective ET_A/ET_B receptor antagonist blocks proliferation of rat vascular smooth-muscle cells after balloon angioplasty. *Life Sci* 56: PL449–PL454
155 Jenkins-West T, Valentine M, Moreland S, Ferrer P (1995) Intimal lesion development in balloon-injured rat carotid arteries is suppressed by the endothelin receptor antagonist BMS-182874. *FASEB J* 9: A343
156 Marsault R, Feolde E, Frelin C (1993) Receptor externalization determines sustained contractile responses to endothelin-1 in the rat aorta. *Am J Physiol* 264: C687–C693
157 Waggoner WG, Genova SL, Rash VA (1992) Kinetic analyses demonstrate that the equilibrium assumption does not apply to [^{125}I]endothelin-1 binding data. *Life Sci* 51: 1869–1876
158 Warner TD, Allcock GH, Vane JR (1994) Reversal of established responses to endothelin-1 *in vivo* and *in vitro* by the endothelin receptor antagonists BQ-123 and PD 145065. *Br J Pharmacol* 112: 207–213
159 Franco-Cereceda A, Matran R, Lou Y-P, Lundberg JM (1990) Occurrence and effects of endothelin in guinea pig cardiopulmonary tissue. *Acta Physiol Scand* 138: 539–547
160 Battistini B, Warner TD, Fournier A, Vane JR (1994) Characterization of ETB receptors mediating contractions induced by endothelin-1 or IRL 1620 in guinea pig isolated airways: effects of BQ-123, FR 139317 or PD 145065. *Br J Pharmacol* 111: 1009–1016

161 Battistini B, Warner TD, Fournier A, Vane JR (1994) Comparison of PD 145065 and Ro 46-2005 as antagonists of contractions of guinea pig airways induced by endothelin-1 or IRL 1620. *Eur J Pharmacol* 252: 341–345

162 Maggi CA, Giuliani S, Patachini R, Santicioli P, Giachetti A, Meli A (1990) Further studies on the response of the guinea pig isolated bronchus to endothelins and sarafotoxin S6b. *Eur J Pharmacol* 176: 1–9

163 Hay DWP, Hubbard WC, Undem BJ (1993) Endothelin-induced contraction and mediator release in human bronchus. *Br J Pharmacol* 110: 392–398

164 Candenas ML, Naline E, Sarria B, Advenier C (1992) Effects of epithelium removal and of enkephalin inhibition on the bronchoconstrictor response to three endothelins of the human isolated bronchus. *Eur J Pharmacol* 210: 291–297

165 Goldie RG, Grayson PS, Knott PG, Self GJ, Henry PJ (1994) Predominance of endothelin$_A$ (ET$_A$) receptors in ovine airway smooth muscle and their mediation of ET-1-induced contraction. *Br J Pharmacol* 112: 749–756

166 Nakamichi K, Ihara M, Kobayishi M, Saeki T, Ishikawa M, Yano M (1992) Different distribution of endothelin receptor subtypes in pulmonary tissues revealed by the novel selective ligands BQ-123 and [Ala^{1,3,11,15}]ET-1. *Biochem Biophys Res Commun* 182: 144–150

167 Pons F, Loquet I, Touvay C, Roubert P, Chabrier P-E, Mencia-Huerta JM, Braquet P (1991) Comparison of the bronchopulmonary and pressor activities of endothelin isoforms ET-1, ET-2, and ET-3 and characterization of their binding sites in guinea pig lung. *Am Rev Respir Dis* 143: 294–300

168 Hemsen A, Larsen O, Lundberg JM (1991) Characteristics of endothelin A and B binding sites and their vascular effects in pig peripheral tissues. *Eur J Pharmacol* 208: 313–322

169 D'Orleans-Juste P, Télémaque S, Claing A, Ihara M, Yano M (1992) Human big endothelin-1 and endothelin-1 release prostacyclin via activation of ET1 receptors in the rat perfused lung. *Br J Pharmacol* 105: 773–775

170 Kent A, Keenan AK (1994) Evidence for signaling by endothelin ET-A and ET-B receptors in bovine pulmonary artery smooth muscle cells. *Br J Pharmacol* 112: 549P

171 Cardell LO, Uddman R, Edvinsson L (1993) A novel ET$_A$-receptor antagonist, FR139317, inhibits endothelin-induced contractions of guinea pig pulmonary arteries, but not trachea. *Br J Pharmacol* 108: 448–452

172 Hemsen A, Franco-Cereceda A, Matran R, Rudehill A, Lundberg JM (1990) Occurrence, specific binding sites and functional effects of endothelin in human cardiopulmonary tissue. *Eur J Pharmacol* 191: 219–228

173 Brink C, Gillard V, Roubert P, Mencia-Huerta JM, Chabrier PE, Braquet P, Verley J (1991) Effects and specific binding sites of endothelin in human lung preparations. *Pulm Pharmacol* 4: 54–59

174 Zamora MA, Dempsey EC, Walchak S, Stelzner TJ (1993) BQ123, an ET$_A$ receptor antagonist, inhibits endothelin-1-mediated proliferation of human pulmonary artery smooth muscle cells. *Am J Respir Cell Mol Biol* 9: 429–433

175 Bonvallet ST, Oka M, Yano M, Zamora MR, McMurtry IF, Seltzer TJ (1993) *Cardiovasc Pharmacol* 22: 39–43

176 Yang CM, Yo YL, Ong R, Hsieh JT (1994) Endothelin- and sarafotoxin-induced phosphoinositide hydrolysis in cultured canine tracheal smooth muscle cells. *J Neurochem* 62: 1440–1448

177 Inui T, James AF, Fujitani Y, Takimoto M, Okada T, Yamamura T, Urade Y (1994) ET$_A$ and ET$_B$ receptors on single smooth muscle cells cooperate in mediating guinea pig tracheal contraction. *Am J Physiol* 266: L113–L124

178 Henry PJ (1993) Endothelin-1 (ET-1)-induced contraction in rat isolated trachea: involvement of ET$_A$ and ET$_B$ receptors and multiple signal transduction systems. *Br J Pharmacol* 110: 435–441

179 Abraham WM, Ahmed A, Cortes A, Spinella MJ, Malik AB, Andersen TT (1993) A specific endothelin-1 antagenist blocks inhaled endothelin-1-induced bronchoconstriction in sheep. *J Appl Physiol* 74: 2537–2542

180 Filep JG, Sirois MG, Foldes-Filep E, Rousseau A, Plante GE, Fournier A, Yano M, Sirois P (1993) Enhancement by endothelin-1 of microvascular permeability via the activation of ET$_A$ receptors. *Br J Pharmacol* 109: 880–886

181 Lippton HL, Hauth TA, Cohen GA, Hyman AL (1993) Functional evidence for different receptors in the lung. *J Appl Physiol* 75: 38–48

182 Spinella MJ, Malik AB, Everitt J, Andersen TT (1991) Design and synthesis of a specific endothelin 1 antagonist: effects on pulmonary vasoconstriction. *Proc Natl Acad Sci USA* 88: 7443–7446

183 Fabregat I, Rozengurt E (1990) [D-Arg^1, D-Phe^5, D-Trp^{7,9}, Leu^{11}] Substance P, a neuropeptide antagonist, blocks binding, Ca^{2+}-mobilizing, and mitogenic effects of endothelin and vasoactive intestinal contractor in mouse 3T3 cells. *J Cell Physiol* 145: 88–94

184 Itoh S, Sasaki T, Ide K, Ishikawa K, Nishikibe M, Yano M (1993) A novel endothelin ET$_A$ receptor antagonist, BQ-485, and its preventive effect on experimental cerebral vasospasm in dogs. *Biochem Biophys Res Commun* 195: 969–975

185 Webb ML, Bird JE, Liu ECK, Rose PM, Serafino R, Stein PD, Moreland S (1995) BMS-182874 is a selective, nonpeptide endothelin ET(A) receptor antagonist. *J Pharmacol Exp Ther* 272: 1124–1134

186 Bunker AM, Edmunds JJ, Berryman KA, Walker DM, Flynn MA, Welch KM, Doherty AM (1996) 1-Benyzl-3-thioaryl-2-carboxyindoles as potent nonpeptide endothelin antagonists. *Bioorganic Med Chem Lett* 6: 1367–1370

187 Balwierczak JL, Bruseo CW, Delgrande D, Jeng AY, Savage P, Shetty SS (1995) Characterization of a potent and selective endothelin-B receptor antagonist, IRL-2500. *J Cardiovasc Pharmacol* 26 (Suppl 3): S393–S396

188 Battistini B, Botting RM, Warner TD (1995) Endothelin: a knockout in London. *Trends Pharmacol Sci* 16: 217–222

189 Oshima T, Hirata M, Oda T, Sasaki A, Shiratsuchi M (1994) Pheorphorbide A, a potent endothelin receptor antagonist for both ET(A) and ET(B) subtypes. *Chem Pharm Bull* 42: 2174–2176

Pulmonary Actions of the Endothelins
ed. by R. G. Goldie and D. W. P. Hay
© 1999 Birkhäuser Verlag Basel/Switzerland

CHAPTER 3
Cellular Localization of the Endothelin System in the Lung

Nicholas W. Morrell*, Carlos Orte, John Wharton and Julia M. Polak

Division of Medicine and Department of Histochemistry, Imperial College School of Medicine, Hammersmith Campus, Du Cane Road, London W12 0NN*

1   Introduction
1.1 The Endothelin System
2   Lung Development
3   The Airways
3.1 Sites of Endothelin Synthesis
3.2 Endothelin Receptors
4   The Parenchyma
4.1 Sites of Endothelin Synthesis
4.2 Endothelin Receptors
5   The Pulmonary Circulation
5.1 Sites of Endothelin Synthesis
5.2 Endothelin Receptors
6   Summary
     References

1. Introduction

Although the endothelium is considered the main site of production of the endothelins, many other cell types in the lung are capable of endothelin synthesis, implying numerous potential roles for endothelins in lung biology. The endothelins were characterised originally as powerful vasoconstrictors, however, it is becoming clear that their effects may vary depending on the predominant receptor subtype present and the initial tone of the vascular bed. In addition, endothelins may be mitogenic and chemotactic for lung fibroblasts and smooth muscle cells derived from the airways or pulmonary circulation. Furthermore, their presence during development raises the possibility that these peptides may contribute to growth of the normal lung in addition to pathological remodelling of the airways and lung vasculature.

This chapter aims to describe the precise anatomical localisation of sites of synthesis, storage and action of the endothelins. This information is required in order to define the role of the endothelin system in the normal and diseased lung. The use of morphological, immunohistochemical, and molecular biological techniques in tissue sections have suggested an autocrine, paracrine and possibly endocrine role for these important peptides.

1.1. The endothelin system

To date, three structurally similar endothelin peptides have been designated: endothelin-1, endothelin-2, and endothelin-3 (ET-1, ET-2 and ET-3), each encoded by a separate gene [1]. Endothelin-1 is the best characterized of the ET family and is formed within the endothelial cell by a three-stage process. Preproendothelin-1, a 212 aminoacid precursor, is cleaved by one or more dibasic pair-specific endopeptidases. Secondly, a carboxypeptidase cleaves two amino acids from the carboxy terminus to form pro-endothelin-1, or big ET-1, a 38 aminoacid peptide. Finally, big ET-1 is then converted to ET-1 (21 amino acids) by endothelin converting enzyme (ECE) and released mainly at the abluminal surface of the endothelium [2, 3].

At least two distinct isoforms of ECE have been described, ECE-1 [4] and ECE-2 [5]. Both enzymes are present in bovine aortic endothelial cells. The localisation of ECE-1 has recently been studied in the human lung [6], but the distribution of ECE-2 in the lung has not yet been reported.

Endothelin receptors exist in at least two separate isoforms (ET_A and ET_B). These receptors belong to the seven-transmembrane, G protein-coupled rhodopsin superfamily. The ET_A receptor displays high affinity for ET-1 and ET-2, is generally present on smooth muscle cells, and mediates vasoconstriction. The ET_B receptor was initially thought to be located mainly on endothelial cells, but now is also realised to be present on vascular and airway smooth muscle, binds all three endothelins and may mediate vasodilatation *via* release of nitric oxide, prostacyclin, or *via* ATP-gated K channels [7], or vasoconstriction. There is also controversial evidence for at least two different subtypes of the ET_A (ET_{A1} and ET_{A2}) and ET_B (ET_{B1} and ET_{B2}) receptors [8, 9], although their existence is not clearly established.

2. Lung Development

High levels of ET-1 and ET-3 mRNA are expressed during lung development [10]. In the rat fetus at 19 days of gestation, *in situ* hybridization studies for ET mRNA reveal that most of the signal is localized to bronchial epithelial cells, with lower levels over small vessels. Endothelin mRNA and peptide expression has been found in the human fetal lung as early as 12 weeks of gestation [11]. Recent studies evaluating homozygous ET-1 null mutation mice (Edn-1) confirm that ET-1 is essential to normal development of the heart and great vessels [12, 13]. Edn-1 −/− mice displayed cardiovascular malformations including interrupted aortic arch, tubular hypoplasia of the aortic arch, aberrant right subclavian artery, and ventricular septal defect with abnormalities of the outflow tract. Although the homozygous −/− mice appear to have no structural abnormality of the lungs, it remains to be determined whether ET-1 is important in the functional maturation of the lung.

3. The Airways

3.1. Sites of Endothelin Synthesis

In the human airway, cells that have been shown to synthesize endothelins include bronchial epithelial cells [14,15], endothelial cells [16], macrophages [17], and pulmonary neuroendocrine cells [11]. Ovine tracheal smooth muscle cells in culture also express preproendothelin-1 (ppET-1) mRNA suggesting a further source of endothelin production [18]. In rodents we have shown previously that the airway epithelium is a rich source of endothelins, with immunostaining observed in mucous, serous and Clara cells; no staining was observed in basal cells and ciliated cells [19]. Indeed, levels of immunoreactive ET-1 in the bronchial epithelium is amongst the highest noted for any tissue [20]. In situ hybridisation studies have identified abundant ET-1 mRNA expression in the bronchiolar epithelial cells of the developing and adult rat lung [10]. In man we have also found that airway epithelium and endocrine cells are a rich source of ET-1 [11].

Of the three isoforms, ET-1 appears to be the most abundant in the human lung [21, 22], though all three isoforms have been detected by immunohistochemistry and in situ hybridization [11]. Proendothelins-1 and -3 are detectable, using polyclonal antibodies, in airway epithelium, whereas all three proendothelins are detectable in submucosal glands [22].

Patients with symptomatic asthma have elevated levels of endothelin-1 in bronchoalveolar lavage fluid [23]. This correlates with the observation of increased immunoreactive ET-1 in bronchial biopsy specimens from patients with asthma compared with controls [24, 25], particularly in airway epithelial and vascular endothelial cells (Fig. 1). In addition, asthmatic airway epithelium demonstrates increased ppET-1 mRNA expression by in situ hybridisation [14]. It has been suggested that the presence of increased numbers of airway epithelial cells staining positively for ET-1 may identify a subset of asthmatics with more poorly controlled asthma. These patients show an increased number of inflammatory cells, especially mast cells and eosinophils, in bronchial biopsies and exhibit increased bronchoconstriction to nebulised distilled water [26]. Moreover, bronchial biopsies from these patients show that approximately 65% of their bronchial epithelial cells have ET-1 immunostaining compared to 8% of cells from patients who do not bronchoconstrict to distilled water [26].

The cellular localization of ECE-1 in airways has been reported by Saleh et al. [6]. In normal lungs, focal moderate expression of ECE-1 is localized

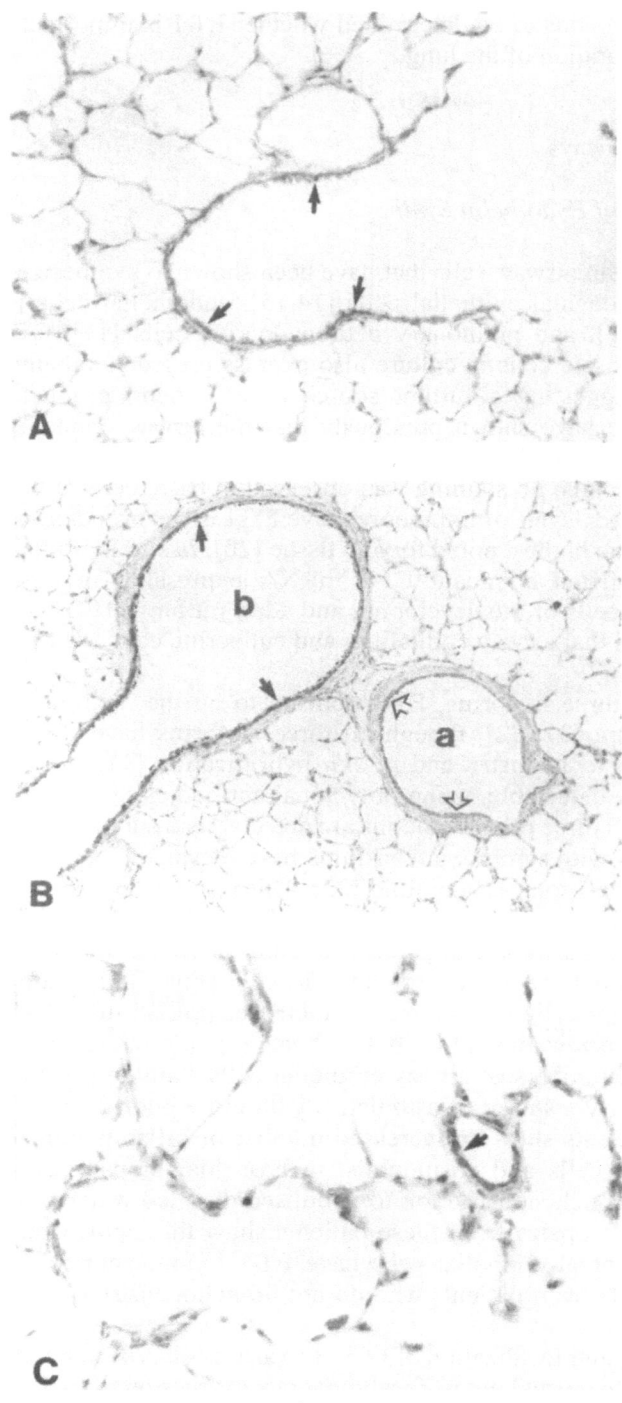

to airway epithelium and airway smooth muscle cells. Serous bronchial glands also expressed ET-1 and ECE-1.

3.2. Endothelin Receptors

All three endothelin peptides have been shown to contract isolated human bronchi [27], with ET-1 being the most potent. In man, the ET_B receptor directly mediates bronchoconstriction in response to ET-1 [28], and in human bronchial smooth muscle 80–85% of ET-1 binding sites are ET_B receptors [29]. In other animals, such as sheep, the ET_A receptor predominates and is responsible for bronchoconstriction [30]. Interestingly, in the airway smooth muscle of mice and rats, ET_A and ET_B receptors exist in almost equal proportions and contraction is mediated by either receptor. As well as in larger airways, binding sites for ET-1 and ET-2 also exist on human peripheral airways [31], alveolar septae [32, 33], nasal mucosa [34], and airway parasympathetic ganglia and nerves [32, 33].

The endothelin system has been implicated in the airway oedema, structural remodelling [35], and airway hyperreactivity that characterise asthma. Studies have shown that bronchial smooth muscle from subjects with asthma are less sensitive to an ET_B receptor agonist than tissues obtained from normal subjects. Since receptor subtype density is similar in the peripheral lung of asthmatic and normal subjects, this decreased sensitivity to ET_B receptor activation is thought to be due to a defect in postreceptor signalling in asthmatics [29].

Endothelins may be involved in the structural remodelling of asthmatic airways which leads to relatively fixed airflow limitation. Endothelin-1 acts as a co-mitogen for airway smooth muscle cells [35, 36] and may contribute to fibrosis by stimulating myofibroblast proliferation [37].

Viral respiratory tract infections have been shown to alter expression of ET_A and ET_B receptors. Influenza A/PR-8/34 infection in mice decreases the total number of ET receptors by 20%. However, the ratio of ET_A/ET_B is dramatically shifted: ET_A receptors are increased by 40% and ET_B receptors decreased by 70% [38], resulting in altered ET-mediated responses.

Figure 1. Photomicrographs of formalin-fixed, paraffin-embedded, sections of lung from control (A) and chronically hypoxic (B and C) Wistar-Kyoto rats immunostained with a monoclonal anti-endothelin-1 antibody (ABR, INC), visualised by the avidin-biotinylated-peroxidase method, and counterstained with haematoxylin. In the control animals (A), immunostaining was confined to occasional rway epithelial cells (arrowed) and endothelial cells. In contrast, in chronically hypoxic animals (B), most of the epithelial cells (closed arrows) of small bronchi (b) were intensely stained, in addition to more marked staining in the endothelium (open arrows) of accompanying arteries (a). Furthermore, intense staining of the endothelium was noted in very small peripheral arteries in hypoxic animals (C).

4. The Parenchyma

4.1. Sites of Endothelin Synthesis

Many cell types have been shown to be sites of endothelin production in human lung parenchyma including pulmonary neuroendocrine cells [11], alveolar epithelial cells [39], endothelial cells [16] and alveolar macrophages [17]. Rat visceral pleural mesothelial cells in culture also make ET-1, suggesting another potential site of production in man [40].

All three ET isopeptides are found in pulmonary endocrine cells of newborn mammals [41], including man [11]. Immunohistochemistry has demonstrated alveolar macrophages and epithelium as sites of ET-1 and ET-3 synthesis in adult human lung [17].

During development, intense immunostaining for ET-1 and ppET-1 mRNA levels in large and small airways, and vascular endothelium and smooth muscle, during the first and second trimester suggest a possible role for ET-1 in lung maturation [11]. The precise role, however, remains uncertain since mice deficient in the ET-1 gene have apparently normal lungs and tracheas, though they die of respiratory failure soon after birth, probably due to abnormal development of pharyngeal anatomy [12].

A number of studies have suggested a role for the endothelins in the pathogenesis of pulmonary fibrosis [39, 42, 43]. Patients with idiopathic fibrosis have elevated levels of ET-1 in plasma [43] and bronchoalveolar lavage fluid [42]. Endothelin-1 immunoreactivity in the normal human lung is mainly confined to conducting airway epithelium, pulmonary vascular endothelium and neuroendocrine cells [39]. In contrast, patients with idiopathic pulmonary fibrosis have increased ppET-1 mRNA expression and ET-1 immunoreactivity in airway epithelial cells and type II pneumocytes [39]. The ET immunoreactivity occurs mainly in hyperplastic type II pneumocytes, undifferentiated epithelial cells, and small airway epithelial cells [39]. Moreover, ET-1 immunostaining directly correlates with the extent of immature granulation tissue and type II cell proliferation. In the fibrotic lung, ET-1 release may stimulate proliferation of interstitial lung fibroblasts [44], or myofibroblasts, promote fibronectin expression and fibroblast chemotaxis [45, 46] and increase fibroblast collagen synthesis [47].

Saleh et al. [6] have recently investigated the expression of ECE-1, big ET-1, and ET-1, using immunohistochemistry and *in situ* hybridization, in the lungs of patients with idiopathic pulmonary fibrosis compared with normal subjects. In normal lungs, focal moderate expression of all three molecules was localized to airway epithelium, pulmonary endothelium, and airway and vascular smooth muscle cells. Serous bronchial glands also expressed ET-1 and ECE-1. In pulmonary fibrosis, strong diffuse expression of ECE-1 was seen in airway epithelium, proliferating type II pneumocytes, and in endothelial and inflammatory cells. Furthermore, ECE-1 immunostaining was colocalized to big ET-1 and ET-1 immunostaining, and correlated with disease activity.

The regulation of ET production by parenchymal cells *in vivo* is unknown, but in cell culture various cytokines such as interleukin-1 (IL-1), IL-2, lipopolysaccharide (LPS), tumour necrosis factor alpha (TNFα) and transforming growth factor beta (TGFβ) have been shown to stimulate ET-1 production from epithelial cell lines [48, 49], whereas platelet-derived growth factor (PDGF) and interferon gamma (IFNγ), and glucocorticoids decrease ET-1 secretion [48, 49]. TNFα may also have a role in the upregulation of ECE-1 expression by epithelial cells [6].

4.2. Endothelin Receptors

In the human lung parenchyma, ET receptors are found in the alveolar septum [32, 33, 50], fetal lung fibroblasts [51], and nerves [32]. As in airway smooth muscle, the predominant receptor subtype in the normal human alveolar wall is ET_B (approximately 70%) rather than ETA (approximately 30%) [50]. *In situ* hybridization studies in rat and rabbit lung have identified ET_B receptor mRNA in bronchial associated lymphoid tissue, ciliated and non-ciliated bronchial epithelial cells, alveolar type II pneumocytes, and visceral pleural mesothelium [52]. Autoradiographic studies have demonstrated that the ET_A receptor subtype predominates in submucosal glands and pulmonary vessels in animals and man [29].

The functional significance of ET receptor expression on alveolar cells may be that autocrine production of ET-1 by alveolar epithelial cells is involved in the regulation of surfactant, since ET-1 increases surfactant synthesis in cultured rat alveolar type II cells [53]. In addition, endogenous ET-1 production may also modulate epithelial cell proliferation and iNOS expression [54]. Furthermore, autocrine production of ET-1 by malignant, especially squamous cell and adenocarcinoma cells [55], may contribute to epithelial cell tumour growth [56].

5. The Pulmonary Circulation

5.1. Sites of Endothelin Synthesis

Pulmonary arterial endothelial cells contain ppET-1 mRNA and appear to release only ET-1, the majority (approximately 80%) of which is released from the basolateral aspect of the cell [57]. ECE has also been localised to the pulmonary endothelium by immunohistochemistry in rats [58] and man [6]. Among the stimuli which have been shown to stimulate ET-1 secretion from pulmonary artery endothelial cells are TGFβ, TNFα, thrombin and an endogenous digitalis-like factor [59, 60]. These effects seem to be regulated at the level of increased transcription of ppET-1 mRNA [61]. There is evidence that ET-1 release may be differentially regulated in endothelial cells derived from the systemic and pulmonary circulations. For example it has been reported that angiotensin II and vasopressin do not stimulate ET-1 release from pulmonary endothelium, but do in systemic endothelium [59, 62]. Evidence from systemic endothelial cells also indicates that prostacyclin [63] and heparin [62] may reduce endothelial ET-1 synthesis and release. Interestingly, mechanical forces such as shear stress and stretch [64] also regulate ET-1 synthesis in human cultured umbilical vein endothelial cells. In addition to endothelial cells, smooth muscle cells (SMC) of the tunica media may be capable of endothelin production. Endothelin-1 and ET-3 production has been reported in human systemic vascular SMC [65], though no similar studies have been reported in SMC derived from the pulmonary circulation.

Endothelin-1 has been implicated in the development of pulmonary hypertension and has become an attractive potential target for therapeutic intervention in this condition. Endothelin-1 acts as a co-mitogen to stimulate proliferation of cultured human pulmonary artery smooth muscle cells [66]. Increased ET-1 and ppET-1 mRNA levels have been reported in lung homogenates in the chronically hypoxic rat model of pulmonary hypertension after 4 weeks [67]. However, two other studies have reported no change in ET-1 or ppET-1 mRNA levels after 3 weeks of hypoxia [68, 69]. Perhaps more convincing evidence for a role for endothelin in the process of pulmonary vascular remodelling comes from studies using ET receptor antagonists: treatment with BQ-123, an ET_A receptor antagonist, almost completely prevented the rise in pulmonary artery pressure, vascular remodelling and right ventricular hypertrophy associated with 2 weeks exposure to hypoxia [70].

Few studies have investigated whether there are changes in the levels or distribution of ET-1 expression during the development of pulmonary hypertension. One group have studied ET-1 expression by immunostaining in a rat model of chronic pulmonary artery ligation [71]. Increased ET-1 expression was seen in the endothelium of bronchial collateral arteries and elastic and muscular pulmonary arteries. Using the same antibody we have examined the pattern of ET-1 immunoreactivity in rat lungs exposed to hypoxia for 3 weeks compared with normoxic rats. In normoxic animals ET-1 immunoreactivity was restricted to occassional endothelial cells and less than 10% of bronchial epithelial cells in extra-acinar airways (Fig. 1 A). In contrast, in chronically hypoxic rats, the bronchial epithelium was intensely and uniformly stained (Fig. 1 B), and small pulmonary arterioles showed enhanced staining of the endothelium (Fig. 1 C). Interestingly, stainingiof the epithelium was consistently more conspicuous than that of the endothelium.

There has been much interest in the regulation of ET-1 expression by hypoxia, particularly by endothelial cells. Following the initial report that hypoxia stimulates ET-1 release from human umbilical vein endothelial cells (HUVECs) [72], it has been found that the hypoxic regulation of the ET-1 gene is highly variable between species and endothelial cells from different vascular beds. In pulmonary artery endothelial cells, studies have found that hypoxia increases [60], does not change [73] or decreases ET-1 release [74, 75]. These differences may be explained by the factors mentioned above, culture differences and by heterogeneity in endothelial cells derived from different parts of the pulmonary circulation.

Patients with pulmonary hypertension, either primary or secondary to underlying cardiopulmonary disease, have elevated plasma ET-1 levels [76, 77]. It is likely that the increased circulating levels of ET-1 are due to increased lung production, decreased lung clearance, or both, since the transpulmonary gradient of ET-1 is increased in these patients [77]. In addition, the whole lung ET-1 content is increased in primary pulmonary hypertension (PPH) patients compared with control subjects [76]. The main site of increased ET-1 synthesis is probably the pulmonary endothelium [78]. Immunohistochemical studies in normal subjects show, perhaps surprisingly, minimal staining for ET-1 in the vascular endothelium [78]. In contrast, patients with PPH demonstrate increased ppET-1 mRNA expression and abundant immunostaining for ET-1 in almost all pulmonary vessels [78], especially elastic and muscular pulmonary arteries. Endothelial cells from arteries with medial thickening or intimal fibrosis also possess intense ET-1 immunostaining. This increased staining seems organ specific in PPH since myocardial or renal blood vessels contain little or no staining. Intense immunostaining is also seen in the plexiform lesion which is the hallmark of primary and severe forms of secondary pulmonary hypertension [78].

5.2. Endothelin Receptors

In the adult lung, endothelin binding sites are widely distributed throughout the bronchial and vascular systems and the lung parenchyma [79]. It is likely that human pulmonary artery endothelial cells, like their counterparts in the systemic circulation, express both ET_A and ET_B receptors. The ET_B receptor predominates on endothelial cells, suggesting a role for endothelin in modulation of endothelial cell function [80].

The media of human pulmonary arteries contain both ET_A and ET_B receptor mRNAs, and autoradiographic studies suggest ET_A (97%) receptors predominate over ET_B (3%) [81, 82]. The receptor subtypes have not been reported in human pulmonary resistance arteries, with most studies concentrating on large elastic extrapulmonary vessels. However, pharmacological studies in human pulmonary resistance arteries suggest that the ET_B receptor, as well as the ET_A receptor mediate vasoconstriction to ET-1 in man [83], though there appear to be species differences in ET_B receptor-mediated contraction [84]. ET binding sites have also been described in human pulmonary veins and in nerves accompanying arteries [32]. In the chronic hypoxic rat model of pulmonary hypertension, hypoxia has been shown to increase the level of ET_A receptor mRNA in whole lung and the level of ET_B receptor mRNA in the main pulmonary artery, but hypoxia did not change ET_A mRNA expression in these vessels [85]. Autoradiographic studies, on the other hand, indicate that chronic hypoxia in adult animals does not change total or subtype-specific ET-1 receptor binding in pulmonary arteries compared with normoxic animals [86]. However, there is evidence that ET_A binding sites are increased in elastic and muscular arteries, and veins, during hypoxic exposure in neonatal pigs [87], concurrent with a rise in plasma ET-1. Whether changes occur in receptor distribution, density, and affinity during the development of pulmonary hypertension in man has not been examined.

6. Summary

This review has summarised what is known regarding the cellular localisation of endothelin synthesis and their receptors and actions in the normal and diseased lung. We have emphasised the widespread expression of endothelins by lung cells which points to numerous diverse functions, from a key modulator of vascular and airway tone, to growth mediator and cytokine. Much work remains to be done regarding the cellular expression of the isoforms of endothelin and ECE, and the subcellular localisation of these molecules. In addition, the recognition of marked endothelin receptor subtype heterogeneity demands further detailed studies, especially in human lung. It is becoming clear that the endothelin system provides an exciting new potential therapeutic target in the treatment of some important pulmonary diseases such as asthma, pulmonary fibrosis and pulmonary hypertension.

References

1 Inoue A, Yanagisawa M, Kimura S, Kasuya Y, Miyauchi T, Goto K (1989) The human endothelin family: three structurally and pharmacologically distinct isopeptides predicted by three separate genes. *Proc Natl Acad Sci USA* 86: 2863–2867
2 Yanagisawa M, Kurihara H, Kimura S, Tomobe Y, Kobayashi M, Mitsui Y (1988) A novel potent vasoconstrictor peptide produced by vascular endothelial cells. *Nature* 332: 411–415
3 Barnes K, Murphy LJ, Takahashi M, Tanzawa K, Turner AJ (1995) Localization and biochemical characterisation of endothelin-converting enzyme. *J Cardiovasc Pharmacol* 26(Suppl 3): S37–39
4 Xu D, Emoto N, Giaid A, Slaughter C, Kaw S, deWit D (1994) ECE-1: a membrane bound metalloproteinase that catalyses the proteolytic activation of big endothelin-1. *Cell* 78: 473–85
5 Emoto N, Yanagisawa M (1995) Endothelin converting enzyme-2 is a membrane-bound phosphoramidon-sensitive metalloproteinase with acidic pH optimum. *J Biol Chem* 270: 15262–15268

6 Saleh D, Furukawa K, Tsao MS, Maghazachi A, Corrin B, Yanagisawa M, Barnes PJ, Giaid A (1997) Elevated expression of endothelin-1 and endothelin-converting enzyme-1 in idiopathic pulmonary fibrosis: possible involvement of proinflammatory cytokines. *Am J Respir Cell Mol Biol* 16: 187–193

7 Ziegler JW, Ivy DD, Kinsella JP, Abman SH (1995) The role of nitric oxide, endothelin, and prostaglandins in the transition of the pulmonary circulation. *Clin Perinatol* 22: 387–403

8 Shyamala V, Moulthrop THM, Stratton-Thomas J, Tekamp-Olson P (1994) Two distinct human endothelin B receptors generated by alternative splicing from a single gene. *Cell Mol Biol Res* 40: 285–296

9 Elshourbagy NA, Adamou JE, Gagnon AW, Wu HL, Pullen M, Nambi P (1996) Molecular characterization of a novel human endothelin receptor splice variant. *J Biol Chem* 271: 25300–25307

10 MacCumber MW, Ross CA, Glaser BM, Snyder SH (1989) Endothelin: visualization of mRNAs by *in situ* hybridization provides evidence for local action. *Proc Natl Acad Sci USA* 86: 7285–7289

11 Giaid A, Polak JM, Gaitonde V, Hamid QA, Moscoso G, Legon S, Uwanogho D, Roncalli M, Shinmi O, Sawamura T (1991) Distribution of endothelin-like immunoreactivity and mRNA in the developing and adult human lung. *Am J Respir Cell Mol Biol* 4: 50–58

12 Kurihara Y, Kurihara H, Suzuki H, Kodama T, Maemura K, Nagai R, Oda H, Kuwaki T, Cao WH, Kamada N (1994) Elevated blood pressure and craniofacial abnormalities in mice deficient in endothelin-1. *Nature* 368: 703–710

13 Kurihara Y, Kurihara H, Oda H, Maemura K, Nagai R, Ishikawa T, Yazaki Y (1995) Aortic arch malformations and ventricular septal defect in mice deficient in endothelin-1. *J Clin Invest* 96: 293–300

14 Vittori E, Marini M, Fasoli A, De Franchis R, Mattoli S (1992) Increased expression of endothelin in bronchial epithelial cells of asthmatic patients and effects of corticosteroids. *Am Rev Respir Dis* 146: 1320–1325

15 Shokeir MO, Pare P, Wright JL (1994) Relation of smoking to immunoreactive endothelin in the bronchiolar epithelial cells. *Thorax* 49: 786–789

16 Visner GA, Staples ED, Chesrown SE, Block ER, Zander DS, Nick HS (1994) Isolation and maintenance of human pulmonary artery endothelial cells in culture isolated from transplant donors. *Am J Physiol* 267: L406–413

17 Ehrenreich H, Anderson RW, Fox CH, Rieckmann P, Hoffman GS, Travis WD (1990) Endothelins, peptides with potent vasoactive properties, are produced by human macrophages. *J Exp Med* 172: 1741–1748

18 Ergul A, Glassberg MK, Wanner A, Puett D (1995) Characterization of endothelin receptor subtypes on airway smooth muscle cells. *Exp Lung Res* 21: 453–468

19 Rozengurt N, Springall DR, Polak JM (1990) Localization of endothelin-like immunoreactivity in airway epithelium of rats and mice. *J Pathol* 160: 5–8

20 Yoshimi H, Hirata Y, Fukuda Y, Kawano Y, Emori T, Kuramochi M (1989) Regional distribution of immunoreactive endothelin in rats. *Peptides* 10: 805–808

21 Hemsen A, Franco-Cereceda A, Matran R, Rudehill A, Lundberg JM (1990) Occurrence, specific binding sites and functional effects of endothelin in human cardiopulmonary tissue. *Eur J Pharmacol* 191: 319–328

22 Marciniak SJ, Plumpton C, Barker PJ, Huskisson NS, Davenport AP (1992) Localization of immunoreactive endothelin and proendothelin in the human lung. *Pulm Pharmacol* 5: 175–182

23 Mattoli S, Soloperto M, Marini M, Fasoli A (1991) Levels of endothelin in the bronchoalveolar lavage fluid of patients with symptomatic asthma and reversible airflow obstruction. *J Allergy Clin Immunol* 88: 376–384

24 Springall DR, Howarth PH, Counihan H, Djukanovic R, Holgate ST, Polak JM (1991) Endothelin immunoreactivity of airway epithelium in asthmatic patients. *Lancet* 337: 697–701

25 Ackerman V, Carpi S, Bellini A, Vassali G, Marini M, Mattoli S (1995) Constitutive expression of endothelin in bronchial epithelial cells of patients with symptomatic and asymptomatic asthma nd modulation by histamine and interleukin-1. *J Allergy Clin Immunol* 96: 618–627

26 Carpi S, Marini M, Vittori E, Vassali G, Mattoli S (1993) Bronchoconstrictive responses to inhaled ultrasonically nebulised distilled water and airway inflammation in asthma. *Chest* 104: 1346–1351

27 Advenier C, Sarria B, Naline E, Puybasset L, Lagente V (1990) Contractile activity of three endothelins (ET-1, ET-2, ET-3) on the human isolated bronchus. *Br J Pharmacol* 100: 168–172

28 Hay DWP, Luttman MA, Hubbard WC, Undem BJ (1993) Endothelin receptor subtypes in human and guinea pig pulmonary tissues. *Br J Pharmacol* 110: 1175–1183

29 Goldie RG, Henry PJ, Knott PG, Self GJ, Luttman MA, Hay DWP (1995) Endothelin-1 receptor density, distribution, and function in human isolated asthmatic airways. *Am J Respir Crit Care Med* 152: 1653–1658

30 Goldie RG, Grayson PS, Knott PG, Self GJ, Henry PJ (1994) Predominance of endothelin A (ET_A) receptors in ovine airway smooth muscle and their mediation of ET-1-induced contraction. *Br J Pharmacol* 112: 749–756

31 Goldie RG, Henry PJ, Paterson JW, Preuss JMH, Rigby PJ (1990) Contractile effects and receptor distributions for endothelin-1 (ET-1) in human and animal airways. *Agents Actions* Suppl 31: 229–232

32 Power RF, Wharton J, Zhao Y, Bloom SR, Polak JM (1989) Autoradiographic localization of endothelin-1 binding sites in the cardiovascular and respiratory systems. *J Cardiovasc Pharmacol* 13 (Suppl 5): S50–56

33 McKay KO, Black JL, Diment LM, Armour CL (1991) Functional and autoradiographic studies of endothelin-1 and endothelin-2 in human bronchi, pulmonary arteries, and airway parasympathetic ganglia. *J Cardiovasc Pharmacol* 17 (Suppl 7): S206–209

34 Wu T, Mullol J, Rieves RD, Logun C, Hausfield J, Kaliner MA, Shelhamer JH (1992) Endothelin-1 stimulates eicosanoid production in cultured human nasal mucosa. *Am J Respir Cell Mol Biol* 6: 168–174

35 Glassberg MK, Ergul A, Wanner A, Puett D (1994) Endothelin-1 promotes mitogenesis in airway smooth muscle cells. *Am J Respir Cell Mol Biol* 10: 316–321

36 Panettieri RA Jr, Goldie RG, Rigby PJ, Eszterhas AJ, Hay DW (1996) Endothelin-1-induced potentiation of human airway smooth muscle proliferation: an ET_A receptor-mediated phenomenon. *Br J Pharmacol* 118: 191–197

37 Brewster CEP, Howarth PH, Djukanovic R, Wilson J, Holgate ST, Roche WR (1990) Myofibroblasts and subepithelial fibrosis in bronchial asthma. *Am J Respir Cell Mol Biol* 3: 507–511

38 Henry PJ, Goldie RG (1994) ET_B but not ET_A receptor-mediated contractions to endothelin-1 attenuated by respiratory tract viral infection in mouse airways. *Br J Pharmacol* 112: 1188–1194

39 Giaid A, Michel RP, Stewart DJ, Sheppard M, Corrin B, Hamid Q (1993) Expression of endothelin-1 in lungs of patients with fibrosing cryptogenic alveolitis. *Lancet* 341: 1550–4

40 Kuwahara M, Kuwahara M, Suzuki N (1992) Production of endothelin-1 and big-endothelin-1 by pleural mesothelial cells. *FEBS Lett* 298: 21–24

41 Seldeslagh KA, Lauweryns JM (1993) Endothelin in normal lung tissue of newborn mammals: immunocytochemical distribution and co-localization with serotonin and calcitonin-gene related peptide. *J Histochem Cytochem* 41: 1495–1502

42 Sofia M, Mormile M, Faraone S, Alifano M, Zofra S, Romano L, Carratu L (1993) Increased endothelin-like immunoreactive material in bronchoalveolar lavage fluid from patients with bronchial asthma and patients with interstitial lung disease. *Respiration* 60: 89–95

43 Uguccioni M, Pulsatelli L, Grigolo B, Facchini A, Fasano L, Cinti C, Fabbri M, Gasbarrini G, Meliconi R (1995) Endothelin-1 in idiopathic pulmonary fibrosis. *J Clin Pathol* 48: 330–334

44 MacNulty EE, Plevin R, Wakelam MJ (1990) Stimulation of the hydrolysis of phosphatidylinositol 4,5-biphosphate and phosphatidylcholine by endothelin, a complete mitogen for rat fibroblasts. *Biochem J* 272: 761–766

45 Peacock AJ, Dawes KE, Shock A, Gray AJ, Reeves JT, Laurent GJ (1992) Endothelin-1 and endothelin-3 induce chemotaxis and replication of pulmonary artery fibroblasts. *Am J Respir Cell Mol Biol* 7: 492–499

46 Marini M, Carpi S, Bellini A, Patalano F, Mattoli S (1996) Endothelin-1 induces increased fibronectin expression in human bronchial epithelial cells. *Biochem Biophys Res Commun* 220: 896–899

47 Guarda E, Katwa LC, Myers PR, Tyagi SC, Weber KT (1993) Effects of endothelins on collagen turnover in cardiac fibroblasts. *Cardiovasc Res* 27: 2130–2134

48 Calderon E, Gomez-Sanchez CE, Cozza EN, Zhou M, Coffey RG, Lockey RF, Prockop LD, Szentivanyi A (1994) Modulation of endothelin-1 production by a pulmonary epithelial cell line, I: regulation by glucocorticoids. *Biochem Pharmacol* 48: 2065–2071

49 Markewitz BA, Kohan DE, Michael JR (1995) Endothelin-1 synthesis, receptors and signal transduction in a cloned rat epithelial cell line: evidence for an autocrine role. *Am J Physiol* 268: L192–L200

50 Knott PG, D'Aprile AC, Henry PJ, Hay DWP, Goldie RG (1995) Receptors for endothelin-1 in asthmatic human peripheral lung. *Br J Pharmacol* 114: 1–3

51 Cambrey AD, Harrison NK, Dawes KE, Southcott AM, Black CM, du Bois RM, Laurent GJ, McAnulty RJ (1994) Increased levels of endothelin-1 in bronchoalveolar lavage fluid from patients with systemic sclerosis contribute to fibroblast mitogenic activity *in vitro*. *Am J respir Cell Mol Biol* 11: 439–445

52 Durham SK, Goller NL, Lynch JS, Fisher SM, Rose PM (1993) Endothelin receptor B expression in the rat and rabbit lung as determined by *in situ* hybridization using nonisotopic probes. *J Cardiovasc Pharmacol* 22 (Suppl 8): S1–S3

53 Sen N, Grunstein MM, Chander A (1994) Stimulation of lung surfactant secretion by endothelin-1 from rat alveolar type II cells. *Am J Physiol* 266: L255–262

54 Markewitz BA, Michael JR, Kohan DE (1997) Endothelin-1 inhibits the expression of inducible nitric oxide synthase. *Am J Physiol* 272: L1078–1083

55 Giaid A, Hamid QA, Springall DR, Yanagisawa M, Shinmi O, Sawamura T, Masaki T, Kimura S, Corrin B, Polak JM (1990) Detection of endothelin immunoreactivity and mRNA in pulmonary tumours. *J Pathol* 162: 15–22

56 Shichiri M, Hirata Y, Nakajima T, Ando K, Imai T, Yanagisawa M, Masaki T, Marumo F (1991) Endothelin-1 is an autocrine/paracrine growth factor for human cancer cell lines. *J Clin Invest* 87: 1867–1871

57 Wagner OF, Christ G, Wojta J, Vierhapper H, Parzer S, Nowotny PJ, Schneider B, Waldhäusl W, Binder BR (1992) Polar secretion of endothelin-1 by cultured endothelial cells. *J Biol Chem* 267: 16066–16068

58 Takahashi M, Fukuda K, Shimada K, Barnes K, Turner AJ, Ikeda M, Koike H, Yamamoto Y, Tanzawa K (1995) Localization of rat endothelin converting enzyme to vascular endothelial cells and some secretory cells. *Biochem J* 311: 657–665

59 Ohlstein EH, Arleth A, Ezekiel M, Horohonich S, Ator MA, Caltabiano MM, Sung C-P (1990) Biosynthesis and modulation of endothelin from bovine pulmonary artery endothelial cells. *Life Sci* 46: 181–188

60 Golden CL, Kohler JP, Nick HS, Visner GA (1995) Effects of vasoactive and inflammatory mediators on endothelin-1 expression in pulmonary endothelial cells. *Am J Respir Cell Mol Biol* 12: 503–12

61 Ryan US, Zhong R, Hayes BA, Visner G, Sauther ML (1993) Regulation of endothelin-1 expression in normal and transfected endothelial cells. *J Cardiovasc Pharmacol* 22 (Suppl 8): S38–S41

62 Imai T, Hirata Y, Emori T, Yanagisawa M, Masaki T, Marumo F (1992) Induction of endthelin-1 gene by angiotensin and vasopressin in endothelial cells. *Hypertension* 19: 753–757

63 Prins BA, Hu R-M, Nazario B, Pedram A, Frank HJL, Weber MA, Levin ER (1994) Prostaglandin E2 and prostacyclin inhibit the production and secretion of endothelin from cultured endothelial cells. *J Biol Chem* 269: 11938–11944

64 Yoshizumi M, Kurihara H, Sugiyama T, Takaku F, Yanagisawa M, Masaki T, Yazaki Y (1989) Hemodynamic shear stress stimulates endothelin production by cultured endothelial cells. *Biochem Biophys Res Commun* 161: 859–864

65 Yu JCM, Davenport AP (1995) Secretion of endothelin-1 and endothelin-3 by human cultured vascular smooth muscle cells. *Br J Pharmacol* 114: 551–557

66 Zamora MA, Dempsey EC, Walchak SJ, Stelzner TJ (1993) BQ123, an ET$_A$ receptor antagonist, inhibits endothelin-1-mediated proliferation of human pulmonary artery smooth muscle cells. *Am J Respir Cell Mol Biol* 9: 429–433

67 Li H, Chen SJ, Chen Y-F, Meng QC, Durand J, Oparil S, Elton TS (1994) Enhanced endothelin-1 and endothelinal rceptor gene expression in chronic hypoxia. *J Appl Physiol* 77: 1451–1459

68 Stelzner TJ, O'Brien RF, Yanagisawa M, Sakurai T, Sato K, Webb S, Zamora M, McMurtry IF, Fisher JH (1992) Increased lung endothelin-1 production in rats with idiopathic pulmonary hypertension. *Am J Physiol* L614–620

69 Ono S, Westcott JY, Voelkel NF (1992) PAF antagonists inhibit pulmonary vascular re-modelling induced by hypobaric hypoxia in rats. *J Appl Physiol* 73: 1084–1092

70 Bonvallet ST, Zamora MR, Hasunuma K, Sato K, Hanasato N, Anderson D, Sato K, Stelz-ner TJ (1994) BQ123, an ET_A-receptor antagonist, attenuates hypoxic pulmonary hyperten-sion in rats. *Am J Physiol* 266: H1327–331

71 Giaid A, Stewart DJ, Michel RP (1993) Endothelin-1-like immunoreactivity in postob-structive pulmonary vasculopathy. *J Vasc Res* 30: 333–338

72 Kourembanas S, Marsden PA, McQuillan LP, Faller DV (1991) Hypoxia induces endothelin gene expression and secretion in cultured human endothelium. *J Clin Invest* 88: 1054–1057

73 Hassoun PM, Thappa V, Landman MJ, Fanburg BL (1992) Endothelin-1: mitogenic activi-ty on pulmonary artery smooth muscle cells and release from hypoxic endothelial cells. *Proc Soc Exp Biol Med* 199: 165–170

74 Wiebke JL, Montrose-Rafizadeh C, Zeitlein PL, Guggino WB (1992) Effect of hypoxia on endothelin-1 production by pulmonary vascular endothelial cells. *Biochim Biophys Acta* 1134: 105–111

75 Markewitz BA, Kohan DE, Michael JR (1995) Hypoxia decreases endothelin-1 synthesis by rat lung endothelial cells. *Am J Physiol* 269: L215–220

76 Cacoub P, Dorent R, Nataf P, Carayon A (1993) Endothelin-1 in pulmonary hypertension. *N Engl J Med* 329: 1967–1968

77 Stewart DJ, Levy RD, Cernacek P, Langleben D (1991) Increased plasma endothelin-1 in pulmonary hypertension: marker or mediator of disease? *Ann Intern Med* 114: 464–469

78 Giaid A, Yanagisawa M, Langleben D, Michel RP, Levy R, Shennib H, Kimura S, Masaki T, Duguid WP, Stewart DJ (1993) Expression of endothelin-1 in the lungs of patients with pulmonary hypertension. *N Engl J Med* 328: 1732–1739

79 Zhao YD, Springall DR, Hamid Q, Yacoub MH, Levene M, Polak JM (1995) Localization and characterization of endothelin-1 binding sites in the transplanted human lung. *J Car-diovasc Pharmacol* 26 Suppl 3: S336–340

80 Hagiwara H, Nagasawa T, Yamamoto T, Lodhi KM, Ito T, Takemura N, Hirose S (1993) Immu-nochemical characterization and localization of endothelin ET_B receptor. *Am J Physiol* 264: R777–783

81 Davenport AP, O'Reilly G, Molenaar P, Maguire JJ, Kuc RE, Sharkey A, Bacon CR, Ferro A (1993) Human endothelin receptors characterised using reverse transcriptase-polymerase chain reaction, *in situ* hybridisation, and subtype selective ligands BQ123 and BQ3020: evi-dence for expression of ET_B receptors in human vascular smooth muscle. *J Cardiovasc Pharmacol* 22 (Suppl 8): S22–25

82 Fukuroda T, Kobayashi M, Ozaki S, Yano M, Miyauchi T, Onizuka M, Sugishita Y, Goto K, Nishikibe M (1994) Endothelin receptor subtypes in human versus rabbit pulmonary ar-teries. *J Appl Physiol* 76: 1976–1982

83 McCulloch KM, Docherty CC, Morecroft I, MacLean MR (1996) Endothelin B receptor-mediated contraction in human pulmonary resistance arteries. *Br J Pharmacol* 119: 1125–1130

84 Bialecki RA, Fisher CS, Murdoch WW, Barthlow HG, Bertelsen DL (1997) Functional comparison of endothelin receptors in human and rat pulmonary artery smooth muscle. *Am J Physiol* 272: L211–218

85 Li H, Elton TS, Chen YF, Oparil S (1994) Increased endothelin receptor gene expression in hypoxic rat lung. *Am J Physiol* 266: L553–560

86 Eddahibi S, Springall DR, Mannan M, Carville C, Chabrier P-E, Levanne M, Raffestin B, Polak JM, Adnot S (1993) Dilator effect of endothelins in the pulmonary circulation: changes associated with chronic hypoxia. *Am J Physiol* 265: L571–580

87 Noguchi Y, Hislop AA, Haworth SG (1997) Influence of hypoxia on endothelin-1 binding sites in neonatal porcine pulmonary vasculature. *Am J Physiol* 272: H669–678

Pulmonary Actions of the Endothelins
ed. by R. G. Goldie and D. W. P. Hay
© 1999 Birkhäuser Verlag Basel/Switzerland

CHAPTER 4
Activity and Distribution of the Endothelin-Converting Enzyme in the Lung

Pedro D'Orléans-Juste[1], Jean-Philippe Gratton[1], Ghassan Bkaily[2] and Adel Giaid[3]

[1] *Department of Pharmacology and* [2] *Department of Anatomy and Cell Biology, Medical School, Université de Sherbrooke, Sherbrooke (Québec) J1H 5N4, Canada*
[3] *Department of Pathology, Montréal General Hospital, 1650 Cedar Avenue, Montréal (Québec) H3G 1A4, Canada*

1   Introduction
2   Endothelin Biosynthesis: General Role of ECE
3   ECE in the Lungs: Molecular Biology and Biochemistry
3.1   Molecular Biology
3.2   Biochemistry
3.2.1 Pulmonary Vascular ECE
3.2.2 Pulmonary Non-Vascular ECE
4   Monitoring ECE Activity in the Lungs
4.1   *In vitro* Pharmacology
4.2   *In vivo* Pharmacology
5   Pulmonary Modulation of Endothelin Plasma Levels
5.1   NO- or PGI_2-dependent Modulation
5.2   Clearance
6   Development of Selective ECE Inhibitors
7   Physiopathological Considerations
7.1   Transgenic Knockout Animals
7.2   Human Pathologies
8   Conclusion
      References

1. Introduction

Within the intricate capillary and bronchial network of the lungs are found important cellular sources for eicosanoids, cytokines, surfactant and the potent endothelium derived-vasoconstrictor, endothelin-1 [1–3]. In large mammals, the lungs contain only 12% of total blood volume at any given time yet the cross-sectional area of the pulmonary circulation represents 40% of that of the entire vasculature [4]. This would suggest that the pulmonary vascular endothelium has the potential capacity of producing and concentrating in the lungs a significant portion of the total vascular-derived endothelin.

Due to its endothelial origin and potent pressor effects [5], intense research activites have been devoted to unraveling a role for endothelin-1 in vascular dysfunctions [6–8]. However, many other cell types have now been reported to secrete the mature peptide. For example, epithelial cells derived from different parts of the respiratory tree, polymorphonuclear cells and smooth muscle have been shown to release measurable quantities of the peptide [9–12].

Because of the anatomical localization of two major cell types producing endothelin-1, namely endothelial and epithelial cells, it is not surprising that both the blood and the bronchial alveolar lavage fluids are normally targeted to assess an increase in the production of endothelin in vascular or airway pathological conditions (for review, see Battistini et al. [13]). It has been suggested that endothelin-1 may be involved in the etiology of debilitating diseases, such as primary pulmonary hypertension. Giaid and co-workers have demonstrated that at the stage of the disease where pulmonary blood vessels show morphological abnormalities (i.e., plexogenic pulmonary arteriopathy), a significant downregulation of nitric oxide synthase and an upregulation of endothelin-1 synthesis and/or endothelin receptors occur [14, 15]. Furthermore, research on endothelin receptors has led to the development of potent and orally-available antagonists, which have been shown to be beneficial in specific cardiovascular dysfunctions in man such as chronic congestive heart failure [16].

2. Endothelin Biosynthesis: General Role of ECE

Although less studied, of potential importance is the contribution of endothelin-converting enzyme (ECE) in cardiovascular and lung diseases. Endothelin-1, 2 and 3 are produced through the specific cleavage of three distinct precursors, namely big-endothelin-1, 2 and 3 [17]. The conversion of big-endothelin-1 to endothelin-1 following the hydrolysis of the Trp^{21}-Val^{22} bond results in the production of the 21 amino acid endothelin-1 and an inactive fragment big-endothelin-1 (22–38) [5]. Similar processing events occur in the biosynthesis of endothelin-2 and endothelin-3 [17]. However, the scissile bond Trp^{21}-Ile^{22} present in big-endothelin-3 primary structure is relatively less susceptible to hydrolysis, suggesting that ECE may show a relative selectivity for the endothelin-1 and 2 precursors. This latter point will be discussed further in Section 4 below. Figure 1 illustrates the possible sites of ECE activity in pulmonary vascular and non-vascular compartments. Although endothelin-1 is clearly identified in the various pulmonary structures and ultrastructures [3], the presence and role of the two other isoforms, endothelin-2 and 3, remain to be clarified in those tissues.

In the last 8 years, less than 50 papers have been devoted specifically to the biochemical, molecular and functional characteristics of ECE in the respiratory and vascular components of the pulmonary system. The relatively sparse literature on the putative roles of ECE in physiopathological

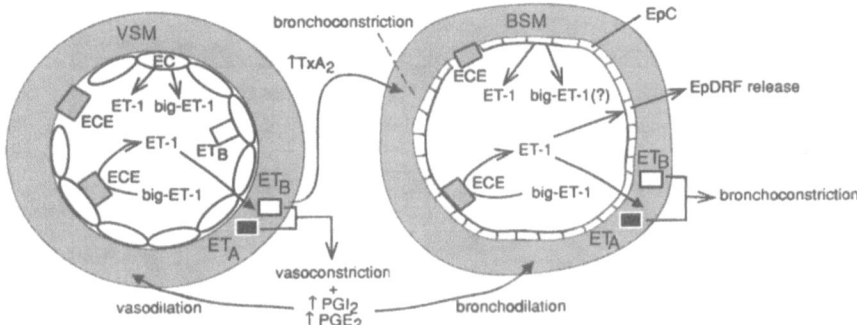

Figure 1   Schematic representation of the localization and role of ECE in the microcirculation (VSM) and the bronchioles (BSM) Endothelin-1 and big-endothelin-1 are both released by the endothelium and epithelium The secreted big-endothelin-1 is converted extracellularly by an ECE (membrane-bound ectopeptidase) Endothelin-1 may activate directly receptors on the endothelium (EC) or epithelium (Epi) to trigger the release of endothelium-derived relaxing factor (EDRF) or epithelium-derived relaxing factor (EpDRF) Furthermore, eicosanoids (PGI_2, PGE_2 or TxA_2) may be stimulated by endothelin-1 through specific receptors and act as modulators of both the microvascular tone or airway resistance

conditions may be explained, in part, by the fact that the lack of availability of potent and orally-available ECE inhibitors. Most of the compounds described to date have low potencies and/or have limited selectivity, affecting other enzymes such as neutral endopeptidase-24.11 [18].

3. ECE in the Lungs: Molecular Biology and Biochemistry

3.1. Molecular Biology

The first biochemical characterization of a purified ECE activity was reported by Takahashi et al. in 1993 from rat lung microsomes [19]. Since then, cDNA cloning studies revealed the existence of two types of ECE, namely ECE-1 and ECE-2 [20, 21]. Figure 2 illustrates the cDNA sequences of the three currently cloned isoforms, in addition to the relative affinities of the different ECEs for the precursors of the three isoforms of endothelins (big-endothelin-1, 2 and 3). ECE-1 exists in two different isoforms, ECE-1α and ECE-1β, according to their NH_2 terminal sequence resulting from alternative promoter controls [22]. ECE-1 has now been shown, using SDS-PAGE, to be a highly glycosylated membrane-bound protein of 130 kDa with one transmembrane region near the NH_2 terminus; rat ECE-1 shares 37% structural homology with rat neutral endopeptidase NEP-24.11 [20]. Current evidence suggests that ECE-1 exists in a dimeric disulfide linked form. The molecular weight of ECE-1 under reducing conditions is estimated to be 130 kDa but migrates at a 300 kDa band under non-reducing conditions [23, 24]. Interestingly, the K_m of the 300 kD dimer is not

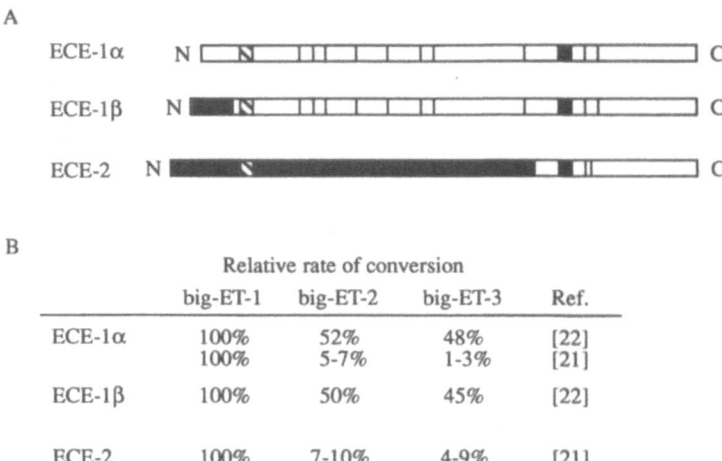

Figure 2. Schematic representation of the primary structures of the three isoforms for human ECE (ECE-1α, ECE-1β, ECE-2). Lines show potential N-glycosilation sites. Cross-hatched boxes show transmembrane domain and closed boxes show zinc-binding motif. Shaded areas represent regions of low sequence similarities. Table B illustrates, in percent of specific activity (big-endothelin-1 set at 100%), the relative efficacy of the different ECEs in converting the three isoforms of endothelin precursors, namely big-endothelin-1 (big-ET-1), big-endothelin-2 (big-ET-2) and big-endothelin-3 (big-ET-3).

significantly different from that of each of the disulphide linked monomers for their preferential substrate, big-endothelin-1 (1.6 and 1.2 µM, respectively), yet differ in terms of K_{cat} (1.01 and 3.21 min, respectively) (A. Jeng et al., personal communication).

ECE-2 is derived from a separate gene motif [21], although it retains 59% structural homology with ECE-1. However, within the COOH-terminal catalytic domain, the homology of ECE-2 with ECE-1 reaches 71%. The estimated amount of ECE-2 is much less than that of ECE-1, at least in endothelial cells where it was reported that the mRNA for ECE-2 represents 1 to 2% of that of ECE-1 [21]. Interestingly, ECE-2 shows optimal activity at acidic pH, in contrast to ECE-1 which is active preferentially at neutral pH. This suggests that ECE-2 might be located in an intracellular acidic compartment such as the *trans*-Golgi network [21]. The physiological role of ECE-2 in endothelin production remains unclear. On the other hand, ECE-1 appears to be the physiologically relevant entity in most endothelin-producing tissues including the lung. It has also been suggested that ECE-1α is the constitutively expressed isoform in contrast to ECE-1β which may act as the predominant enzymatic form expressed in pathological states [25]. The exact location of the various ECE isoforms in different pulmonary cell types remains to be elucidated.

3.2. Biochemistry

Approaches using lung homogenates have allowed the characterization of phosphoramidon-sensitive activity and partial characterization of ECE in cellular membranes and organelles of various species (Table 1). Characteristically, the dynamic conversion of big-endothelin-1 to its active metabolite is normally markedly reduced or abolished by phosphoramidon. Because of different methodological approaches used to obtain ECEs, variable molecular weights have been appointed to each of these entities. It is of interest that the crude ECE preparation reported by Sawamura et al. [26] possesses a molecular weight of around 300 kDa similar to that derived from the cloned ECE-1 in host carrier cells.

Very few studies have addressed directly the capacity of the lungs to produce endothelin-1. Some groups have also demonstrated the contribution of moieties other than the phosphoramidon-sensitive ECE in the production of endothelin in rat lungs. Indeed, Wypij et al. [27] demonstrated that rat perfused lungs release endothelin when stimulated by the mast cell degranulator, compound 48/80, via a phosphoramidon-insensitive but chymostatin-sensitive mechanism.

One of the main difficulties in biochemically studying the production of endothelin and the activity of ECE in intact lungs resides in the fact that many cell types (macrophages, endothelial cells, epithelial cells, airway and vascular smooth muscle cells) generate endothelin-1. Because of the approximately 40 different cell types found in the pulmonary system, immunohistochemical evidence is required to assess the predominant ECE-rich sites in the lung. Figure 3 illustrates the results obtained from various pulmonary structures of healthy donors. It has also been demonstrated that in idiopathic pulmonary fibrosis (IPF), there is a marked increase in the density of endothelin-1 and of ECE in the endothelium and

Table 1. *In vitro* lung preparations with phosphoramidon-sensitive ECE activity

Source	Species	Preparation	Approximate M.W. (kDa)	Inhibitors	Ref.
Lung homogenates	Rat	Membranes	–	Phosphoramidon	[75]
		Microsomes	130	Phosphoramidon	[32]
		Golgi apparatus	–	Phosphoramidon	[76]
	Guinea pig	Cytosol	38	None	[77]
		Membranes	162	Phosphoramidon	[78]
	Bovine	Membranes	–	Phosphoramidon	[79]
	Porcine	Membranes	300	Phosphoramidon	[26]
			65	Phosphoramidon	[26]
Lung cells	Guinea pig	Clara cells	–	Phosphoramidon	[11]
		Epithelial cells	–	Phosphoramidon	[77]
	Rat	Endothelial	130	Phosphoramidon	[80]

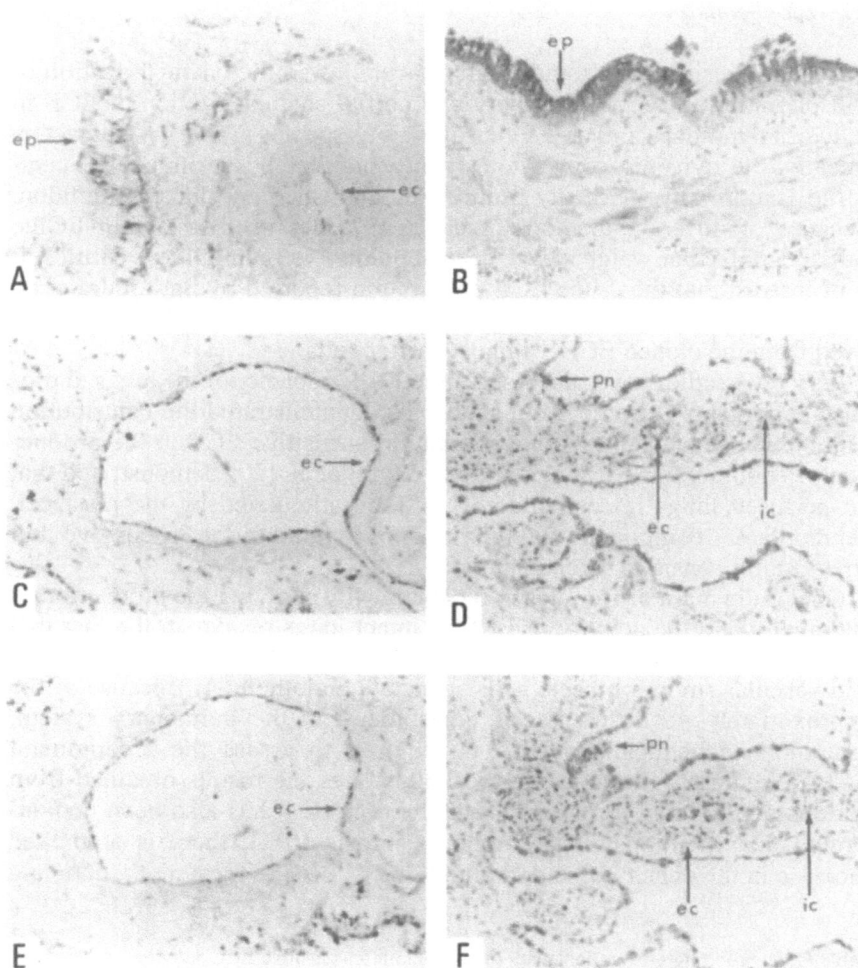

Figure 3. Immunostaining of pulmonary ultrastructures illustrating the presence of immunore-active (IR) ECE-1 or endothelin in epithelium (ep), endothelium (ec) pneumocytes (pn) and inflammatory cells (ic) in lung tissues of healthy subjects or IPF patients. Stronger immunore-activity for ECE-1 is shown in the epithelium (Fig. 3 B), microvascular endothelium, type II pneumocytes and inflammatory cells (Fig. 3 D) of IPF patients than in tissues of healthy sub-jects (Figs. 3 A and 3 C). Figure 3 F shows the high IR-endothelin density in microvascular endothelium, pneumocytes or inflammatory cells of IPF patients when compared to the endo-thelium of healthy tissues (Figure 3 E) (magnification Figs. 3 A, 3 B, 3 C and 3 E: × 450; Figs. 3 D and 3 F: × 500). Figure modified from [28], with permission.

smooth muscle of pulmonary microvasculature as well as in the epithelium of lower bronchioles, when compared with tissues obtained from healthy individuals [28]. Monitoring ECE activity may indeed be correlated with the immunostaining for endothelin-1 in both normal and pathological tissues. Interestingly, Saleh et al. [28] also reported an increase in immunostaining for ECE-1 and IR-endothelin in other cell types, such as macrophages and type II pneumocytes. An increase in the production of endothelin-1 or in ECE activity in these cell types may contribute to inflammatory states or cellular disorders leading to pulmonary fibrosis. Interestingly, enhanced activity of ECE, perhaps contributing to the development of fibrotic processes, has been demonstrated recently in patients with IPF (A. Giaid, personal communication).

3.2.1. Pulmonary vascular ECE: In the lungs, there seems to be a delicate equilibrium between the production of hormones and autacoids generated through the activity of ECE, nitric oxide synthase and cyclooxygenase [3, 29]. Indeed, interfering with the production of nitric oxide or prostaglandins will affect the pulmonary synthesis of endothelin as demonstrated in endotoxin shock and hypertensive animal models [30, 31].

Characteristically, the conversion of the intermediate precursor, big-endothelin-1, occurs through enzymatic activity of ECE-1 which is sensitive to phosphoramidon and has been shown to be a membrane-bound ectopeptidase located at the luminal surface of blood vessels [32]. It has been suggested that ECE-1 shows a relatively low conversion efficacy, as it is responsible for only 35% of the conversion of the intravenously injected precursor, big-endothelin-1 in the pig [33]. However, the kinetics of the purified enzyme may not be indicative of the true affinity of the native ECE in its isosteric conformation on the intact cell membrane. Although big-endothelin-1 is a significantly less potent constrictor than endothelin-1 in isolated vascular tissues (about 150 times less potent), the two peptides are almost equipotent *in vivo* [34]. The near equipressor properties of the precursor big-endothelin-1 and endothelin-1 *in vivo* may give a better indication of the true pharmacological efficacy of ECE. However the physiological role of ECE is best determined through the activity of the enzyme on the endogeneous production of endothelin-1. One must also take into consideration that elements other than the basic activity of endothelial and epithelial ECEs are involved in the control of plasma levels of endothelin. For example, shear stress or endothelial stretching [35] and other ECE-rich blood-borne cells (i.e. macrophages; [9]) are known to be involved in the generation of endothelins.

Early studies have also indirectly shown a functional role for the ECE in the pulmonary vasculature. For example, Vemulapalli et al. [36] demonstrated that ischemia/hypoxia induces a phosphoramidon-sensitive release of endothelin-1 and related increases in pulmonary insufflation pressure in perfused lungs of the guinea-pig.

3.2.2. Pulmonary non-vascular ECE: Characteristically, endothelin-1 is a more potent constrictor (at low concentrations) of human bronchi than big-endothelin-1, although both peptides possess the same intrinsic activity [37]. This would suggest that the conversion of big-endothelin-1 in non-vascular smooth muscle of the lung by the phosphoramidon-sensitive ECE-1 would be relatively more efficient than in vascular preparations *in vitro* [34]. Interestingly, although the pulmonary vascular ECE activity has been identified following i.v. administration of big-endothelin-1 [38], the direct bonchoconstrictive effect of the peptide administered by aerosol has not been studied. It is, therefore, not possible to confirm at the present time if there is a functional ECE on the epithelium or airway smooth muscle *in vivo*.

Conversely, some groups have demonstrated that non-specified epithelial cells were able to generate endothelin-1 [10, 89]. More recent work has identified that guinea pig Clara cells (non-ciliated epithelial cells) not only release endothelin-1, but seem also to possess a strong ECE activity which is sensitive to phosphoramidon (Table 1). These Clara cells, located in the epithelium of the trachea, upper and lower bronchi, preferentially convert exogenous big-endothelin-1 and 2 but not big-endothelin-3 [11]. In addition, Clara cells have, as another interesting characteristic, the capacity to release almost as much endothelin-1 as endothelial cells; this later property may suggest an important contribution of Clara cells in the overall epithelial-derived secretion (or excretion) of endothelin-1 in bronchoalveolar lavage fluids.

The glucocorticoid dexamethasone is a potent inhibitor of the endothelin-1 release from Clara and other lung epithelial cells [39, 40]. Interestingly, dexamethasone also reduces the concentration of immunoreactive endothelin in bronchoalveolar lavage fluids of asthmatic patients [41, 42]. The mechanism of the inhibitory effects of that glucocorticoid has yet to be clarified. It is, however, currently believed that glucocorticoids, such as budesonide and dexamethasone, enhance the NEP-dependent degrading activity in human epithelial cells [43]. This activity would reduce the actual concentration of endothelin in the bronchoalveolar lavage fluid. However, it is not known if glucocorticoids have a direct effect on the expression or function of the ECE *per se*. As the increased presence of big-endothelin-1 is a better marker of the reduced activity of ECE than is endothelin-1 [33], it would be useful to assess the effect of dexamethasone or other glucocorticoids in bronchoalveolar fluid levels of the precursor rather than the intermediate metabolite.

4. Monitoring ECE Activity in the Lungs

The putative role of ECE in the airways may be indirectly monitored through the phosphoramidon-sensitive pharmacological effects of big-endothelin-1. When injected intra-arterially in isolated perfused lungs of various animal species, big-endothelin-1 can either generate the release of prostacyclin, induce vasoconstriction or increase the resistance of the airways to mechanical ventilation. As previously mentioned, *in vivo* the intra-arterial administration of endothelin-1 or big-endothelin-1 may also have a marked effect on the pulmonary insufflation pressure in anaesthetized animals, such as the guinea pig [38, 44, 45]. Such approaches, summarized in Tables 2 and 3,

Table 2. Isolated preparations used to investigate the pharmacology of the phosphoramidon-sensitive endothelin-converting enzyme

Tissues	Species	Big-endothelin activity	Effect	Ref.
Pulmonary artery	Bovine (cultured smooth muscle cells)	Big-ET-1	$\uparrow$ IP$_3$	[81]
Parenchyma	Guinea pig	Big-ET-1 > 2 ▶ 3	Contraction	[82] [83]
	Rat	Big-ET-1	$\uparrow$ Vascular permeability	[84]
Bronchus	Guinea pig	Big-ET-1 > 2 ▶ 3	Contraction	[85]
	Rat	Big-ET-1	$\uparrow$ Vascular permeability	[84]
	Human	Big-ET-1	Contraction	[37]
Trachea	Rat	Big-ET-1	$\uparrow$ Vascular permeability	[84]
Isolated lung	Guinea pig	Big-ET-1 ▶ 2,3	$\uparrow$ PG,Tx	[44]
	Rabbit	Big-ET-1 > 2	Pressor effect	[46]
		Big-ET-1	$\uparrow$ PG	[86]
	Rat	Big-ET-1 ▶ 2,3	$\uparrow$ PG	[47]

IP$_3$: Inositol triphosphate; PG: prostacyclin; Tx: Thromboxane A$_2$.

Table 3. Qualitative summary of the *in vivo* effects of endothelins and their precursors

Peptides	Pressor Effect[1]			Bronchoconstrictor effect[2]			Increase in Plasma prostacyclin[3]	
	Rat [47]	Guinea pig [44]	Rabbit [46]	Rat [47]	Guinea pig [44]	Rabbit [46]	Rat [87]	Rabbit[4]
Endothelin-1	+++	+++	+++	ne	+++	ne	+++	+++
Endothelin-2	+++	+++	+++	ne	+++	ne	++	+++
Endothelin-3	++	++	ND	ne	+++	ne	++	ND
Big-endothelin-1	+++	+++	+++	ne	+++	ne	+	++
Big-endothelin-2	+++	+++	++	ne	+	ne	+	++
Big-endothelin-3	ne	–	ND	ne	–	ND	+	ND

[1] Sustained increase in mean arterial pressure.
[2] Expressed in change in pulmonary insufflation pressure.
[3] Expressed in ng/ml of 6-keto-PGF$_{1\alpha}$ platelet-poor plasma as detected by radioimmunoassay.
[4] Gratton et al. (unpublished results).
ne: no effect; ND: not determined.

have been exploited to assess the specificity characteristic of ECE for the various endothelin precursors, as described in the following two sections.

4.1. In vitro *Pharmacology*

When injected in the pulmonary circulation via the pulmonary artery, endothelin-1 triggers the release of eicosanoids from the rat, guinea pig and rabbit perfused lungs [29, 45, 46]. It was of interest to assess whether the precursors would trigger the release of these eicosanoids in the pulmonary circuits and whether that later effect would be sensitive to phosphoramidon. Table 2 illustrates the effects of the three endothelins and their respective precursors in the isolated perfused lungs of three animal species, namely the rabbit, rat and guinea pig, in terms of eicosanoid-releasing and/or constrictor effects. Characteristically, big-endothelin-1 induces a phosphoramidon-sensitive release of eicosanoids in the lungs of the rat and guinea pig [44, 47]. In contrast, although big-endothelin-2 is active as a pressor agent in the rat and guinea pig, big-endothelin-2 does not stimulate the release of either PGI_2 or thromboxane A_2 in the isolated perfused lungs of the two above-mentioned species [44, 47]. Furthermore, big-endothelin-2 is significantly less potent than big-endothelin-1 in inducing a phosphoramidon-sensitive increase in vascular resistance in the rabbit perfused lung [46] (Tab. 2).

4.2. In vivo *Pharmacology*

Table 3 summarizes the contribution of phosphoramidon-sensitive ECE in the *in vivo* pharmacological properties of big-endothelins in various animal species. Intravenous administration of big-endothelin-1 induces a phosphoramidon and indomethacin-sensitive increase in pulmonary insufflation pressure (PIP) in the anaesthetized guinea pig [38]. It was suggested that big-endothelin-1 may increase airway resistance through the release of thromboxane following conversion by the phosphoramidon-sensitive ECE. We demonstrated later that in the guinea pig lung, the ECE involved in the effect of big-endothelin-1 on PIP must be localized at or near the sites responsible for the generation of bronchoconstrictive eicosanoids [44].

In addition to its lack of eicosanoid-releasing effects in perfused lungs, big-endothelin-2 has also been shown to be inactive in increasing pulmonary insufflation pressure in the anaesthetized guinea pig [44]. It was suggested that big-endothelin-2, in contrast to big-endothelin-1, is converted at a site other than in the pulmonary vasculature since it was unable to induce an indomethacin-sensitive bronchoconstriction *in vivo* and to release either prostacyclin or thromboxane from the guinea pig pulmonary circulation. This profile suggests the presence of a different type of converting enzyme for big-endothelin-2, which remains to be characterized. Interestingly, most of the ECEs which have been recently cloned and functionally expressed show relatively low conversion efficacy against big-endothelin-2 when compared to big-endothelin-1 [20–22]. We have also recently shown that human big-endothelin-2 is a less potent vasoconstrictor than big-endothelin-1 in the pulmonary but not the renal vasculature of the rabbit [48]. On the other hand, big-endothelin-3, is inactive as a pressor agent or in the pulmonary circulation of rat guinea pig and rabbit [29, 44, 47], suggesting that the ECE present in those models does not recognize the precursor of endothelin-3.

As initially shown by Whittle et al. [49], endothelin-1 induces a direct bronchoconstrictive effect in the guinea pig when administered by aerosol. In contrast, the parenteral administration of the potent vasoactive peptide induces an increase in PIP which is mediated completely by constrictive eicosanoids [45]. The precursor, big-endothelin-1, when injected intravenously or by aerosol, possesses the same bronchoconstrictive characteristics as endothelin-1 [38]. Albeit currently unconfirmed, we suspect that, as with endothelin-1 [49], indomethacin would have no effect on the bronchoconstriction induced by aerosolized big-endothelin-1 in the guinea pig. This would provide evidence for the presence of a converting enzyme for endothelin, which would be localized in anatomically distinct areas in the airways and which would be responsible for a local (BAL fluid) increase in the production of endothelin. Hence, in physiological conditions, plasma and BAL fluid endothelin levels are generated separately via a similar ECE, predominantly localized as an ectopeptidase at the surfaces of pulmonary capillary endothelial cells and bronchial epithelium, respectively. However, in conditions where there is structural damage at the microcirculatory or alveolar levels, such as in IPF, cystic fibrosis or primary pulmonary hypertension, an exchange between the plasma and BAL fluid endothelin levels may occur.

5. Pulmonary Modulation of Endothelin Plasma Levels

Although the lungs contain both endothelin-1-producing endothelium and epithelium, they are not only an important site for the production of endothelin but also possess counter-regulatory mechanisms which may control the blood and BAL fluid levels of this potent vasoactive peptide.

5.1. NO- or PGI$_2$-dependent Modulation

As discussed earlier, shifts in the NO/endothelin-1 balance may play an important role in the etiology of primary pulmonary hypertension. It is established that NO released from the endothelium will increase its own intracellular cGMP levels [50]. This enhanced activity of guanylate cyclase will interfere with post-transcriptional (endothelin-1 mRNA) and secretory processes, resulting in a reduced production of endothelin-1 [51]. In conditions of NO-synthase impairment, endothelin-1 message and production are increased [51]. Although clearly illustrated for primary pulmonary hypertension (PPH), the mechanisms described above have not been determined in pulmonary hypertension secondary to degenerative processes occurring in various respiratory afflictions (cyctic fibrosis, congestive heart failure, etc).

Recent observations from our laboratories confirm the important role of NO in the control of circulating endothelin-1. Indeed, in the anaesthetized rabbit, an NO synthase inhibitor, L-NAME, induced both a phosphoramidon-sensitive pressor effect and increases in the plasma level of big-endothelin-1 but not endothelin-1 [52]. As the circulating plasma levels of big-endothelin-1 (pg/ml) are lower than the general K_m of the cloned ECE-1α, an NO-induced reduction in ECE activity is unlikely to explain the raised plasma levels of the precursor. In addition, the L-NAME-induced pressor response is sensitive to an ET_A receptor selective antagonist, BQ-123, suggesting a dynamic conversion of the precursor to endothelin-1 in the basolateral rather than luminal portion of the endothelium [52]. On the other hand, Langleben and co-workers [88] have demonstrated that in primary pulmonary hypertensive patients, the net endothelin-1 clearance improved following chronic treatment with prostacyclin. This would suggest that this prostanoid might be involved in the modulation of endothelin-1 release, perhaps via the increase in $[cAMP]_i$ in the endothelium and/or bronchiolar epithelium. It would be expected that both NO and PGI_2 would synergistically repress the production of endothelial cell-derived endothelin-1. It remains to be clarified if the activity of ECE *per se* is directly modulated by NO or PGI_2 *in vitro* and *in vivo*.

5.2. Clearance

Although not directly related to ECE activity, the contribution of the clearance receptors in the overall control of endothelin-1 plasma levels in the lung, is of importance. Because of the abundance of endothelial cells in the pulmonary circulation, the lungs have been frequently suggested as an important organ both in terms of synthesis and clearance of endothelin-1 [53]. With respect to the ratio of ET_A/ET_B receptors in human pulmonary vasculature, evidence to date suggests that the endothelium exclusively contains the latter and the smooth muscle the former receptor population [3, 54]. In the vascular circuits of the lungs, ECE-1 has been predominantly localized in the endothelium.

Fukuroda et al. [55] were the first to suggest a clearance of circulating endothelin-1 by ET_B receptors in rats. Intravascular administration of radiolabeled endothelin-1 was rapidly cleared from the circulation and retained in the lungs, among other tissues. An ET_B receptor antagonist, BQ-788 [56], markedly reduces the clearance rate at the pulmonary vascular level in the rat and the dog [55, 57]. These ET_B receptors have a very high affinity for their natural ligands, endothelin-1 and endothelin-3 [58]. Dupuis et al. [57] have also recently shown that the lungs are the major sites for the removal of endothelin-1 from the dog circulation. Using a well-defined dilution curve method, this group has demonstrated a significant contribution ($\approx 40\%$) of the lungs in retaining endothelin-1. It was suggested that pulmonary clearance of endothelin-1 was determined by ET_B but not ET_A receptors. This postulate for pulmonary clearance of endothelin-1 has been confirmed by Dupuis et al. [53] in healthy human subjects. However, interference with ET_B receptors affects plasma levels of not only endothelin-1, but also of big-endothelin-1 [52]. As big-endothelin-1 only poorly binds to rat lung tissues [59], we suggest that ET_B receptor antagonists, such as BQ-788, reduce not only the clearance of endothelin-1, but also the ET_B receptor-dependent release of NO [60], which consequently will reduce the retro-inhibition of big-endothelin-1 release or synthesis [52]. Interestingly, experimental pulmonary hypertension was recently associated to a reduced expression of ET_B clearance receptors [73, 90].

6. Development of Selective ECE Inhibitors

Of the currently used inhibitors, the best known is phosphoramidon [18, 25, 26, 37, 44] which is also a neutral endopeptidase and thermolysine inhibitor. More recently, molecules developed by Novartis (e.g., CGS 26303) have been shown to act as dual neutral endopeptidase and ECE inhibitors [61]. In addition, highly selective NEP inhibitors such as SQ-28603 [62] have also been shown to inhibit the big-endothelin-1-induced pressor effects in rats *in vivo* [62]. Besides the above-mentioned inhibitors, some groups have also shown that metal chelators [63], will interfere with the zinc-containing site of ECE. Although non-specific for ECE (metal chelators may affect other metalloendoproteases as well [18]), these compounds inhibit the conversion of big-endothelin-1 from semi-purified ECE in the low micromolar concentration range [63], and prevented big-endothelin-1-induced sudden death in mice [64].

An alternative approach is to design big-endothelin-1 analogs which occupy the ECE and renders the enzyme unavailable to process the precursors. Such false substrates include N-terminal truncated compounds which possess a Trp[21]-Phe[22] [65] or Trp[21]-D-Val[22] [66] hydrolysis site and the 15–16 amino acid C-terminal portion of big-endothelin-1. We suggest that these false substrates are hydrolized by ECE and generate a short N-terminal inactive tripeptide as well as an inactive C-terminal fragment (22–37) [65, 67]. We have also shown recently that such an analog of big-endothelin-1, [Phe^{22}]big-endothelin-1(19−37), was a potent inhibitor of big-endothelin-1-induced renal vasoconstriction in the rabbit as well as an inhibitor of big-endothelin-1-induced IP$_3$ hydrolysis in rat spinal cord slices [65, 67].

Development of these analogs is of interest since they may act preferentially in the systemic ECE without affecting the delicate balance between nitric oxide and endothelin in the pulmonary vascular circuit [48]. In the future, such compounds may be developed from truncated analogs of big-endothelin-2. The latter precursor, as described above, appears to be poorly converted in the pulmonary as opposed to the systemic circuit. However, peptidic false substrates of the ECE may have disadvantages compared to the more practical non-peptidic molecules in interfering with the enzymatic entity, especially with respect to pharmacodynamic profile and oral activity. Nonetheless, the development of truncated analogs of big-endothelin-2 may give us some interesting tool compounds and lead to the development of an ECE inhibitor which will more efficiently act in the systemic and renal circulation than at the pulmonary levels.

7. Physiopathological Considerations

7.1. Transgenic Knockout Animals

Full knockout of a single gene for endothelin-1 or the ET$_A$ receptor induces major craniofacial and cardiac defects in mice [68, 69]. Knockouts of ECE-1 generate the same cranofacial defects [69]. This latter observation points out to the vital contribution of ECE in the generation of endothelin from its precursor in mice. There is, however, no indication of important alterations of the lung structure in these homozygous knockout mice, either for endothelin-1, ET$_A$ or the ECE-1 gene repression.

On the other hand, Yanagisawa et al. [69] have recently disclosed that a complete knockout of the single gene for ECE-2 in transgenic mice was devoid of any detectable or life-threatening alterations in these animals. This would suggest that ECE-2 (localized intracellularly) unlike ECE-1, does not appear to be involved in the developmental properties of endothelin-1. It further suggests that of the ECEs which have been currently identified, only the ECE-1 has a vital contribution in the cardiovascular system and developmental mechanisms.

7.2. Human Pathologies

Increased expression of endothelin-1 has been described in IPF and cryptogenic fibrosis alveolitis (CFA) [28, 70]. However, the issue as to whether the processing of the precursor to its active metabolite is enhanced in these debilitating pulmonary diseases had to await the recent study by Saleh and co-workers which demonstrated that an elevated expression of ECE-1 was found in IPF. ECE-1 expression was markedly enhanced in airway epithelium, vascular endothelium, pneumocytes, macrophages and neutrophils, and in tissues from IPF patients [28].

In addition, there is an increasing body of evidence suggesting that proinflammatory cytokines, such as TNFα, TGFβ and IL-1, may play an important role in the production and/or expression of endothelins (for review, see Battistini et al. [71]). Interestingly, in human umbilical vein endothelial cells, Stankova et al. [72] have shown that endothelin-1 and interleukin-1β act synergistically in increasing the expression and secretion of another inflammatory cytokine, interleukin-6. The recent study by Saleh et al. [28] demonstrated that TNFα stimulates not only endothelin-1, but also ECE-1 mRNA and protein production in normal bronchial epithelial cells. As the same cell types show an elevated production of inflammatory cytokines in IPF, it was suggested by Saleh et al. [28] that there may be a positive cross-talk between proinflammatory factors and the expression and function of ECE-1 in the same cell types. Whether TNFα upregulates not only the expression but also the activity of ECE-1 in cells from the pulmonary vasculature remains to be determined.

It is also of interest that there appears to be cross-talk between the expression of endothelin and nitric oxide synthase in lungs of primary pulmonary hypertensive (PPH) patients [14, 15]. Indeed, an upregulation of endothelin expression and production occurs in vascular tissues of PPH patients, whereas NOS in the same tissues (in particular in the endothelial cells) is down-regulated. This would suggest that, as initially reported by Boulanger and Luscher [51], nitric oxide, through the increase of intracellular cyclic GMP, will interfere with the expression and/or release of endothelin-1 from endothelial cells, not only in the systemic but in the pulmonary circulation as well. Furthermore, in patients with severe PPH, the impairment in the NO pathway may result in the upregulation of endothelin-1 which, as a potent vasoconstrictor, would enhance vascular resistance in those subjects.

It remains to be confirmed if there is a cross-talk between the vascular and non-vascular compartments in the production of endothelin-1. Indeed, in asthma or pulmonary hypertension, it has to be clarified if the overproduction of endothelin (through the enhanced activity of the ECE) will result in a cross-overspilling of the potent vasoactive factor within the airway or vascular compartments. This could be investigated by injecting intravenously relatively high concentrations of big-endothelin and subsequently monitoring the BAL fluid level of the precursor or its active metabolite in normoxic or hypoxic conditions. One possibility is that any cross-overspilling which occurs may be due to deterioration of the hemato-alveolar barrier rather than an active diffusion of the peptide from one compartment to the other in the pulmonary system of IPF, PPH or CFA patients.

8. Conclusion

Due to the high structural complexity of the pulmonary system and the large number of different cell types which may be involved in the production of endothelin, the contribution of ECE in homeostasis or in physiopathological conditions remains a difficult topic to study. ECE (in particular, ECE-1α) is predominantly localized as an ectopeptidase oriented towards the luminal surface of either the pulmonary microvessels (endothelial cells) or bronchioles (epithelial cells). Because of the large quantity of endothelial cells (i.e., high density of ET_B clearance receptors), one must take into account the possible modulatory role of endothelial-derived vasoactive and growth factors (EDRF, NO, PGI_2, IL-6) as possible modulators of the activity of ECE. It is, however, not yet known if these other vasoactive factors directly interfere with ECE activity. On the other hand, it is worthy of mention that patients with anti-endothelial cell antibodies show an increase in circulating ir-endothelin-1 and thus a reduction in the overall ET_B receptor-dependent clearance of the vasoactive peptide [74]. Thus, in some pathological conditions, an increase in plasma endothelin levels may not be necessarily directly related to an enhancement of the ECE activity.

Nonetheless, evidence of an increase in production of endothelin-1 and in expression of the ECE, in pulmonary disorders, supports the continued exploration of the role of ECE in pulmonary diseases as an important area of research. Another interesting point to address would be to compare the efficacies of selective or mixed endothelin receptor antagonists against ECE inhibitors in animal models reproducing some of the pulmonary dysfunctions discussed in this chapter. Ultimately, clinical trials for these pathological conditions will determine if there is a clear advantage in using one class of compounds over the other two.

Acknowledgements

The authors wish to acknowledge H. Morin for secretarial assistance. P.D.J. and A.G. are scholars of the Fonds de la recherche en santé du Québec (F.R.S.Q.). J.P.G. is in receipt of a joint studentship from the Fonds des chercheurs et aide à la recherche (F.C.A.R.) and F.R.S.Q. This project was financially supported by the Medical Research Council of Canada (M.R.C.C.) and the Heart and Stroke Foundation of Canada (H.S.F.C.).

References

1 Blackwell GJ, Flower RJ, Nijkamp FP, Vane JR (1978) Phospholipase A_2 activity of guinea-pig isolated perfused lungs: stimulation, and inhibition by anti-inflammatory steroids. *Br J Pharmacol* 62: 79–89

2 Bry K, Lappalainen U, Hallman M (1996) Cytokines and production of surfactant components. *Semin Perinatol* 20: 194–205

3 Barnes PJ (1994) Endothelins and pulmonary diseases. *J Appl Physiol* 77: 1051–1059

4 Milner WR (1982) Hemodynamics. *Williams & Wilkins*, Baltimore.

5 Yanagisawa M, Kurihara H, Kimura S, Tomobe Y, Kobayashi M, Mitsui Y, Yazaki Y, Goto K, Masaki T (1988) A novel potent vasoconstrictor peptide is produced by vascular endothelial cells. *Nature* 332: 411–415

6 Ferri C, Bellini C, de Angelis C, de Siati L, Perrone A, Properzi G, Santucci A (1995) Circulating endothelin-1 concentrations in patients with chronic hypoxia. *J Clin Pathol* 48: 519–524

7 Feng CS, Huang NQ, Zeng Q (1993) Determination of plasma endothelin-1 in aged patients with chronic obstructive pulmonary disease in its clinical significance. *Chung Hua Chieh Ho Ho Hu Hsi Tsa Chih* 16: 287–289

8 Schersten H, Aarnio P, Burnett JC jr, McGregor CG, Miller VM (1994) Endothelin-1 in bronchoalveolar lavage during rejection of allotransplanted lungs. *Transplantation* 57: 159–161

9 Sessa WC, Kaw S, Hecker M, Vane JR (1991) The biosynthesis of endothelin-1 by human polymorphonuclear leukocytes. *Biochem Biophys Res Commun* 174: 613–618

10 Black PN, Ghatei MA, Takahashi K, Bretherton-Watt D, Krausz T, Dollery CT, Bloom SR (1989) Formation of endothelin by cultured airway epithelial cells. *FEBS Lett* 255: 129–132

11 Laporte J, D'Orléans-Juste P, Sirois P (1996) Guinea pig clara cells secrete endothelin-1 through a phosphoramidon-sensitive pathway. *Am J Respir Cell Mol Biol* 14: 356–362

12 Takeda Y, Itoh Y, Yoneda T, Miyamori I, Takeda R (1993) Cyclosporine A induces endothelin-1 release from cultured rat vascular smooth muscle cells. *Eur J Pharmacol* 223: 299–301

13 Battistini B, D'Orléans-Juste P, Sirois P (1993) Endothelin involvement in physiopathologies: Circulating plasma levels and presence in biological fluids. *Lab Invest* 68: 600–628

14 Giaid A, Saleh D (1995) Reduced expression of endothelial nitric oxide synthase in the lungs of patients with pulmonary hypertension. *N Engl J Med* 333: 214–221

15 Giaid A, Yanagisawa M, Langleben D, Michel RP, Levy R, Shennib H, Kimura S, Masaki T, Duguid WP, Stewart DJ (1993) Expression of endothelin-1 in the lungs of patients with pulmonary hypertension. *N Engl J Med* 328: 1732–1739

16 Sutsch G, Bertel O, Kiowski W (1996) Acute and short-term effects of the nonpeptide endothelin-1 receptor antagonist bosentan in humans. *Cardiovasc Drugs Ther* 10: 717–725

17 Inoue A, Yanagisawa M, Kimura S, Kasuya Y, Miyauchi T, Goto K, Masaki T (1989) The human endothelin family: Three structurally and pharmacologically distinct isopeptides predicted by three separate genes. *Proc Natl Acad Sci USA* 86: 2863–2867

18 McMahon EG, Palomo MA, Moore WM, McDonald JF, Stern MK (1991) Phosphoramidon blocks the pressor activity of porcine big endothelin-1-(1-39) *in vivo* and conversion of big endothelin-1-(1-39) to endothelin-1-(1-21) *in vitro*. *Proc Natl Acad Sci USA* 88: 703–707

19 Takahashi M, Matsumura Y, Iijma Y, Tanzawa K (1993) Purification and characterization of endothelin converting enzyme from rat lung. *J Biol Chem* 268: 21394–21398

20 Xu D, Emoto N, Giaid A, Slaughter CKS, deWit D, Yanagisawa M (1994) ECE-1: A membrane-bound metalloprotease that catalyzes the proteolytic activation of big-endothelin-1. *Cell* 78: 473–485

21 Emoto N, Yanagisawa M (1995) Endothelin-converting enzyme-2 is a membrane-bound, phosphoramidon-sensitive metalloprotease with acidic pH optimum. *J Biol Chem* 270: 15262–15268

22 Shimada K, Takahashi M, Ikeda M, Tanzawa K (1995) Identification and characterization of two isoforms of an endothelin-converting enzyme-1. *FEBS Lett* 371: 140–144

23 Barnes K, Murphy LJ, Takahashi M, Tanzawa K, Turner AJ (1995) Localization and biochemical characterization of endothelin-converting enzyme. *J Cardiovasc Pharmacol* 29: S37–S39

24 Shimada K, Takahashi M, Turner AJ, Tanzawa K (1996) Rat endothelin-converting enzyme-1 forms a dimer through Cys^{412} with a similar catalytic mechanism a distinct substrate binding mechanism compared with neutral endopeptidase 24-11. *Biochem J* 315: 863–867

25 Turner AJ, Tanzawa K (1997) Mammalian membrane metallopeptidases: NEP, ECE, KELL, and PEX. *FASEB J* 11: 355–364

26 Sawamura T, Shinmi O, Kishi N, Sugita Y, Yanagisawa M, Goto K, Masaki T, Kimura S (1993) Characterization of phosphoramidon-sensitive metalloproteinase with endothelin-converting enzyme activity in porcine lung membrane. *Biochem Biophys Acta* 1161: 295–302

27 Wypij DM, Nichols JS, Novak PJ, Stacy DL, Berman J, Wiseman JS (1992) Role of mast cell chymase in the extracellular processing of big-endothelin-1 to endothelin-1 in the perfused lung. *Biochem Pharmacol* 43: 845–853

28 Saleh D, Furukawa K, Tsao MS, Maghazachi A, Corrin B, Yanagisawa M, Barnes PJ, Giaid A (1997) Elevated expression of endothelin-1 and endothelin-converting enzyme-1 in idiopathic pulmonary fibrosis: possible involvement of proinflammatory cytokines. *Am J Respir Cell Mol Biol* 16: 187–193

29 de Nucci G, Thomas R, D'Orléans-Juste P, Antunes E, Walder C, Warner TD, Vane JR (1988) Pressor effects of circulating endothelin are limited by its removal in the pulmonary circulation and by the release of prostacyclin and endothelium-derived relaxing factor. *Proc Natl Acad Sci USA* 85: 9797–9800

30 Weitzberg E, Lundberg JM, Rudehill A (1995) Inhibitory effects of diclofenac on the endotoxin shock response in relation to endothelin turnover in the pig. *Acta Anaesthesiol Scand* 39: 50–59

31 Oka M, Hasunuma K, Webb, Stelzner TJ, Rodman DM, McMurty IF (1993) EDRF suppresses an unidentified vasoconstrictor mechanism in hypertensive rat lungs. *Am J Physiol* 264: L587–L597

32 Takahashi M, Fukuda K, Shimada K, Barnes K, Turner AJ, Ikeda M, Koike H, Yamamoto Y, Tanzawa K (1995) Localization of rat endothelin-converting enzyme to vascular endothelial cells and some secretory cells. *Biochem J* 311: 657–665

33 Hemsen A, Pernow J, Lundberg JM (1991) Regional extraction of endothelins and conversion of big endothelin to endothelin-1 in the pig. *Acta Physiol Scand* 141: 325–334

34 Kashiwabara T, Inagaki Y, Ohta H, Iwamatsu A, Nozimu M, Morita A, Nishikori K (1989) Putative precursors of endothelin have less vasoconstrictor activity *in vitro* but a potent pressor effect *in vivo*. *FEBS Lett* 247: 73–76

35 Macarthur H, Warner TD, Wood EG, Corder R, Vane JR (1994) Endothelin-1 release from endothelial cells in culture is elevated both acutely and chronically by short periods of mechanical stretch. *Biochem Biophys Res Commun* 200: 395–400

36 Vemulapalli S, Rivelli M, Chiu PJ, delPrado M, Hey JA (1992) Phosphoramidon abolishes the increase in endothelin-1 release induced by ischemia-hypoxia in isolated perfused guinea-pig lungs. *J Pharmacol Exp Ther* 262: 1062–1069

37 Advenier C, Lagente V, Zhang Y, Naline E (1992) Contractile activity of big endothelin-1 on the human isolated bronchus. *Br J Pharmacol* 106: 883–887

38 Pons F, Touvay C, Lagente V, Mencia-Huerta JM, Braquet P (1992) Involvement of phosphoramidon-sensitive endopeptidase in the processing of big-endothelin-1 in the guinea-pig. *Eur J Pharmacol* 217: 65–70

39 Laporte J, D'Orléans-Juste P, Singh G, Sirois P (1995) Dexamethasone and phosphoramidon inhibit endothelin release by cultured nonciliated bronchiolar epithelial (Clara) cells. *J Cardiovasc Pharmacol* 26: S53–S55

40 Calderon E, Gomez-Sanchez CE, Cozza EN, Zhou M, Coffey RG, Lockey RF, Prockop LD, Szentivanyi A (1994) Modulation of endothelin-1 production by a pulmonary epithelial cell line. I. Regulation by glucocorticoids. *Biochem Pharmacol* 48: 2065–2071

41 Mattoli S, Soloperto M, Marini M, Fasoli A (1991) Levels of endothelin in the bronchoalveolar lavage fluid of patients with symptomatic asthma and reversible airflow obstruction. *J Allergy Clin Immunol* 88: 376–384

42 Redington AE, Springall DR, Ghatei MA, Lau LC, Bloom SR, Holgate ST, Polak JM, Howart PH (1995) Endothelin in bronchoalveolar lavage fluid and its relation to airflow obstruction in asthma. *Am J Resp Crit Care Med* 151: 1034–1039.

43 Borson DB, Gruenert DC (1991) Glucocorticoids induce neutral endopeptidase in transformed human tracheal epithelial cells. *Am J Physiol* 260: L83–L89

44 Gratton JP, Rae GA, Claing A, Télémaque S, D'Orléans-Juste, P (1995) Different pressor and bronchoconstrictor properties of human big-endothelin-1, 2 (1-38) and 3 in ketamine/xylazine-anaesthetized guinea-pigs. *Br J Pharmacol* 114: 720–726

45 Pons F, Touvay C, Lagente V, Mencia-Huerta JM, Braquet P (1991) Comparison of the effect of intra-arterial and aerosol administration of endothelin-1 (ET-1) in the guinea-pig isolated lung. *Br J Pharmacol* 102: 791–796

46 Gratton JP, Maurice MC, D'Orléans-Juste P (1995) Characterization of endothelin receptors and endothelin-converting enzyme activity in the rabbit lung. *J Cardiovasc Pharmacol* 26: S88–S90

47 Gratton JP, Maurice MC, Rae GA, D'Orléans-Juste P (1995) Pharmacological properties of endothelins and big-endothelins in ketamine/xylazine or urethane anesthetized rats. *Am J Hypertens* 8: 1121–1127

48 D'Orléans-Juste P, Gratton JP, Bkaily G, Claing A (1996) L'endothéline: de la pharmacologie moléculaire à ses implications en physiopathologie. *Méd Sci* 12: 563–574

49 Whittle BJ, Payne AN, Esplugues JV (1989) Cardiopulmonary and gastric ulcerogenic actions of endothelin-1 in the guinea pig and rat. *J Cardiovasc Pharmacol* 13: S103–S107

50 Ignarro LJ (1996) Physiology and pathophysiology of nitric oxide. *Kidney Int* 55: S2–S5

51 Boulanger C, Luscher TF (1990) Release of endothelin form the porcine aorta. Inhibition by endothelium-derived nitric oxide. *J Clin Invest* 85: 587–590

52 Gratton JP, Cournoyer G, Löffler BM, Sirois P, D'Orléans-Juste P (1997) ET_B receptor and nitric oxide synthase blockades induce BQ-123-sensitive pressor effects in the anaesthetized rabbit. *Hypertension* 30: 1204–1209

53 Dupuis J, Stewart DJ, Cernacek P, Gosselin G (1996) Human pulmonary circulation is an important site for both clearance and production of endothelin-1. *Circulation* 94: 1578–1584

54 Russell FD, Davenport AP (1995) Characterization of endothelin receptors in the human pulmonary vasculature using bosentan, SB 209670 and 97–139. *J Cardiovasc Pharmacol* 26: S346–S347

55 Fukuroda T, Fujikawa T, Ozaki S, Ishikawa K, Yano M, Nishikibe M (1994) Clearance of circulating endothelin-1 by ET_B receptors in rats. *Biochem Biophys Res Commun* 199: 1461–1465

56 Ishikawa K, Ihara M, Noguchi K, Mase T, Mino N, Saeki T, Fukuroda T, Fukami T, Ozaki S, Nagase T et al (1994) Biochemical and pharmacological profile of a potent and selective endothelin B-receptor antagonist, BQ-788. *Proc Natl Acad Sci USA* 91: 4892–4896

57 Dupuis J, Goresky CA, Fournier A (1996) Pulmonary clearance of circulating endothelin-1 in dogs *in vivo*: exclusive role of ET_B receptors. *J Appl Physiol* 81: 1510–1515

58 Sakurai T, Yanagisawa M, Takuwa Y, Miyazaki H, Kimura S, Goto K, Masaki T (1990) Cloning of a cDNA encoding a non-peptide-selective subtype of the endothelin receptor. *Nature* 348: 732–735

59 Neuser D, Steinke W, Dellweg H, Kazda S, Stasch JP (1991) [125I]-endothelin-1 and [125I]-big endothelin-1 in rat tissues: autoradiographic localization and receptor binding. *Histochemistry* 95: 621–628

60 Warner TD, Mitchell JA, de Nucci G, Vane JR (1989) Endothelin-1 and endothelin-3 release EDRF from isolated arterial vessels of the rat and rabbit. *J Cardiovasc Pharmacol* 13: S85–S88

61 De Lombaert S, Ghai RD, Jeng AY, Trapani AJ, Webb RL (1994) Pharmacological profile of a non-peptidic dual inhibitor of neutral endopeptidase 24.11 and endothelin-converting enzyme. *Biochem Biophys Res Communun* 204: 407–412

62 Gardiner SM, Kemp PA, Bennett T (1992) Effects of the neutral endopeptidase inhibitor, SQ 28,603, on regional haemodynamic responses to atrial natriuretic peptide or proendothelin-1 [1–38] in concious rats. *Br J Pharmacol* 106: 180–186

63 Ashizawa N, Okumura H, Kobayashi F, Aotsuka T, Takahashi M, Asakura R, Arai K, Matsuura A (1994) Inhibitory activities of metal chelators on endothelin-converting enzyme. I. *In vitro* studies. *Biol Pharm Bull* 17: 207–211

64 Ashizawa N, Okumura H, Kobayashi F, Aotsuka T, Asakura R, Arai K, Ashikawa N, Matsuura A (1994) Inhibitory activities of metal chelators on endothelin-converting enzyme. II. *In vivo* studies. *Biol Pharm Bull* 17: 212–216

65 Claing A, Neugebauer W, Yano M, Rae G, D'Orléans-Juste P (1995) [Phe22]-big endothelin-1 [19–37]: a new and potent inhibitor of the endothelin-converting enzyme. *J Cardiovasc Pharmacol* 26: S72–S74

66 Morita A, Nomizu M, Okitsu M, Horie K, Yokogoshi H, Roller PP (1994) D-Val22 containing human big endothelin-1 analog, [D-Val22]Big ET-1 [16–38], inhibits the endothelin converting enzyme. *FEBS Lett* 353: 84–88

67 Poulat P, de Champlain J, D'Orléans-Juste P, Couture R (1996) Receptor and mechanism that mediate endothelin- and big endothelin-1-induced phosphoinositide hydrolysis in the rat spinal cord. *Eur J Pharmacol* 315: 327–334

68 Kurihara K, Kurihara H, Suzuki H, Kodama T, Maemura K, Nagai R et al (1994) Elevated blood pressure in craniofacial abnormalities in mice deficient in endothelin-1. *Nature* 368: 703–710

69 Yanagisawa M (1995) Molecular genetic dissection of the endothelin system. *C.O.E. International Symposium* (Osaka, Japan) (Abstract 0–21)

70 Giaid A, Michel RP, Stewart DJ, Sheppard M, Corrin B, Hamid Q (1993) Expression of endothelin-1 in lungs of patients with cryptogenic fibrosing alveolitis. *Lancet* 341: 1550–1554
71 Battistini B, Forget MA, Laight D (1996) Potential roles for endothelins in systemic inflammatory response syndrome with particular relationship to cytokines. *Shock* 5: 167–183
72 Stankova J, D'Orléans-Juste P, Rola-Pleszczynski M (1996) ET-1 induces IL-6 gene expression in human umbilical vein endothelial cells – synergistic effect of IL-1. *Am J Physiol* 271: C1073–C1078.
73 Ivy DD, Le Cras, Horan MP, Abman SH (1998) Increased lung prepro ET-1 and decreased ET_B-receptor gene expression in fetal pulmonary hypertension. *Am J Physiol* 274: L535–L541
74 Filep JG, Bodolay E, Sipka S, Gyimesi E, Csipo I, Szegedi G (1995) Plasma endothelin correlates with antiendothelial antibodies in patients with mixed connective tissue disease. *Circulation* 92: 2969–2974
75 Matsumura Y, Umekawa T, Kawamura H, Takaoka M, Robinson PS, Cook ND, Morimoto S (1992) A simple method for measurement of phosphoramidon-sensitive endothelin converting enzyme activity. *Life Sci* 51: 1603–1611
76 Gui G, Xu D, Emoto N, Yanagisawa M (1993) Intracellular localization of membrane-bound endothelin-converting enzyme from rat lung. *J Cardiovasc Pharmacol* 22: S53–S56
77 Shima M, Yamanouchi M, Omori K, Sugiura M, Kawashima K, Sato T (1995) Endothelin-1 production and endothelin converting enzyme expression by guinea pig airway epithelial cells. *Biochem Mol Biol Int* 37: 1001–1010
78 Shima H, Kawashima Y, Ohmori K, Sugiura M, Kawashima K (1994) Endothelin-converting enzymes in guinea-pig lung membrane fractions: purifications and characterizations. *Biochem Mol Biol Int* 34: 1227–1234
79 Kundu GC, Wilson IB (1992) Identification of endothelin converting enzyme in bovine lung membranes using a new fluorogenic substrate. *Life Sci* 50: 965–970
80 Shimada K, Takahashi M, Tanzawa K (1994) Cloning and functional expression of endothelin-converting enzyme from rat endothelial cells. *J Biol Chem* 269: 18275–18278
81 Kent A, Keenan AK (1995) Evidence for signalling by big endothelin-1 via conversion to endothelin-1 in pulmonary artery smooth muscle cells. *Life Sci* 57: 1191–1196
82 Lebel N, D'Orléans-Juste P, Fournier A, Sirois P (1996) Role of neutral endopeptidase 24.11 in the conversion of big-endothelins in guinea pig lung parenchyma. *Br J Pharmacol* 117: 184–188
83 Battistini B, Brown M, Vane JR (1995) Selective proteolitic activation and degradation of ETs and Big-ETs in parenchymal strips of the guinea-pig lung. *Biochem Biophys Res Commun* 207: 675–681
84 Lehoux S, Plante GE, Sirois MG, Sirois P, D'Orléans-Juste P (1992) Phosphoramidon blocks big-endothelin-1 but not endothelin-1 enhancement of vascular permeability in the rat. *Br J Pharmacol* 107: 996–1000
85 Lebel N, D'Orléans-Juste P, Fournier A, Sirois P (1995) Characterization of the endothelin-converting enzyme in guinea-pig upper bronchus. *J Cardiovasc Pharmacol* 26: S81–S83
86 D'Orléans-Juste P, Lidbury PS, Télémaque S, Warner TD, Vane JR (1991) Human big-endothelin releases prostacyclin *in vivo* and *in vitro* through a phosphoramidon-sensitive conversion to endothelin-1. *J Cardiovasc Pharmacol* 17: S251–S255
87 Mattera GG, Catalioto RM, Criscuoli M, Subissi A (1994) Endothelins induce prostacyclin release in both vascular and non-vascular tissues. *Nauyn Schmied Arch Pharmacol* 350: 410–415
88 Langleben D, Long WA, Barst RJ, Tapson VF, Groves M, Badesch DB, Bourge RC, Murali S, Ettinger N, Shalit E, Stewart DJ & The North American Primary Pulmonary Hypertension Study Group (1995) Prostacyclin infusion improves the balance between pulmonary clearance and production of endothelin-1 in patients with primary pulmonary hypertension. *Circulation* 92: I–241
89 Mattoli S, Mezzetti M, Riva G, Allegra L, Fasoli A (1990) Specific binding of endothelin on human bronchial smooth muscle cells in culture and secretion of endothelin-like material from bronchial epithelial cells. *Am J Respir Cell Mol Biol* 3 (2) 145–151
90 Gosselin R, Gutkowska J, Baribeau J, Penneault T (1997) Endothelin receptor changes in hypoxia-induced pulmonary hypertension in the newborn Piglet. *Am J Physiol* 273: L72–L79

Pulmonary Actions of the Endothelins
ed. by R. G. Goldie and D. N. P. Hay
© 1999 Birkhäuser Verlag Basel/Switzerland

CHAPTER 5
ET Receptor-Linked Signal Transduction Processes in the Airway Wall

Peter J. Henry

Department of Pharmacology, University of Western Australia, Stirling Hwy, Nedlands 6907, Western Australia, Australia

1   Introduction
2   Airway Smooth Muscle
2.1 Contraction
2.1.1 Direct increases in cytosolic calcium
2.1.2 Diacylglycerol and protein kinase C
2.1.3 Sodium-hydrogen antiporter
2.1.4 Phospholipase A_2 and arachidonic acid metabolites
2.1.5 Chloride channels
2.2 Relaxation
2.3 Mitogenesis
3   Epithelial Cells
4   Submucosal Glands
5   Bronchial Circulation, Microvascular Leakage and Oedema
6   Cholinergic Nerves
7   Summary
     References

1. Introduction

The vascular endothelium and airway epithelium are well-documented cellular sources of ET-1. The release of ET-1 from epithelial cells appears to be towards the submucosal surface and into the airway wall, since the ET-1 content of the basal side of airway epithelial cells is many fold higher than that of the apical side [1]. Due to its close proximity to the airway epithelium, the underlying airway smooth muscle is likely to be an important target within the airway wall for epithelium-derived ET-1. Additional studies [2], which have assessed the topographical location of vascular endothelial cells and the ability of these cells to generate ET-1, suggest that endothelium cell-derived ET-1 might also exert significant effects on airway smooth muscle. The induction of a strong and long-lasting contraction is the most extensivefy studied action of ET-1 on airway smooth muscle, although it has been reported to bring about other actions including relaxation and mitogenesis. Furthermore, ET-1 significantly affects the functions of other important structures within the airway wall, causing microvascular leakage and oedema, potentiation of cholinergic nerve-mediated contraction, stimulation of mucous gland secretion, as well as exerting wide-ranging effects on the airway epithelium [3]. Each of these effects is produced via the interaction of ET-1 with specific ET receptors and the activation of regulatory G proteins and intracellular signal transduction processes. Interestingly, recent studies suggest that ET-1, via ET receptors, can activate a wide range of quite different signal transduction pathways. By way of illustration, ET-1 has been shown in many cell types to stimulate the phosphoinositide pathway leading to the release of intracellular calcium, as well as to promote the influx of extracellular calcium through plasma membrane channels. In other instances, the signal transduction pathway activated by ET-1 has been shown to involve enzyme systems (e.g., protein kinase C, phospholipase A_2, phospholipase D, protein tyrosine kinase, adenylate cyclase and guanylate cyclase), ion transporters (e.g., Na^+-H^+ exchange) and ion channels (e.g., chloride channels). This chapter provides an overview of our current, albeit incomplete, understanding of the signal transduction processes which link ET receptors to cellular responses in the airway wall.

2. Airway Smooth Muscle

2.1. Contraction

2.1.1. Direct increases in cytosolic calcium: Isometric tension recording studies demonstrate that ET-1 is a potent spasmogen of airway smooth muscle from the human and many animal species. Contraction of airway smooth muscle induced by ET-1, like several other spasmogens, is accompanied by an increase in intracellular Ca^{2+} concentration ($[Ca^{2+}]_i$), usually consisting of two phases. Firstly, there is a rapid initial increase in $[Ca^{2+}]_i$ resulting from the mobilisation of Ca^{2+} from intracellular stores, presumably the sarcoplasmic reticulum. The second phase is dependent on the influx of extracellular Ca^{2+} through transmembrane calcium channels. An important consequence of increases in $[Ca^{2+}]_i$ in smooth muscle cells is the activation of Ca^{2+}/calmodulin-dependent myosin light chain kinase (MLCK) and phosphorylation of the 20-kDa myosin light chains (LC_{20}), resulting in an increase in myosin ATPase activity and cross-bridge cycling (for review see Horowitz [4]). Thus, changes in intracellular Ca^{2+} play a central role in the regulation of contraction in smooth muscle, and here we review ET-1-induced activation of signal transduction pathways which lead to increased levels of $[Ca^{2+}]_i$ in airway smooth muscle (Fig. 1).

Interestingly, a study in human isolated bronchial preparations [5] observed that the magnitude of ET-1-induced contractions obtained in media containing physiological concentrations of Ca^{2+} was similar to that obtained in nominally Ca^{2+}-free solutions. The presence of a strong contractile

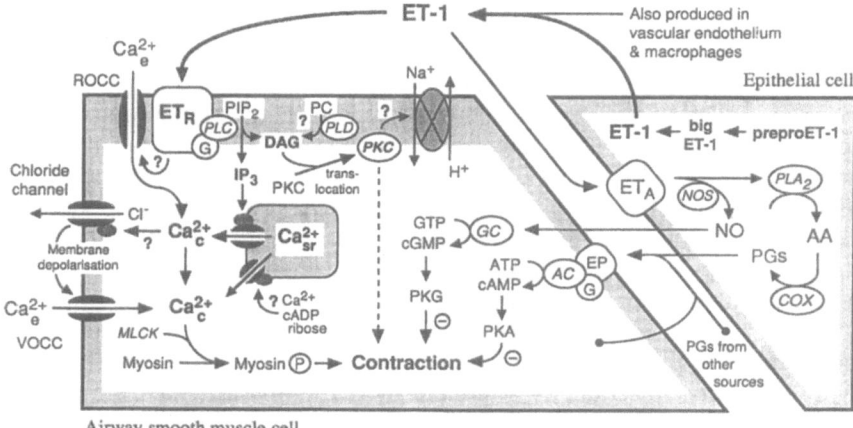

Figure 1. Schematic representation of signal transduction pathways present in airway epithelial and smooth muscle cells.

response to ET-1 in the absence of extracellular Ca^{2+} is indicative of the contractile response being dependent upon the release of Ca^{2+} from intracellular stores. Consistent with this, ET-1 caused a rapid transient increase in $[Ca^{2+}]_i$ in cultured bronchial smooth muscle cells which was insensitive to removal of extracellular Ca^{2+}, but was attenuated by treatments that reduced the intracellular stores of Ca^{2+} [6]. A major intracellular store of Ca^{2+} is the sarcoplasmic reticulum and Ca^{2+} release occurs through both inositol 1,4,5 trisphosphate ($Ins(1,4,5)P_3$)-sensitive and an $Ins(1,4,5)P_3$-insensitive mechanisms.

Phospholipase C and Ins (1,4,5)-P_3: Agonist-induced activation of phospholipase C and facilitation of phosphoinositide turnover is one of the major signal transduction pathways utilized by cells. Phospholipase C catalyses the phosphodiesteric cleavage of a minor phospholipid in the plasma membrane, phosphatidylinositol 4,5-bisphosphate, which results in the generation of $Ins(1,4,5)P_3$ and diacylglycerol (Fig. 1). $Ins(1,4,5)P_3$ diffuses from the cell membrane into the cytosol where it binds to an $InsP_3$ receptor on the sarcoplasmic reticulum. The $InsP_3$ receptor has binding domains for $Ins(1,4,5)P_3$ and Ca^{2+}, and has a central aqueous channel which, when activated permits the passage of Ca^{2+} down its concentration gradient into the cytosol, causing a transient increase in $[Ca^{2+}]_i$, activation of MLCK and initiation of smooth muscle contraction. Diacylglycerol exerts its second messenger action through activation of Ca^{2+} and phospholipid-dependent protein kinases (protein kinase C). Phosphoinositide turnover is a pathway activated by a wide variety of agonists to increase $[Ca^{2+}]_i$ levels and produce contraction within airway smooth muscle [7, 8].

The first report of ET-1-induced activation of phosphoinositide turnover in airway smooth muscle came from studies measuring [³H]-InsP accumulation in guinea-pig trachea by Hay (1990) [9]. Since then, ET-1-induced increases in InsPs have been demonstrated in intact preparations of rat trachea [10], and human [11] and bovine [12] bronchus, as well as in airway smooth muscle cell cultures derived from human bronchus [6] and rabbit [13], canine [14, 15] and bovine [16] trachea. As expected, ET-1-induced increases in Ins(1,4,5)P$_3$ were inhibited by neomycin, an inhibitor of phospholipase C [13]. Furthermore, ET-1-induced increases in InsP levels in canine cultured airway smooth muscle cells were inhibited in the presence of an activator of protein kinase C, phorbol 12-myristate 13-acetate [14], indicating that protein kinase C activation may inhibit phosphoinositide hydrolysis and consequently attenuate increases in [Ca^{2+}]$_i$. A similar response has been observed to other agonists in airway smooth muscle [17, 18] and to ET-1 in other smooth muscle cell types.

Interestingly, Nally and coworkers [11, 12] reported that the nonselective agonist ET-1, but not ET$_B$ receptor-preferring agonist ET-3, evoked a rise in the levels of Ins(1,4,5)P$_3$ in human and bovine bronchi, even though both agonists induced a contractile response in these preparations. A possible explanation for these findings is that contraction involves the activation of at least two signal transduction pathways, with ET$_A$, but not ET$_B$ receptors being linked to InsP generation and mobilisation of [Ca^{2+}]$_i$. However, in stark contrast, another study of human isolated bronchus [19] has proposed that contractions induced by an ET$_B$ receptor-selective agonist, sarafotoxin S6c occurred via stimulation of phosphoinositide turnover and [Ca^{2+}]$_i$ mobilisation, whereas ET-1 elicited contraction through ET$_B$ receptor activation and also ET$_A$ receptor stimulation, which increased Ca^{2+} influx through dihydropyridine-sensitive voltage-dependent channels.

Ryanodine receptor: In smooth muscle, Ca^{2+} from intracellular stores is mobilised through two Ca^{2+} release channels; the Ins(1,4,5)P$_3$ receptor and the ryanodine receptor. Although ryanodine receptors are activated by the binding of ryanodine and caffiene, and blocked by ruthenium red, the identity of the intracellular activators is uncertain. Only two physiological agonists have been recognised so far, Ca^{2+} itself and cyclic ADP ribose, a metabolite of nicotinamide adenine dinucleotide generated by ADP-ribosyl cyclase [20]. Several contractile agonists (e. g., cholecystokinin octapeptide) have been shown to stimulate ADP-ribosyl cyclase, and its product, cyclic ADP ribose acts as an endogenous modulator of Ca^{2+}-induced Ca^{2+} release in intestinal longitudinal muscle [21]. However, recent studies in permeabilized porcine coronary artery smooth muscle cells provided evidence that cyclic ADP ribose stimulates Ca^{2+} release from the sarcoplasmic reticulum through a mechanism that is independent of activation of either Ins(1,4,5)P$_3$ or the ryanodine receptor [22]. The physiological role of the Ca^{2+}-gated ryanodine receptor is unclear and controversial, although in some cell types it may be involved in the Ca^{2+}-induced Ca^{2+} release response that is activated by the influx of extracellular Ca^{2+}.

The possibility that ET-3 induces the release of Ca^{2+} from intracellular stores through an $Ins(1,4,5)P_3$-insensitive pathway has been raised in studies of rat aortic vascular smooth muscle cells in culture [23]. The prospect that an influx of extracellular Ca^{2+} may trigger Ca^{2+} release from intracellular stores has been generated from studies in which L-type Ca^{2+} channel blockers were found to inhibit both the transient and sustained increases in $[Ca^{2+}]_i$ in cultured vascular smooth muscle cells [24, 25]. Whether ET-1 and related peptides stimulate the release of Ca^{2+} from the sarcoplasmic reticulum via these ryanodine receptor-linked Ca^{2+} channels in airway smooth muscle cells, remains to be established. Nevertheless, studies in intact rat isolated tracheal smooth muscle have found some striking similarities in the characteristics of the contractions induced by ET-1 through the activation of ET_A receptors, and by ryanodine in K^+-pre-contracted preparations [26]. Both contractions were selectively inhibited by nordihydroguaiaretic acid, possibly via an inhibitory action on Ca^{2+}-channel activity [27] and were slow to develop. An intriguing, but untested, possibility is that the characteristically slowly developing and sustained contraction induced by ET-1 in airway smooth muscle occurs via the release of Ca^{2+} through ryanodine-sensitive pathways.

Extracellular calcium and plasma membrane Ca^{2+}-channels: In airway smooth muscle, ET-1-induced increases in cytosolic $[Ca^{2+}]$ and tension generation are frequently attenuated in the absence of extracellular Ca^{2+} (Ca^{2+}-free solutions), indicative of the need for Ca^{2+} influx in producing these effects (Fig. 1). By way of example, removal of extracellular Ca^{2+} abolished ET-1-induced contractions in ovine trachea, indicating an absolute dependence of ET-1-induced contraction in ovine tracheal smooth muscle on extracellular Ca^{2+} [28]. Furthermore, in human and canine cultured airway smooth muscle cells, removal of extracellular Ca^{2+} abolished the sustained increase in Ca^{2+} observed in the presence of ET-1 [6, 29]. Clearly though, the influx of extracellular Ca^{2+} may not be obligatory for ET-1-induced contractions in airway smooth muscle from all species. Several studies, including one in human bronchus, have reported that ET-1-induced contractions are not reduced in Ca^{2+}-free solution, indicating that when present, extracellular Ca^{2+} does not contribute significantly to the contractile response [5, 9].

Interestingly, studies in human isolated bronchial preparations indicate that the influence of extracellular Ca^{2+} on ET receptor-mediated contractions is dependent on the choice of ET receptor agonist, such that the L-type Ca^{2+}-channel blocker nifedipine had no effect on contractions induced by ET-1, but attenuated those induced by ET-3 [11]. Other studies in human bronchus reveal that the influence of extracellular Ca^{2+} on ET receptor-mediated contractions is also dependent on the concentration range of agonist used. Whereas Ca^{2+}-free solution and nicardipine each significantly reduced contractions induced by low concentrations of ET-1 (subnanomolar range), these treatments did not affect the contractile responses induced by higher concentrations of ET-1 (nanomolar to micromolar range) [30].

An exciting prospect to emerge from the initial report of the discovery of ET-1 by Yanagisawa and coworkers was that this peptide might be an endogenous agonist of voltage-dependent, L-type plasma-membrane Ca^{2+} channels. Indeed, in ferret [31], rabbit [32, 33], dog [33] and sheep [28] airway smooth muscle, ET-1-induced contractions were significantly inhibited by verapamil, nicardipine and/or nifedipine, suggesting that Ca^{2+}-influx occurred through L-type, voltage-dependent Ca^{2+}-channels and contributed to the contractile response. On the other hand, in many instances the extracellular Ca^{2+}-dependent component of ET-1-induced contraction in airway smooth muscle was not inhibited by L-type Ca^{2+}-channel blockers [11, 34–36], but were blocked by cadmium and lanthanum ions [37, 38]. These latter findings are consistent with numerous other studies in a variety of tissues demonstrating that ET-1 does not compete or displace the binding of ligands at L-type Ca^{2+}-channels. Thus, despite the earlier attractive postulate, ET-1 is not a ligand of L-type Ca^{2+} channels and the observed dependence in some preparations of ET-1-induced contraction on voltage-dependent Ca^{2+} channels must be a consequence of indirect gating [39]. Possible indirect mechanisms through which ET-1 might induce membrane depolarization and promote Ca^{2+} influx through voltage-dependent channels include the activation of chloride channels or by stimulation of the Na^+/H^+ antiporter (see below).

In addition to flow through voltage-dependent Ca^{2+}-channels, channel-mediated influx of extracellular Ca^{2+} may occur through voltage-insensitive receptor-operated channels or non-selective cation channels. Although the existence of one or other of these channels is supported by the findings that ET-1-induced contractions are dependent upon the influx of extracellular Ca^{2+} via non-L-type channels, little is known of these channels in airway smooth muscle.

Endothelin receptor subtypes and links to activator Ca^{2+}: Recent studies indicate that ET-1-induced contractions in airway smooth muscle involve both ET_A and ET_B receptors and multiple signal transduction systems. In intact preparations of rat isolated tracheal smooth muscle, ET_A receptor-mediated contraction involved the mobilisation of intracellular Ca^{2+} through both $InsP_3$-dependent and $InsP_3$-independent pathways [26, 36], whereas ET_B receptor-mediated contractions appeared to result from the influx of extracellular Ca^{2+} via non L-type Ca^{2+} channels. ET-1 activates both pathways. Subsequently, Inui and coworkers have established in single smooth muscle cells from guinea-pig trachea that ET_A and ET_B receptors coexist in a major population of cells and cooperate in mediating smooth muscle contraction [40].

Autoradiographic studies indicate that, depending on the species and region of airway studied, airway smooth muscle contained a predominance of ET_A or ET_B receptors or a mixture of both receptor subtypes [3]. For example, whereas human bronchial smooth muscle contained a predominance of ET_B receptors [41, 42], sheep tracheal smooth muscle contained a homogeneous population of ET_A receptors [28] and rat and murine tracheal smooth muscle possessed similar proportions of ET_A and ET_B receptors [26, 43]. Evidence of marked regional differences in the proportions of ET_A and ET_B receptors in airway smooth muscle have been reported in both the pig [44] and guinea-pig [45] respiratory tract.

Thus, there are significant interspecies differences in both the endothelin receptor subtype present on airway smooth muscle and the source of activator Ca^{2+} that mediates ET-1-induced contraction. However, there is no supportive evidence that, across the species, one or other of the ET receptor subtypes is linked to a particular source of activator Ca^{2+} or signal transduction pathway. Studies of rat, human and sheep airways provide a stark example of the apparent lack of a positive relationship between binding to ET receptor subtype and activation of a specific signal transduction system leading to airway smooth muscle contraction. ET_A receptor-mediated contraction in rat tracheal smooth muscle is due to stimulation of phosphoinositide turnover and the release of intracellular Ca^{2+} [26, 36], but in sheep tracheal smooth muscle is almost entirely attributable to the influx of extracelluar Ca^{2+} [28]. Similar differences have been observed with respect to the ET_B receptor-efffector systems mediating airway smooth muscle contraction. ET_B receptor-mediated contraction was mediated via the influx of extracellular Ca^{2+} in rat trachea [36], whereas in human bronchus it appears to be mediated by activation of the phosphoinositide pathway and release of intracellular Ca^{2+} [19]. Recent studies using a range of different cell lines have also demonstrated that ET receptors are not always coupled to the same signal transduction system [46].

2.1.2. Diacylglycerol and protein kinase C: Numerous contractile agonists have been demonstrated to generate diacylglycerol in smooth muscle through the hydrolysis of either phosphatidylinositol 4,5-bisphophate by phospholipase C or of phosphatidylcholine by phospholipase D. Studies in vascular smooth muscle demonstrated that agonist-induced diacylglycerol accumulation is biphasic; phosphatidylinositol 4,5-bisphosphate hydrolysis by phospholipase C produced a short-lived increase in diacylglycerol levels, which was followed by a sustained accumulation of diacylglycerol from phosphatidylcholine hydrolysis. Diacylglycerol is a well-characterized physiological activator of protein kinase C, causing the sustained translocation of protein kinase C from the cytosolic to the membrane fraction that is commonly associated with sustained contractile responses (for review see Orallo, 1996 [47]). Unlike phosphatidylinositol 4,5-bisphosphate, the hydrolysis of phosphatidylcholine is not associated with the generation of Ins(1,4,5)P$_3$, and thus the protein kinase C pathway can be stimulated independently of changes in [Ca^{2+}]$_i$ (Fig. 1).

Several lines of evidence suggest that the activation of protein kinase C is involved in ET-1-induced contraction in vascular smooth muscle. Firstly, inhibitors of protein kinase C such as staurosporine and calphostin C, have been shown to inhibit ET-1-induced contractions. Secondly, activators of protein kinase C such as the phorbol esters evoke slowly developing and long-lasting contractions similar to those induced by ET-1. Thirdly, ET-1 caused the translocation of protein kinase C with a direct correlation between the time course of tension induced and kinase translocated [48].

Studies of airway smooth muscle have indicated that activation of protein kinase C might contribute to ET-1-induced contraction in rat [26], rabbit [32, 33] and dog airways [33], but not in guinea-pig [9] or bovine [12] airways. In human bronchus, protein kinase C activation does not appear to contribute to the magnitude of an ET-1-induced contraction [11, 33] but may contribute to the generation of a sustained phase of contraction [33]. These conclusions were based on the ability or otherwise of inhibitors of protein kinase C (staurosporine, H-7) to inhibit ET-1-induced contractions. However, these compounds act by competing at the ATP binding site of protein kinase C that shares substantial homology with other protein kinases, and are therefore likely to lack specificity for inhibiting protein kinase C. Of particular interest would be an investigation into the effects on ET-1-induced contraction of more selective inhibitors of protein kinase C, such as calphostin C, which compete at the diacylglycerol binding site. Moreover, direct measures of ET-1-induced diacylglycerol generation and protein kinase C activity, and information as to the temporal relationship between protein kinase C activity and ET-1-induced contraction are currently lacking in airway smooth muscle. Until these and related studies have been completed, the role of protein kinase C activation on ET-1-induced contraction in airway smooth muscle must remain unresolved.

2.1.3. Sodium-hydrogen antiporter: Activation of the Na$^+$/H$^+$ antiporter in the plasmalemma of smooth muscle cells will increase the extrusion of H$^+$ and intracellular concentration of Na$^+$, thereby inhibiting the Na$^+$/Ca^{2+} exchanger and preventing extrusion of Ca^{2+} from the cell [49] (Fig. 1). Battistini and coworkers [50] raised the possibility that ET-1-induced activation of Na$^+$/H$^+$ exchange plays a role in ET-1-induced contraction of guinea-pig airways. These workers reported that ET-1-induced contractions of guinea-pig tracheal and bronchial strips were dose-dependently attenuated by several analogues of amiloride, established inhibitors of the Na$^+$/H$^+$ antiporter [50]. Whether this ET-1-induced effect in airway smooth muscle is mediated by the activation of protein kinase C, as has been proposed for a variety of agonists, including ET-1, in vascular smooth muscle [51–53] remains to be determined.

2.1.4. Phospholipase A_2 *and arachidonic acid metabolites:* The lypolytic enzyme phospholipase A_2 functions intracellularly to catalyse the hydrolysis of membrane bound phospholipids to yield arachidonic acid and a lysophosphatide. Arachidonic acid can be further metabolised to prostaglandins, thromboxanes and leukotrienes. Several studies have indicated that these latter eicosanoid metabolites of arachidonic acid may play an important role as second messengers in mediating some of the biological actions of ET-1, including constriction of airway smooth muscle.

There is general agreement that the *in vivo* bronchoconstrictor response to ET-1 involves the actions of cyclooxygenase products such as thromboxane A_2. By way of example, the ET-1-induced bronchoconstrictor response in guinea-pigs was attenuated by indomethacin and meclofenamate (inhibitors of cyclooxygenase) [54–56], as well as by thromboxane A_2 synthesis inhibitors [57] and receptor antagonists [58, 59]. A similar dependence of ET-1-induced bronchoconstriction on thromboxane A_2 has been reported in cats [60] and rats [61]. Consistent with this, ET-1 was a potent releaser of cyclooxygenase-derived eicosanoids in isolated perfused lungs from guinea-pig or rat [62, 63]. In human isolated bronchus, ET-1 evoked the release of an array of prostanoids, but there was no evidence that cyclooxygenase products mediated ET-1-induced contractions in this tissue [64]. In other studies of isolated airway smooth muscle, cyclooxygenase inhibitors have been shown to inhibit, to potentiate or to have no effect on contractions induced by ET-1 (for review see Goldie et al., 1996 [3]). The reasons for the conflicting data are not known.

As indicated above, ET-1 is a potent releaser of cyclooxygenase-derived eicosanoids in the lung, however the precise mechanisms involved in ET-1-induced activation of phospholipase A_2 and the primary cellular source of phospholipase A_2, (airway smooth muscle?) remains to be determined.

2.1.5. Chloride channels: It is now evident that in many types of smooth muscle, agonist-induced activation of the phosphoinositide pathway generates $Ins(1,4,5)P_3$, which leads to the opening of Ca^{2+}-activated Cl^- channels (for review see [65]). The opening of these Cl^- channels and the subsequent efflux of Cl^- will drive the membrane potential towards E_{Cl} (between -20 and -30 mV) and hence produce depolarisation. This powerful depolarizing mechanism will increase the probability of voltage-dependent Ca^{2+}-channels becoming open and of Ca^{2+} flowing into the cell and producing contraction. Therefore, agonists that release Ca^{2+} from intracellular stores can produce contraction directly via the action of release Ca^{2+} on contractile proteins or indirectly by stimulating the opening of Ca^{2+}-activated Cl^- channels, causing depolarization and Ca^{2+} influx through voltage-dependent channels [65] (Fig. 1).

Agonist-induced Ca^{2+}-activated Cl^- currents have been measured in airway smooth muscle cells in response to excitatory neurotransmitters including acetylcholine [66, 67] and substance P [68] and other mediators of contraction including histamine [69] and neurokinin A [70]. These conductances can be blocked by Cl⁻ channel blockers such as A-9-C (anthracene-9-carboxylic acid), niflumic acid and various stilbene derivatives (e.g., SITS; 4-acetamido-4′-isothiocyanostilbene-2,2′disulphonic acid). To date, the influence of ET-1 and related peptides on Cl⁻ channels in airway smooth muscle cells has yet to be determined, although such a mechanism has been demonstrated in various vascular smooth muscle cell types including porcine coronary artery [71], human mesenteric artery [71], rat renal resistance arteries [72], rat pulmonary artery [73] and A7r5 cells [74]. Thus, the activation of Cl⁻ channels currently offers a convincing explanation for the ionic basis of ET-1-induced membrane depolarization in vascular smooth muscle [39], but remains to be established in airway smooth muscle.

2.2. Relaxation

In several studies, low concentrations of ET-1 have been reported to induce a transient relaxation of airway smooth muscle prior to contraction [32, 75–79]. In isolated human and guinea-pig airways, ET-1-induced relaxation is due at least partly to the release of epithelium-derived nitric oxide (NO), leading to the activation of guanylate cyclase and the production of cyclic GMP [77, 79, 80] (Fig. 1). In human bronchial preparations with an intact epithelium, ET-1-induced contractions were potentiated in the presence of an inhibitor of NO synthesis (l-NAME) or by BQ-123, suggesting that ET-1 activated ET_A receptors on the airway epithelium, leading to the release of relaxant factors such as NO [80]. There is also a significant body of evidence in support of the postulate that the relaxant response to ET-1 occurs partly *via* the generation of secondary mediators such as PGE_2. For example, ET-1 stimulated the release of the relaxant prostanoid PGE_2 (and PGD_2) from guinea pig trachea [81], and induced an indomethacin-sensitive increase in cyclic AMP formation [82]. The possibility that ET-1 modulates the cyclic AMP and cyclic GMP cascades has recently been reviewed by Sokolovsky [83]. In addition, the opening of charybdotoxin-sensitive K^+ channels [79], and the generation of lipoxygenase-derived hydroperoxides of arachidonic acid [75] have also been suggested as playing a possible role in ET-1-induced relaxations in airway smooth muscle. Hadj-Kaddour and coworkers (1995) [79] have raised the possibility that ET-1 and ET-3 relax guinea-pig isolated tracheal smooth muscle through different mechanisms. Irrespective of the mechanism of ET-1-induced relaxation, the smooth muscle relaxant responses are likely to be functionally antagonised by both the direct spasmogenic actions of ET-1 and by the simultaneous production of spasmogenic cyclooxygenase products such as thromboxane A_2 [77].

2.3. Mitogenesis

The recent findings in human and animal airway smooth muscle cell cultures that ET-1 promoted [³H]-thymidine incorporation into DNA, stimulated protein synthesis, enhanced the transient expression of proto-oncogenes such as c-fos and increases cell number, indicates that ET-1 has mitogenic actions [13, 84–89]. In human airway smooth muscle cells, the mitogenic action of ET-1 was inhibited by BQ-123 [86, 88] and ET-3 was inactive as a mitogen [90]. Together, these findings are indicative of ET-1-induced mitogenesis being an ET_A receptor-mediated process. However, the mitogenic actions of ET-1 are relatively weak compared to those of the growth factors EGF [85, 88] and PDGF [87, 91]. One possible explanation is that signal transduction pathways, such as phospholipase C activation, $Ins(1,4,5)P_3$ formation and intracellular Ca^{2+} release, which are important in mediating the contractile actions of ET-1 are less critical in producing mitogenesis of airway smooth muscle cells [85, 92]. Consistent with this, (a) other spasmogens of human airway smooth muscle such as bradykinin, thromboxane A_2 and leukotriene D_4 are also weak mitogens and (b) neither inhibition of phospholipase C (PLCβ) nor depletion of PKC affected the synergism between EGF and ET-1 [93]. In this latter study, Fujitani and Bertrand [93] suggested that ET-1 can interact with an EGF-induced mitogenic axis through a pertussis toxin-sensitive G_i protein-dependent pathway which is distinct from its direct mitogenic pathway. In summary, these findings indicate that ET-1 acts as a potent co-mitogen in airway smooth muscle [88], as previously demonstrated in other cell systems [94] (see [95] for review).

The signal transduction pathways that link ET receptors to the transcription of genes in airway smooth muscle have not been fully elucidated, but may involve the activation of mitogen-activated protein (MAP) kinases, tyrosine kinases and phospholipases, and increased $[Ca^{2+}]_c$. Using rabbit cultured airway smooth muscle cells, Noveral and coworkers [13] reported that ET-1-induced mitogenesis was associated with the activation of a pertussis toxin-sensitive G protein coupled to the stimulation of phospholipase A_2, and the generation of thromboxane A_2. Consistent with this, arachidonic acid or its metabolites have been suggested to play a role in mitogenic signalling in vascular smooth muscle cells, linking the hypertrophic (e. g., those induced by endothelin-1) and the hyperplastic (e. g., those induced by PDGF) signal transduction pathways [96].

A major signalling pathway for airway smooth muscle cell growth and division used by growth factors such as PDGF involves the stimulation of receptors with tyrosine kinase activity and the subsequent activation of MAP kinases via a cascade of multiple tyrosine- and serine/threonine-phosphorylations (e. g., Raf-1, a MAP kinase kinase kinase; MEK, a MAP kinase kinase). MAP kinases help relay the signal to the nucleus and may play an important role in regulating several early events in mitogenesis (activation of protein synthesis, phosphorylation of proto-oncogenes and stimulation of glucose transport). MAP kinase activation has also been suggested to dictate the relative efficacies of growth factor and contractile (G protein-coupled receptor) agonists as mitogens [97, 98]. Whelchel and coworkers [99] have recently demonstrated that ET-1 was a potent stimulator of the extracellular regulated kinase 2 (ERK2) subgroup of MAP kinases, and ERK2 activation was tightly correlated with the proliferation of rat cultured airway smooth muscle cells. Furthermore, the endothelin signal transduction pathway that culminated in ERK2 activation and proliferation could be inhibited by PD98059, a small molecular inhibitor of MEK or by depletion of PKC [99]. Together, these data indicate that the endothelin signal transduction pathway leading to airway smooth muscle proliferation involved the progressive activation of PKC, possibly Raf-1, MEK and ERK2.

As indicated above, ET-1 has been shown to activate MAP kinase in airway smooth muscle cell cultures derived from rat [89, 99] and bovine [87, 91] trachea. However, the finding that ET-1 and PDGF induce MAP kinase to a similar extent, but ET-1 is only a fraction as effective as PDGF in stimulating DNA synthesis, suggest that additional, distinct signalling pathways exist for these two agonists [91]. These investigators proposed that the mitogenic efficacy of agonists that generate comparable MAP kinase signals is determined by their relative abilities to induce phosphatidylinositol 3-kinase (PtdIns 3-kinase) activation of p70 ribosomal protein S6 kinase (p70^{s6k}). Activation of p70^{s6k}, which has been shown to be involved in the regulation of both protein and DNA synthesis [100, 101] was stimulated 15-fold by PDGF but only 2-fold by ET-1, similar to their respective 50-fold and two-fold increases in DNA synthesis [91]. These studies indicate an obligatory role for a PtdIns 3-kinase/p70^{s6k} pathway in agonist-stimulated DNA synthesis in airway smooth muscle cells. Thus, ET-1, which failed to stimulate PtdIns 3-kinase, was a very weak mitogen of airway smooth muscle cells.

Recent studies by Shapiro and coworkers [89] indicate that ET-1 (as well as thrombin, another ligand that acts via receptors of the seven-transmembrane receptor superfamily) causes proliferation of airway smooth muscle cells via the activation of a novel intracellular pathway involving Jun kinases, a pathway used by agents that induce cellular stress such as UV light and tumor necrosis factor. These studies also reported that ET-1 and thrombin activated Raf-1-independent mechanisms of MAP kinase. However, at present, the upstream kinases that couple the endothelin receptor to activation of the MAP kinase and Jun kinase pathways have not been clearly defined.

As outline above, ET receptor activation stimulates phospholipase activity, resulting in increased diacylglycerol levels and activation of protein kinase C. The activation of protein kinase C is an important signalling event for the proliferation of many cells, including smooth muscle (see

[102] for review), however the role of protein kinase C in ET-1-induced mitogenesis in airway smooth muscle cells remains unclear. Malarkey and coworkers [87], reported that the protein kinase C inhibitor, Ro-318220, or the down regulation of protein kinase C by chronic phorbol ester pretreatment, substantially reduced ET-1-induced stimulation of MAP kinase, suggesting that a component of the MAP kinase signal involved protein kinase C-mediated activation of an intermediate kinase, possibly Raf-1 or MEK kinase [87]. Protein kinase C is reported to phosphorylate and activate Raf-1 [103], however recent studies by Shapiro and coworkers [89] indicate that ET-1 is a poor activator of Raf-1 in rat tracheal smooth muscle cells. Thus, although a protein kinase C-dependent component of ET-1-induced MAP kinase activation has been identified, the underlying signal transduction pathways have not been clarified. Furthermore, it is unclear which of the various isoforms of protein kinase C are involved in ET-1-induced mitogenesis of airway smooth muscle cells.

Since the ET-1-induced MAP kinase response in airway smooth muscle cells was only partially inhibited by protein kinase C inhibition, a protein kinase C-independent component of the response also probably exists [87]. ET-1-stimulated activation of MAP kinase [87] and of cell proliferation [13] were inhibited by pertussis toxin. Consistent with this, several G protein coupled agonists including thrombin may activate MAP kinase via a pertussis toxin-sensitive, protein kinase C-independent stimulation of $p21^{ras}$, a small molecular weight G protein [87, 97, 104]. Furthermore, Shapiro and coworkers [89] reported that receptor coupling to Jun kinase activation may involve heterotrimeric G proteins since the kinase was activated in cells treated with aluminium fluoride.

In airway smooth muscle, the stimulation of receptors (e. g., β-adrenoceptors) coupled to G_s leads to an increase in adenylyl cyclase activity, elevated levels of cytosolic cAMP and activation of protein kinase A. Stimulation of this pathway appears to have a direct inhibitory effect on the proliferative response to growth factors such as EGF and PDGF, and perhaps to ET-1. For example, forskolin, an activator of G_s, inhibited MAP kinase activation and [³H]-thymidine incorporation into rat airway smooth muscle cells induced by either thrombin or ET-1 [89]. Although protein kinase A activation appears to have pleiotrophic effects on cell growth that are poorly understood [89], a recent study by Scott and coworkers [91] indicates that the ability of cAMP-raising agents to potently inhibit growth factor-stimulated DNA synthesis in airway smooth muscle cells can be attributed to inhibition of the PtdIns 3-kinase/p70^{s6k} pathway, possibly at or above the level of tyrosine phosphorylation of the p85 regulatory subunit of PtdIns 3-kinase. In contrast, studies in human airway smooth muscle cells by Tomlinson and associates [86] demonstrated that salbutamol caused a β_2-adrenoceptor-mediated inhibition of [³H]-thymidine incorporation induced by EGF, thrombin and a thromboxane A$_2$-mimetic U46619, but did not inhibit the small degree of cell proliferation induced by ET-1.

3. Epithelial Cells

In addition to being a major cellular source of ET-1 [105–109], various epithelial cellular functions can also be influenced by ET-1, suggesting that epithelial ET-1 might act as an autocrine, as well as a paracrine hormone within the airways. In epithelial cell cultures, ET-1 alone [110] or in combination with EGF [111] increased [^{3}H]-thymidine incorporation and cell numbers. ET-1 also increased the negativity of the transepithelial potential difference [112, 113] and the short-circuit current caused by stimulation of Cl$^-$ transport across the epithelium [113–115]. Cilia beating frequency [115] and mucociliary activity in the nasal sinus and tracheal mucosae [116] are also stimulated by ET-1. Many of these effects of ET-1 on ion transport and mucociliary transport were reduced in the presence of indomethacin, suggesting that the ET-1-induced effects were being mediated via the generation of cyclooxygenase products [113–115]. Consistent with this, feline and human bronchial epithelial cells have been found to release various prostanoids including PGE$_2$, PGF$_{2\alpha}$ and thromboxane B$_2$ in response to ET-1 [111, 117]. In addition to cyclooxygenase products, ET-1 also has been reported to induce the synthesis and release of the cytokines Il-6, Il-8 and GM-CSF from a human bronchial epithelial cell line (BEAS-2B) [118]. In contrast, ET-1 did not stimulate the release of these cytokines from normal human bronchial epithelial cells [119].

In contrast to airway smooth muscle cells, relatively little is known of the signal transduction pathways that mediate ET-1-induced responses in airway epithelial cells. To date, one of the more extensive evaluations of the signal transduction pathways activated by ET-1 was conducted by Markewitz and coworkers [120] using a cloned rat alveolar epithelial cell line, called L2 cells. In these cells, ET-1 increased PGE$_2$ and cAMP levels through an ET$_A$ receptor-mediated pathway. At least part of the increase in cAMP levels was sensitive to inhibition by indomethacin, suggesting it was due to the actions of endogenous cyclooxygenase product(s), probably PGE$_2$, released by ET-1 [120]. ET-1-induced stimulation of Cl$^-$ secretion from canine tracheal epithelium is thought to involve at least two pathways i.e., a primary phosphorylation pathway involving cAMP/protein kinase A (probably via PGE$_2$ generation) and a second pathway involving a rise in [Ca^{2+}]$_i$ [114]. ET-1 also induced an ET$_A$ receptor-mediated increase in [Ca^{2+}]$_i$ in human bronchial epithelial cells [111]. The finding that ET-1 induced increases in Cl$^-$ secretion and ciliary beating frequency were greatly reduced in Ca^{2+}-free media suggests that the influx of extracellular Ca^{2+} plays an important role in the signal transduction process, perhaps by activating phospholipase A$_2$, which catalyses the release of arachidonic acid from phosphatidylcholine with the resultant production of cyclooxygenase products from arachidonic acid [115]. The mechanisms through which ET-1 induces its other effects in epithelial cells (e.g., mitogenesis, cytokine release) have still to be determined.

4. Submucosal Glands

ET-1, but not ET-2 or ET-3, induced mucus glycoprotein secretion from feline tracheal isolated submucosal glands [121]. Stimulus-secretion coupling in ET-1-evoked mucus secretion was associated with a rise in $[Ca^{2+}]_i$, but no change in cAMP levels [121]. Removal of extracellular Ca^{2+} abolished both ET-1-evoked increases in $[Ca^{2+}]_i$ and glycoprotein secretion, indicating that these processes result from Ca^{2+} influx. Thus, ET-1-induced mucous glycoprotein secretion from airway submucosal glands is dependent moreso on the influx of extracellular Ca^{2+} than on the release of Ca^{2+} from intracellular stores, as with some other secretogogues and in other exocrine cells [121]. Nevertheless, recent studies suggest that secretory processes in cultured submucosal glands, in response to the muscarinic cholinoceptor agonist methacholine, may be mediated by Ca^{2+}-activated Cl^- channels, perhaps via $Ins(1,4,5)P_3$-induced release of intracellular Ca^{2+} [122]. Protein kinase C activation also appears to have a direct stimulatory role in stimulus-secretion coupling in submucosal glands [123]. Together, these findings raise the hitherto untested possibility that a component of ET-1-induced secretion in submucosal glands occurs via activation of the phosphoinositide pathway. Finally, although NO formation in airway submucosal glands appears to be important for bradykinin-induced increases in mucus glycoprotein and electrolyte secretion [124, 125], the role of NO in ET-1-induced secretion is not known.

Interestingly, in tracheal explants or isolated glands with cultured epithelial cells, ET-1 produced significant reductions in glycoconjugate secretion from submucosal glands [121]. This is consistent with the inhibitory effect of ET-1 reported on submucosal gland secretion from ferret whole trachea [126]. Shimura and coworkers suggest that the inhibitory effect of ET-1 is mediated by a combination of a non-cyclooxygenase, inhibitory factor(s) derived from epithelium and by cyclooxygenase product(s) of arachidonic acid released from sources other than epithelium (e. g., PGI_2 from vascular endothelial cells) [121]. A component of the inhibitory effect has been attributed to the activation of dihydropyridine-sensitive Ca^{2+}-channels [126], an unusual mechanism given that influx of Ca^{2+} would be expected to have a stimulatory effect on submucosal gland secretion.

5. Bronchial Circulation, Microvascular Leakage and Oedema

The bronchial circulation, derived from the systemic circulation, supplies blood to structures within the airway wall including airway smooth muscle, epithelium, submucosal glands and nerves. In pigs, the intravenous administration of ET-1 and ET-3 caused a pronounced, concentration-dependent and long-lasting increase in bronchial blood flow [127, 128]. The vasodilator effect of ET-1 did not appear to be mediated by the release of vasodilator prostanoids since the cyclooxygenase inhibitor diclofenac did not modify the ET-1-induced response [127]. Although untested, NO has been proposed as a possible mediator of ET-1-induced vasodilatation in the bronchial circulation [127]. In contrast, administration of ET-1 to dogs reduced both tracheal and main bronchi mucosal blood flow by between 40 and 50% [129]. Although human bronchial blood vessels contain ET receptors [130], neither the influence of ET-1 on human bronchial blood flow nor the signal transduction pathways operating have yet been determined.

In addition to its effects on vascular tone, ET-1 exerts significant effects on microvascular permeability of the tracheobronchial circulation. Filep and coworkers, using tissue accumulation of Evans Blue dye, have demonstrated that the bolus injection of ET-1 enhanced albumin extravasation in the trachea, upper and lower bronchi, but not in the pulmonary parenchyma [131]. The permeability enhancing effects of ET-1 are thought to result from the induction of interendothelial gap formation in post-capillary venules, and mediated through the activation of both ET_A and ET_B receptors [131]. ET-1-induced gap formation may occur directly via activation of endothelial ET_B receptors, which predominate on endothelial cells [132], or through release of secondary mediators such as platelet activating factor [133] or thromboxane A_2 [134]. Whether ET-1-induced oedema in the airways is inhibited by L-type calcium channel blockers (via the inhibiton of secondary mediators) as has recently been demonstrated in the heart [135], remains to be seen. Extravasated plasma causes oedema and swelling within the airway wall, and upon traversing the epithelium may compromise epithelial integrity and mucociliary transport, leading to airways obstruction and bronchial hyperreactivity (for review see Goldie and Pedersen [136]). Developing a greater understanding of the signal transduction pathways through which ET-1 produces microvascular leakage is likely to significantly enhance our chances of combating these pro-inflammatory actions of ET-1.

6. Cholinergic Nerves

Accumulating evidence, obtained from studies performed using airway tissue of both human and animal origin, indicated that ET-1 and related petides exerted significant neuromodulatory actions at parasympathetic postganglionic nerves. In the presence of ET receptor agonists, contractile responses to cholinergic nerve stimulation (induced by electrical field stimulation, EFS), but not to exogenously applied acetylcholine, were augmented in isolated airways preparations from rabbit [137, 138], rat [139], mouse [43, 140] and human [141]. These findings provided compelling, albeit indirect support for the existence of a population of prejunctional ET receptors which are linked to enhanced acetylcholine release. Further support for this mechanism was obtained from a recent study in rat trachea which revealed that EFS-induced release of [^{3}H]-acetylcholine was enhanced by ET-1 [139]. ET-1 does not potentiate cholinergic nerve-me-diated contractions in the airways of all species. Somewhat surprisingly, EFS-induced contractions of sheep isolated tracheal smooth muscle were attenuated by ET-1, and this was associated with an ET_B receptor-mediated reduction in EFS-induced acetylcholine release [142].

Despite the strong probability that stimulation of prejunctional ET recep-tors modulates release of the neurotransmitter acetylcholine from airway cholinergic nerves, nothing is known of the signal transduction pathways involved in this process. Indeed, compared with many other cell types such as smooth muscle, relatively little is known of the signal transduction pro-cesses in nerve terminals. This is due primarily to a variety of practical reasons including the fact that the nerve terminal is a very small unit in an effector tissue that frequently contains many cell types [143]. On the basis of our current knowledge, a most likely mechanism for ET-1-induced neu-romodulation in cholinergic nerves will probably involve ion-channel modulation (Ca^{2+}, K^+), perhaps by direct G protein linking. The influx of Ca^{2+} will probably involve the activation of voltage-dependent Ca^{2+} chan-nels. There is currently little evidence that the Ca^{2+} releasing actions of Ins(1,4,5)P$_3$ alters neurotransmission or is involved in receptor-mediated neurotransmission [144]. It is possible that activation of the cAMP/protein kinase A and/or the phopholipase A$_2$ pathways might be involved in neuro-modulation, however the mechanisms are unclear. Perhaps with the appli-cation of recently developed cholinergic nerve cell cultures from guinea-pig airways [145], some progress in this exciting area of research might be anticipated.

7. Summary

A significant body of knowledge exists in regard to the signal transduction pathways which mediate the powerful spasmogenic actions of ET-1 in air-way smooth muscle. Two signal transduction pathways, Ins(1,4,5)P$_3$-mediated release of Ca^{2+} from intracellular stores and/or the influx of Ca^{2+} from the extracellular space appear central to the generation of ET-1-in-duced contractions in airway smooth muscle. On the other hand, the role of other signal transduction pathways involving the activation of protein ki-nase C, the Na^+/H^+ antiporter, Cl^- channels, phospholipase A$_2$ and ryan-odine receptors, have yet to gain universal acceptance. The pathways that link ET receptors to the effects exerted by ET-1 on non-smooth muscle structures within the airway wall are considerably less well understood. For example, although ET-1 has been shown to exert powerful neuromodul-atory actions within the airway wall, virtually nothing is known of the signal transduction processes involved. Indeed, a major challenge for re-searchers lay in the elucidation of the signal transduction pathways which mediate the wide-ranging and powerful effects induced by ET-1 in the airways, and in the development of therapeutic agents which modulate ET receptor-effector pathways.

Acknowledgements

P.J. Henry is funded by the National Health & Medical Research Council of Australia.

References

1 Noguchi Y, Uchida Y, Endo T, Ninomiya H, Nomura A, Sakamoto T et al (1995) The induction of cell differentiation and polarity of tracheal epithelium cultured on the amniotic membrane. *Biochem Biophys Res Commun* 210: 302–309

2 Mariassy AT, Glassberg MK, Salathe M, Maguire F, Wanner A (1996) Endothelial and epithelial sources of endothelin-1 in sheep bronchi. *Am J Physiol* 270: L54–61

3 Goldie RG, Knott PG, Carr MJ, Hay DW, Henry PJ (1996) The endothelins in the pulmonary system. *Pulm Pharmacol* 9: 69–93

4 Horowitz A, Menice CB, Laporte R, Morgan KG (1996) Mechanisms of smooth muscle contraction. *Physiol Rev* 76: 967–1003

5 McKay KO, Black JL, Armour CL (1991) The mechanism of action of endothelin in human lung. *Br J Pharmacol* 102: 422–428

6 Mattoli S, Soloperto M, Mezzetti M, Fasoli A (1991) Mechanisms of calcium mobilization and phosphoinositide hydrolysis in human bronchial smooth muscle cells by endothelin 1. *Am J Respir Cell Mol Biol* 5: 424–430

7 Chilvers ER, Lynch BJ, Challiss RA (1994) Phosphoinositide metabolism in airway smooth muscle. *Pharmacol Ther* 62: 221–245

8 Chilvers ER, Nahorski SR (1990) Phosphoinositide metabolism in airway smooth muscle. *Am J Respir Dis* 141: S137–S140

9 Hay DW (1990) Mechanism of endothelin-induced contraction in guinea-pig trachea: comparison with rat aorta. *Br J Pharmacol* 100: 383–392

10 Henry PJ, Rigby PJ, Self GJ, Preuss JM, Goldie RG (1992) Endothelin-1-induced [^{3}H]-inositol phosphate accumulation in rat trachea. *Br J Pharmacol* 105: 135–141

11 Nally JE, McCall R, Young LC, Wakelam MJ, Thomson NC, McGrath JC (1994) Mechanical and biochemical responses to endothelin-1 and endothelin-3 in human bronchi. *Eur J Pharmacol* 288: 53–60

12 Nally JE, McCall R, Young LC, Wakelam MJ, Thomson NC, McGrath JC (1994) Mechanical and biochemical responses to endothelin-1 and endothelin-3 in bovine bronchial smooth muscle. *Br J Pharmacol* 111: 1163–1169

13 Noveral JP, Rosenberg SM, Anbar RA, Pawlowski NA, Grunstein MM (1992) Role of endothelin-1 in regulating proliferation of cultured rabbit airway smooth muscle cells. *Am J Physiol* 263: L317–L324

14 Yang CM, Ong R, Chen YC, Hsieh JT, Tsao HL, Tsai CT (1995) Effect of phorbol ester on phosphoinositide hydrolysis and calcium mobilization induced by endothelin-1 in cultured canine tracheal smooth muscle cells. *Cell Calcium* 17: 129–40

15 Yang CM, Yo YL, Ong R, Hsieh JT (1994) Endothelin- and sarafotoxin-induced phosphoinositide hydrolysis in cultured canine tracheal smooth muscle cells. *J Neurochem* 62: 1440–1448

16 Oda K, Fujitani Y, Watakabe T, Inui T, Okada T, Urade Y et al (1992) Endothelin stimulates both cAMP formation and phosphatidylinositol hydrolysis in cultured embryonic bovine tracheal cells. *FEBS Lett* 299: 187–191

17 Murray RK, Bennett CF, Fluharty SJ, Kotlikoff MI (1989) Mechanism of phorbol ester inhibition of histamine-induced IP$_3$ formation in cultured airway smooth muscle. *Am J Physiol* 257: L209–L216

18 Kotlikoff MI, Murray RK, Reynolds EE (1987) Histamine-induced calcium release and phorbol antagonism in cultured airway smooth muscle cells. *Am J Physiol* 253: C561–C566

19 Hay DW, Luttmann MA, Goldie RG (1997) Calcium (Ca^{2+}) translocation mechanisms mediating endothelin-1 (ET-1)- and sarafotoxin S6c (S6c)-induced contractions in isolated human bronchus. *Am J Resp Crit Care Med* 151: A1083

20 Galione A (1992) Ca^{2+}-induced Ca^{2+} release and its modulation by cyclic ADP- ribose. *Trends Pharmacol Sci* 13: 304–306

21 Kuemmerle JF, Makhlouf GM (1995) Agonist-stimulated cyclic ADP ribose. Endogenous modulator of Ca^{2+}-induced Ca^{2+} release in intestinal longitudinal muscle. *J Biol Chem* 270: 25488–25494

22 Kannan MS, Fenton AM, Prakash YS, Sieck GC (1996) Cyclic ADP-ribose stimulates sarcoplasmic reticulum calcium release in porcine coronary artery smooth muscle. *Am J Physiol* 270: H801–H806

23 Little PJ, Neylon CB, Tkachuk VA, Bobik A (1992) Endothelin-1 and endothelin-3 stimulate calcium mobilization by different mechanisms in vascular smooth muscle. *Biochem Biophys Res Commun* 183: 694–700

24 Huang S, Simonson MS, Dunn MJ (1993) Manidipine inhibits endothelin-1-induced $[Ca^{2+}]_i$ signaling but potentiates endothelin's effect on c-fos and c-jun induction in vascular smooth muscle and glomerular mesangial cells. *Am Heart J* 125: 589–597

25 Gardner JP, Tokudome G, Tomonari H, Maher E, Hollander D, Aviv A (1992) Endothelin-induced calcium responses in human vascular smooth muscle cells. *Am J Physiol* 262: C148–C155

26 Henry PJ (1994) Inhibitory effects of nordihydroguaiaretic acid on ETA-receptor- mediated contractions to endothelin-1 in rat trachea. *Br J Pharmacol* 111: 561–569

27 Korn SJ, Horn R (1990) Nordihydroguaiaretic acid inhibits voltage-activated Ca^{2+} currents independently of lipoxygenase inhibition. *Mol Pharmacol* 38: 524–530

28 Goldie RG, Grayson PS, Knott PG, Self GJ, Henry PJ (1994) Predominance of endothelin$_A$ (ET_A) receptors in ovine airway smooth muscle and their mediation of ET-1-induced contraction. *Br J Pharmacol* 112: 749–756

29 Yang CM, Yo YL, Ong R, Hsieh JT, Tsao HL (1994) Calcium mobilization induced by endothelins and sarafotoxin in cultured canine tracheal smooth muscle cells. *Naunyn-Schmeidebergs Arch Pharmacol* 350: 68–76

30 Advenier C, Sarria B, Naline E, Puybasset L, Lagente V (1990) Contractile activity of three endothelins (ET-1, ET-2 and ET-3) on the human isolated bronchus. *Br J Pharmacol* 100: 168–172

31 Lee HK, Leikauf GD, Sperelakis N (1990) Electromechanical effects of endothelin on ferret bronchial and tracheal smooth muscle. *J Appl Physiol* 68: 417–420

32 Grunstein MM, Chuang ST, Schramm CM, Pawlowski NA (1991) Role of endothelin 1 in regulating rabbit airway contractility. *Am J Physiol* 260: L75–L82

33 McKay KO, Armour CL, Black JL (1996) Endothelin receptors and activity differ in human, dog, and rabbit lung. *Am J Physiol* 270: L37–L43

34 Turner NC, Power RF, Polak JM, Bloom SR, Dollery CT (1989) Endothelin-induced contractions of tracheal smooth muscle and identification of specific endothelin binding sites in the trachea of the rat. *Br J Pharmacol* 98: 361–366

35 Ninomiya H, Uchida Y, Saotome M, Nomura A, Ohse H, Matsumoto H et al (1992) Endothelins constrict guinea pig tracheas by multiple mechanisms. *J Pharmacol Exp Ther* 262: 570–576

36 Henry PJ (1993) Endothelin-1 (ET-1)-induced contraction in rat isolated trachea: involvement of ET_A and ET_B receptors and multiple signal transduction systems. *Br J Pharmacol* 110: 435–441

37 Chand N, Diamantis W, Sofia RD (1990) Pharmacologic modulation of endothelin-induced contraction in isolated rat tracheal segments. *Res Commun Chem Pathol Pharmacol* 70: 173–181

38 Sarria B, Naline E, Morcillo E, Cortijo J, Esplugues J, Advenier C (1990) Calcium dependence of the contraction produced by endothelin (ET-1) in isolated guinea-pig trachea. *Eur J Pharmacol* 187: 445–453

39 Rubanyi GM, Polokoff MA (1994) Endothelins: molecular biology, biochemistry, pharmacology, physiology, and pathophysiology. *Pharmacol Rev* 46: 325–415

40 Inui T, James AF, Fujitani Y, Takimoto M, Okada T, Yamamura T et al (1994) ET_A and ET_B receptors on single smooth muscle cells co perate in mediating guinea pig tracheal contraction. *Am J Physiol* 266: L113–L124

41 Goldie RG, Henry PJ, Knott PG, Self GJ, Luttmann MA, Hay DW (1995) Endothelin-1 receptor density, distribution, and function in human isolated asthmatic airways. *Am J Resp Crit Care Med* 152: 1653–1658

42 Fukuroda T, Ozaki S, Ihara M, Ishikawa K, Yano M, Miyauchi T et al (1996) Necessity of dual blockade of endothelin ET_A and ET_B receptor subtypes for antagonism of endothelin-1-induced contraction in human bronchi. *Br J Pharmacol* 117: 995–999

43 Carr MJ, Goldie RG, Henry PJ (1996) Time course of changes in ET_B receptor density and function in tracheal airway smooth muscle during respiratory tract viral infection in mice. *Br J Pharmacol* 117: 1222–1228

44 Goldie RG, D'Aprile AC, Cvetkovski R, Rigby PJ, Henry PJ (1996) Influence of regional differences in ET_A and ET_B receptor subtype proportions on endothelin-1-induced contractions in porcine isolated trachea and bronchus. *Br J Pharmacol* 117: 736–742

45 Hay DW, Luttmann MA, Hubbard WC, Undem BJ (1993) Endothelin receptor subtypes in human and guinea-pig pulmonary tissues. *Br J Pharmacol* 110: 1175–1183

46 Suzaki A, Yamaguchi K, Adachi I, Kimura S (1997) Calcium mobilizing system coupled to endothelinA receptors (ET_A) in Swiss 3T3, A10 and NRK cells. *Biomedical Research* (Tokyo) 18: 221–229

47 Orallo F (1996) Regulation of cytosolic calcium levels in vascular smooth muscle. *Pharmacol Ther* 69: 153–171

48 Haller H, Smallwood JI, Rasmussen H (1990) Protein kinase C translocation in intact vascular smooth muscle strips. *Biochem J* 270: 375–381

49 Pollock DM, Keith TL, Highsmith RF (1995) Endothelin receptors and calcium signaling. *FASEB Journal* 9: 1196–1204

50 Battistini B, Filep JG, Cragoe EJ jr., Fournier A, Sirois P (1991) A role for Na^+/H^+ exchange in contraction of guinea pig airways by endothelin-1 *in vitro. Biochem Biophys Res Commun* 175: 583–588

51 Grinstein S, Rothstein A (1986) Mechanisms of regulation of the Na^+/H^+ exchanger. *J Membrane Biol* 90: 1–12

52 Lonchampt MO, Pinelis S, Goulin J, Chabrier PE, Braquet P (1991) Proliferation and Na^+/H^+ exchange activation by endothelin in vascular smooth muscle cells. *Am J Hypertens* 4: 776–779

53 Danthuluri NR, Brock TA (1990) Endothelin receptor-coupling mechanisms in vascular smooth muscle: a role for protein kinase C. *J Pharmacol Exp Ther* 254: 393–399

54 Payne AN, Whittle BJ (1988) Potent cyclo-oxygenase-mediated bronchoconstrictor effects of endothelin in the guinea-pig *in vivo. Eur J Pharmacol* 158: 303–304

55 Lagente V, Chabrier PE, Mencia-Huerta JM, Braquet P (1989) Pharmacological modulation of the bronchopulmonary action of the vasoactive peptide, endothelin, administered by aerosol in the guinea-pig. *Biochem Biophys Res Commun* 158: 625–632

56 Macquin-Mavier I, Levame M, Istin N, Harf A (1989) Mechanisms of endothelin-mediated bronchoconstriction in the guinea pig. *J Pharmacol Exp Ther* 250: 740–745

57 Nambu F, Yube N, Omawari N, Sawada M, Okegawa T, Kawasaki A et al (1991) Inhibition of endothelin-induced bronchoconstriction by OKY-046, a selective thromboxane A_2 synthetase inhibitor, in guinea pigs. *Adv Prostag Thrombox Leukotr Res* 21: 453–456

58 Noguchi K, Noguchi Y, Hirose H, Nishikibe M, Ihara M, Ishikawa K et al (1993) Role of endothelin ET_B receptors in bronchoconstrictor and vasoconstrictor responses in guinea-pigs. *Eur J Pharmacol* 233: 47–51

59 Lueddeckens G, Bigl H, Sperling J, Becker K, Braquet P, Forster W (1993) Importance of secondary TXA_2 release in mediating of endothelin-1 induced bronchoconstriction and vasopressin in the guinea-pig. *Prostag Leukotr Ess Fatty Acids* 48: 261–263

60 Dyson MC, Kadowitz PJ (1991) Influence of SK&F 96148 on thromboxane-mediated responses in the airways of the cat. *Eur J Pharmacol* 197: 17–25

61 Uhlig S, von Bethmann AN, Featherstone RL, Wendel A (1995) Pharmacologic characterization of endothelin receptor responses in the isolated perfused rat lung. *Am J Resp Crit Care Med* 152:1449–1460

62 de Nucci G, Thomas R, D'Orleans-Juste P, Antunes E, Walder C, Warner TD et al (1988) Pressor effects of circulating endothelin are limited by its removal in the pulmonary circulation and by the release of prostacyclin and endothelium-derived relaxing factor. *P Natn Acad Sci USA* 85: 9797–9800

63 D'Orleans-Juste P, Claing A, Telemaque S, Maurice MC, Yano M, Gratton JP (1994) Block of endothelin-1-induced release of thromboxane A_2 from the guinea pig lung and nitric oxide from the rabbit kidney by a selective ET_B receptor antagonist, BQ-788. *Br J Pharmacol* 113: 1257–1262

64 Hay DW, Hubbard WC, Undem BJ (1993) Endothelin-induced contraction and mediator release in human bronchus. *Br J Pharmacol* 110: 392–398

65 Large WA, Wang Q (1996) Characteristics and physiological role of the Ca^{2+}-activated Cl^- conductance in smooth muscle. *Am J Physiol* 271: C435–C454

66 Janssen LJ, Sims SM (1992) Acetylcholine activates non-selective cation and chloride conductances in canine and guinea-pig tracheal myocytes. *J Physiol* 453: 197–218

67 Daniel EE, Jury J, Bourreau JP, Jager L (1993) Chloride and depolarization by acetylcholine in canine airway smooth muscle. *Can J Physiol Pharmacol* 71: 284–292

68 Janssen LJ, Sims SM (1994) Substance P activates Cl^- and K^+ conductances in guinea-pig tracheal smooth muscle cells. *Can J Physiol Pharmacol* 72: 705–710

69 Janssen LJ, Sims SM (1993) Histamine activates Cl^- and K^+ currents in guinea-pig tracheal myocytes: convergence with muscarinic signalling pathway. *J Physiol* 465: 661–677

70 Nakajima T, Hazama H, Hamada E, Omata M, Kurachi Y (1995) Ionic basis of neurokinin-A-induced depolarization in single smooth muscle cells isolated from guinea-pig trachea. *Pfugers Arch-Eur J Physiol* 430: 552–562

71 Klockner U, Isenberg G (1991) Endothelin depolarizes myocytes from porcine coronary and human mesenteric arteries through a Ca-activated chloride current. *Pfugers Arch-Eur J Physiol* 418: 168–175

72 Gordienko DV, Clausen C, Goligorsky MS (1994) Ionic currents and endothelin signaling in smooth muscle cells from rat renal resistance arteries. *Am J Physiol* 266: F325–F341

73 Salter KJ, Kozlowski RZ (1996) Endothelin receptor coupling to potassium and chloride channels in isolated rat pulmonary arterial myocytes. *J Pharmacol Exp Ther* 279:1053–1062

74 Van Renterghem C, Lazdunski M (1993) Endothelin and vasopressin activate low conductance chloride channels in aortic smooth muscle cells. *Pfugers Arch-Eur J Physiol* 425: 156–163

75 Uchida Y, Saotome M, Nomura A, Ninomiya H, Ohse H, Hirata F et al (1991) Endothelin-1-induced relaxation of guinea pig trachealis muscles. *J Cardiovasc Pharmacol* 17: Suppl 7: S210–S212

76 White SR, Hathaway DP, Umans JG, Tallet J, Abrahams C, Leff AR (1991) Epithelial modulation of airway smooth muscle response to endothelin-1. *Am J Respir Dis* 144: 373–378

77 Filep JG, Battistini B, Sirois P (1993) Induction by endothelin-1 of epithelium-dependent relaxation of guinea-pig trachea *in vitro*: role for nitric oxide. *Br J Pharmacol* 109:637–644

78 Battistini B, Warner TD, Fournier A, Vane JR (1994) Characterization of ET_B receptors mediating contractions induced by endothelin-1 or IRL 1620 in guinea-pig isolated airways: effects of BQ-123, FR139317 or PD 145065. *Br J Pharmacol* 111: 1009–1016

79 Hadj-Kaddour K, Michel A, Chevillard C (1995) Endothelin-1 and endothelin-3 relax isolated guinea pig trachea through different mechanisms. *J Cardiovasc Pharmacol* 26: Suppl 3: S115–S116

80 Naline E, Bertrand C, Biyah K, Okada T, Fujitani Y, Sakaki J et al (1997) Characterization of endothelin-ET_A receptors in epithelium of human isolated airways responsible for a relaxant component of smooth muscle through NO release. *Am J Resp Crit Care Med* 153: A644

81 Hay DW, Hubbard WC, Undem BJ (1993) Relative contributions of direct and indirect mechanisms mediating endothelin-induced contraction of guinea-pig trachea. *Br J Pharmacol* 110: 955–962

82 El Mowafy AM, Abou-Mohamed GA (1996) Endothelins-induce cyclic AMP formation in the guinea-pig trachea through an ET_A receptor- and cyclooxygenase-dependent mechanism. *Br J Pharmacol* 118: 531–536

83 Sokolovsky M (1995) Endothelin receptor heterogeneity, G-proteins, and signaling via cAMP and cGMP cascades. *Cell Mol Neurobiol* 15: 561–571

84 Glassberg MK, Ergul A, Wanner A, Puett D (1994) Endothelin-1 promotes mitogenesis in airway smooth muscle cells. *Am J Respir Cell Mol Biol* 10: 316–321

85 Stewart AG, Grigoriadis G, Harris T (1994) Mitogenic actions of endothelin-1 and epidermal growth factor in cultured airway smooth muscle. *Clin Exp Pharmacol Physiol* 21: 277–285

86 Tomlinson PR, Wilson JW, Stewart AG (1994) Inhibition by salbutamol of the proliferation of human airway smooth muscle cells grown in culture. *Br J Pharmacol* 111: 641–647

87 Malarkey K, Chilvers ER, Lawson MF, Plevin R (1995) Stimulation by endothelin-1 of mitogen-activated protein kinases and DNA synthesis in bovine tracheal smooth muscle cells. *Br J Pharmacol* 116: 2267–2273

88 Panettieri RA jr, Goldie RG, Rigby PJ, Eszterhas AJ, Hay DW (1996) Endothelin-1-induced potentiation of human airway smooth muscle proliferation: an ET_A receptor-mediated phenomenon. *Br J Pharmacol* 118: 191–197

89 Shapiro PS, Evans JN, Davis RJ, Posada JA (1996) The seven-transmembrane-spanning receptors for endothelin and thrombin cause proliferation of airway smooth muscle cells and activation of the extracellular regulated kinase and c-Jun NH2-terminal kinase groups of mitogen-activated protein kinases. *J Biol Chem* 271: 5750–5754

90 Fujitani Y, Bertrand C (1997) Differential role of endothelin A and B receptors in cultured human airway smooth muscle cells. *Am J Resp Crit Care Med* 153: A843

91 Scott PH, Belham CM, al-Hafidh J, Chilvers ER, Peacock AJ, Gould GW et al (1996) A regulatory role for cAMP in phosphatidylinositol 3-kinase/p70 ribosomal S6 kinase-mediated DNA synthesis in platelet-derived-growth-factor-stimulated bovine airway smooth-muscle cells. *Biochem J* 318: 965–971

92 Panettieri RA, Hall IP, Maki CS, Murray RK (1995) Alpha-thrombin increases cytosolic calcium and induces human airway smooth muscle cell proliferation. *Am J Respir Cell Mol Biol* 13: 205–216

93 Fujitani Y, Bertrand C (1997) ET-1 cooperates with EGF to induce mitogenesis via a PTX-sensitive pathway in airway smooth muscle cells. *Am J Physiol* 41: C1492–C1498

94 Weissberg PL, Witchell C, Davenport AP, Hesketh TR, Metcalfe JC (1990) The endothelin peptides ET-1, ET-2, ET-3 and sarafotoxin S6b are co-mitogenic with platelet-derived growth factor for vascular smooth muscle cells. *Atherosclerosis* 85: 257–262

95 Battistini B, Chailler P, D'Orleans-Juste P, Briere N, Sirois P (1993) Growth regulatory properties of endothelins. *Peptides* 14: 385–399

96 Irons CE, Flynn MA, Mok LM, Reynolds EE (1996) Endothelin and PDGF enhance arachidonic acid release and DNA synthesis in vascular smooth muscle cells. *Am J Physiol* 270: C1642–C1646

97 van Corven EJ, Hordijk PL, Medema RH, Bos JL, Moolenaar WH (1993) Pertussis toxin-sensitive activation of p21ras by G protein-coupled receptor agonists in fibroblasts. *P Natn Acad Sci USA* 90: 1257–1261

98 Kahan C, Seuwen K, Meloche S, Pouyssegur J (1992) Coordinate, biphasic activation of p44 mitogen-activated protein kinase and S6 kinase by growth factors in hamster fibroblasts. Evidence for thrombin-induced signals different from phosphoinositide turnover and adenylylcyclase inhibition. *J Biol Chem* 267: 13369–13375

99 Whelchel A, Evans J, Posada J (1997) Inhibition of ERK activation attenuates endothelin-stimulated airway smooth muscle cell proliferation. *Am J Respir Cell Mol Biol* 16: 589–596

100 Lane HA, Fernandez A, Lamb NJ, Thomas G (1993) p70[s6k] function is essential for G1 progression. *Nature* 363: 170–172

101 Reinhard C, Fernandez A, Lamb NJ, Thomas G (1994) Nuclear localization of p85[s6k]: functional requirement for entry into S phase. *EMBO Journal* 13: 1557–1565

102 Panettieri RA (1994) Airways smooth muscle cell growth and proliferation. In: Raeburn D, Giembycz MA (eds). *Airways smooth muscle: Development and regulation of contractility.* Basel/Switzerland: Birkhäuser Verlag, 41–68

103 Kolch W, Heidecker G, Kochs G, Hummel R, Vahidi H, Mischak H, et al (1993) Protein kinase C alpha activates RAF-1 by direct phosphorylation. *Nature* 364: 249–252

104 McLees A, Graham A, Malarkey K, Gould GW, Plevin R (1995) Regulation of lysophosphatidic acid-stimulated tyrosine phosphorylation of mitogen-activated protein kinase by protein kinase C- and pertussis toxin-dependent pathways in the endothelial cell line EAhy 926. *Biochem J* 307: 743–748

105 Black PN, Ghatei MA, Takahashi K, Bretherton-Watt D, Krausz T, Dollery CT et al (1989) Formation of endothelin by cultured airway epithelial cells. *FEBS Lett* 255: 129–132

106 MacCumber MW, Ross CA, Glaser BM, Snyder SH (1989) Endothelin: visualization of mRNAs by *in situ* hybridization provides evidence for local action. *P Natn Acad Sci USA* 86: 7285–7289

107 Rozengurt N, Springall DR, Polak JM (1990) Localization of endothelin-like immunore-activity in airway epithelium of rats and mice. *J Pathol* 160: 5–8

108 Endo T, Uchida Y, Matsumoto H, Suzuki N, Nomura A, Hirata F et al (1992) Regulation of endothelin-1 synthesis in cultured guinea pig airway epithelial cells by various cytokines. *Biochem Biophys Res Commun* 186: 1594–1599

109 Nakano J, Takizawa H, Ohtoshi T, Shoji S, Yamaguchi M, Ishii A et al (1994) Endotoxin and pro-inflammatory cytokines stimulate endothelin-1 expression and release by airway epithelial cells. *Clin Exp Allergy* 24: 330–336

110 Murlas CG, Gulati A, Singh G, Najmabadi F (1995) Endothelin-1 stimulates proliferation of normal airway epithelial cells. *Biochem Biophys Res Commun* 212: 953–959

111 Takimoto M, Oda K, Sasaki Y, Okada T (1996) Endothelin-A receptor-mediated prostanoid secretion via autocrine and deoxyribonucleic acid synthesis via paracrine signaling in human bronchial epithelial cells. *Endocrinology* 137: 4542–4550

112 Webber SE, Yurdakos E, Woods AJ, Widdicombe JG (1992) Effects of endothelin-1 on tracheal submucosal gland secretion and epithelial function in the ferret. *Chest* 101: 63S–67S

113 Satoh M, Shimura S, Ishihara H, Nagaki M, Sasaki H, Takishima T (1992) Endothelin-1 stimulates chloride secretion across canine tracheal epithelium. *Respiration* 59: 145–150

114 Plews PI, Abdel-Malek ZA, Doupnik CA, Leikauf GD (1991) Endothelin stimulates chloride secretion across canine tracheal epithelium. *Am J Physiol* 261: L188–L194

115 Tamaoki J, Kanemura T, Sakai N, Isono K, Kobayashi K, Takizawa T (1991) Endothelin stimulates ciliary beat frequency and chloride secretion in canine cultured tracheal epithelium. *Am J Respir Cell Mol Biol* 4: 426–431

116 Sirvio ML, Metsarinne K, Saijonmaa O, Fyhrquist F (1990) Tissue distribution and half-life of ^{125}I-endothelin in the rat: importance of pulmonary clearance. *Biochem Biophys Res Commun* 167: 1191–1195

117 Wu T, Rieves RD, Larivee P, Logun C, Lawrence MG, Shelhamer JH (1993) Production of eicosanoids in response to endothelin-1 and identification of specific endothelin-1 binding sites in airway epithelial cells. *Am J Respir Cell Mol Biol* 8: 282–290

118 Mullol J, Baraniuk JN, Logun C, Benfield T, Picado C, Shelhamer JH (1996) Endothelin-1 induces CM-CSF, IL-6 and IL-8 but not G-CSF release from a human bronchial epithelial cell line (BEAS-2B). *Neuropeptides* 30: 551–556

119 Takizawa H, Ohtoshi T, Kikutani T, Okazaki H, Akiyama N, Sato M et al (1995) Histamine activates bronchial epithelial cells to release inflammatory cytokines *in vitro*. *Int Arch Allergy Immunol* 108: 260–267

120 Markewitz BA, Kohan DE, Michael JR (1995) Endothelin-1 synthesis, receptors, and signal transduction in alveolar epithelium: evidence for an autocrine role. *Am J Physiol* 268: L192–L200

121 Shimura S, Ishihara H, Satoh M, Masuda T, Nagaki N, Sasaki H et al (1992) Endothelin regulation of mucus glycoprotein secretion from feline tracheal submucosal glands. *Am J Physiol* 262: L208–L213

122 Griffin A, Newman TM, Scott RH (1996) Electrophysiological and ultrastructural events evoked by methacholine and intracellular photolysis of caged compounds in cultured ovine trachea submucosal gland cells. *Exp Physiol* 81: 27–43

123 Shimura S, Ishihara H, Nagaki M, Sasaki H, Takishima T (1993) A stimulatory role of protein kinase C in feline tracheal submucosal gland secretion. *Respir Physiol* 93: 239–247

124 Nagaki M, Shimura S, Irokawa T, Sasaki T, Oshiro T, Nara M et al (1996) Bradykinin regulation of airway submucosal gland secretion: role of bradykinin receptor subtype. *Am J Physiol* 270: L907–L913

125 Nagaki M, Shimura MN, Irokawa T, Sasaki T, Shirato K (1995) Nitric oxide regulation of glycoconjugate secretion from feline and human airways *in vitro*. *Respir Physiol* 102: 89–95

126 Yurdakos E. Webber SE (1991) Endothelin-1 inhibits pre-stimulated tracheal submucosal gland secretion and epithelial albumin transport. *Br J Pharmacol* 104: 1050–1056

127 Matran R, Alving K, Hemsen A, Lundberg JM (1990) Endothelin-1 increases airway mucosa blood flow in the pig. *Agents & Actions – Suppl* 31: 237–241

128 Hemsen A, Larsson O, Lundberg JM (1991) Characteristics of endothelin A and B binding sites and their vascular effects in pig peripheral tissues. *Eur J Pharmacol* 208: 313–322

129 Barman SA, Ardell JL, Taylor AE (1993) Effect of endothelin-1 on canine airway blood flow. *J Cardiovasc Pharmacol* 22:Suppl 8: S274–S277

130 Power RF, Wharton J, Zhao Y, Bloom SR, Polak JM (1989) Autoradiographic localization of endothelin-1 binding sites in the cardiovascular and respiratory systems. *J Cardiovasc Pharmacol* 13: Suppl 5: S50–S56

131 Filep JG, Fournier A, Foldes-Filep E (1995) Acute pro-inflammatory actions of endothelin-1 in the guinea-pig lung: involvement of ET_A and ET_B receptors. *Br J Pharmacol* 115: 227–236

132 Hosoda K, Nakao K, Hiroshi-Arai, Suga S, Ogawa Y, Mukoyama M et al (1991) Cloning and expression of human endothelin-1 receptor cDNA. *FEBS Lett* 287: 23–6

133 Filep JG, Sirois MG, Rousseau A, Fournier A, Sirois P (1991) Effects of endothelin-1 on vascular permeability in the conscious rat: interactions with platelet-activating factor. *Br J Pharmacol* 104: 797–804

134 Filep JG, Fournier A, Foldes-Filep E (1994) Endothelin-1-induced myocardial ischaemia and oedema in the rat: involvement of the ET_A receptor, platelet-activating factor and thromboxane A2. *Br J Pharmacol* 112: 963–971

135 Filep JG, Skrobik Y, Fournier A, Foldes-Filep E (1996) Effects of calcium antagonists on endothelin-1-induced myocardial ischaemia and oedema in the rat. *Br J Pharmacol* 118: 893–900

136 Goldie RG, Pedersen KE (1995) Mechanisms of increased airway microvascular permeability: role in airway inflammation and obstruction. *Clin Exp Pharmacol Physiol* 22: 387–396

137 McKay KO, Armour CL, Black JL (1993) Endothelin-3 increases transmission in the rabbit pulmonary parasympathetic nervous system. *J Cardiovasc Pharmacol* 22:Suppl 8: S181–S184

138 Yoneyama T, Hori M, Tanaka T, Matsuda Y, Karaki H (1995) Endothelin ET_A and ET_B receptors facilitating parasympathetic neurotransmission in the rabbit trachea. *J Pharmacol Exp Ther* 275: 1084–1089

139 Knott PG, Fernandes LB, Henry PJ, Goldie RG (1996) Influence of endothelin-1 on cholinergic nerve-mediated contractions and acetylcholine release in rat isolated tracheal smooth muscle. *J Pharmacol Exp Ther* 279: 1142–1147

140 Henry PJ, Goldie RG (1995) Potentiation by endothelin-1 of cholinergic nerve-mediated contractions in mouse trachea via activation of ET_B receptors. *Br J Pharmacol* 114: 563–569

141 Fernandes LB, Henry PJ, Rigby PJ, Goldie RG (1996) Endothelin$_B$ (ET_B) receptor-activated potentiation of cholinergic nerve-mediated contraction in human bronchus. *Br J Pharmacol* 118: 1873–1874

142 Henry PJ, Shen A, Mitchelson F, Goldie RG (1996) Inhibition by endothelin-1 of cholinergic nerve-mediated acetylcholine release and contraction in sheep isolated trachea. *Br J Pharmacol* 118: 762–768

143 Majewski H, Barrington M (1995) Second messenger pathways in the modulation of neurotransmitter release. In: Powis DA, Bunn SJ (eds) *Neurotransmitter release and its modulation*. Cambridge: Cambridge University Press, 163–181

144 Burgoyne RD, Cheek TR (1995) Mechanisms of exocytosis and the central role of calcium. In: Powis DA, Bunn SJ (eds) *Neurotransmitter release and its modulation*. Cambridge: Cambridge University Press, 7–21

145 Fryer AD, Elbon CL, Kim AL, Xiao HQ, Levey AI, Jacoby DB (1996) Cultures of airway parasympathetic nerves express functional M_2 muscarinic receptors. *Am J Respir Cell Mol Biol* 15: 716–725

Pulmonary Actions of the Endothelins
ed. by R. G. Goldie and D. W. P. Hay
© 1999 Birkhäuser Verlag Basel/Switzerland

CHAPTER 6
In Vitro Effects of the Endothelins on Airway and Vascular Smooth Muscle Tone

Claude Bertrand[1], Emmanuel Naline[2] and Charles Advenier[2]

[1] *Roche Bioscience, Inflammatory Diseases Unit, 3401 Hillview Avenue, Palo Alto, CA 94304-1397, USA*
[2] *Faculté de Médecine Paris-Ouest, Laboratoire de Pharmacologie, 15, rue de l'Ecole de Médecine, F-75270 Paris Cedex 06, France*

1   Introduction
2   Airway Smooth Muscle
2.1   Functional Effects
2.1.1 Contractile activity
2.1.2 Relaxant activity
2.1.3 Indirect effects on neurotransmitter release
2.1.4 Effect of big ET-1
2.1.5 Signal transduction mechanisms
2.2   Receptors Involved
3   Pulmonary Vascular Smooth Muscle
3.1   Functional Effects
3.1.1 Contractile activity
3.1 2 Relaxant activity
3.2   Role of Hypoxia
4   Conclusion
     References

1. Introduction

Endothelin-1 (ET-1) was first isolated from vascular endothelial cells, and susbsequently found in large amount in the lungs, where it is produced mainly by airway epithelial cells and airway macrophages. It is noteworthy that isolated airways and pulmonary vascular preparations, largely because of the abundance of ET receptors in these tissues and the ability to conduct organ bath preparations, have been used extensively in studies investigating ET receptor characterization, subtyping and the different intracellular pathways following activation of these receptors. This field of investigation has rapidly expanded over the last 5 years and it became clear that ET-1 had a large array of functional effects including contraction and relaxation due to direct activation of endothelin receptors, as well as release of mediators and subsequent activation of their corresponding receptors.

In this chapter a comprehensive review of our current knowledge of the influence of the ETs in isolated airway and vascular smooth muscle will be presented with a special emphasis on the receptors involved.

2. Airway Smooth Muscle

2.1. Functional Effects

2.1.1. Contractile activity

Description of the functional effect: The ETs are potent and effective contractile agonists in isolated airway smooth muscle preparations from a variety of species including humans [1–5]; the contraction develops relatively quickly, is well maintained and reversed by washing. Potency values (pD$_2$) of ET-1 in different species (Tab. 1) range from 7.30 to 8.88, demonstrating that ET-1 is one of the most contractile agents described, being more potent than acetylcholine, histamine or leukotriene D$_4$ (Tab. 2). In terms of efficiency (expressed as % of the acetylcholine effect), the relative magnitude of the maximal contractile effect of ET-1 is species dependent and varied from 60% in the pig bronchus [6] to 110% in the rabbit bronchus [7]. In human bronchus, the maximal contractile effect is approximately 80% of the acetylcholine-induced response [1]. However, within the same species, the efficacies vary depending on the size of the airways used in the study. For instance, in the guinea-pig airways, the contractile effect of sarafotoxin 6c, a selective ET$_B$ receptor agonist, was greater in bronchus (maximal effect of 69% of carbachol effect) or distal trachea (48%) than in middle or proximal trachea (14 and 19%, respectively), although these differences were less evident when using ET-1 [8].

Indirect mechanisms: Early on, indirect mechanisms were suspected to be involved in the contractile activity of the ETs in the airways. In this context, the prostanoids was the first family of mediators to be studied, using cyclooxygenase inhibitors. In human bronchus, ET-1-induced contraction is not modulated by cyclooxygenase products [1] or only slightly potentiated [2, 9]. Similar findings were observed using guinea-pig [5, 9–12], ferret [13], bovine [14–16], dog and rabbit airways [17]. Although prostanoids do not appear to contribute significantly to the contractile response, it is noteworthy that ET-1 stimulated the release of various prostanoids, such as thromboxane A$_2$ (TXA$_2$), PGD$_2$, PGE$_2$ and PGI$_2$ or their metabolites, from guinea-pig trachea and human bronchi [8, 9]. In addition, Hay et al. [2] provided strong evidence that ET-1 produced contraction of the human bronchi which did not involve the significant contribution of acetylcholine, PAF, histamine, peptidoleukotrienes or tachykinins. Although this group reached the same conclusion for the guinea-pig trachea, ET-1-induced contraction in this preparation was reported to be mediated in part by PAF, thromboxane A$_2$ and peptidoleukotrienes [18–20]. In addition, a role of histamine was suggested, and it was postulated that ET-1 was able to activate mast cells in the guinea-pig trachea [21].

Table 1. Comparison of the potencies (pD$_2$ or EC$_{50}$) of the contractile effects of endothelin-1 in different species and the receptors involved in mediating these responses

		Endothelin-1 pD$_2$ or EC$_{50}$	Receptor involved (§)	
Human	*Bronchus*	7.90 ± 0.17 [1]	ET$_A$	Epithelial, release of NO and prostanoids, relaxation of ASM
		7.58 ± 0.15 [8]	ET$_A$	ASM, involved in contraction induced by ET-1
		8.88 ± 0.16 [56]	ET$_B$	ASM, contraction
Guinea-pig	*Trachea*	8.15 ± 0.14 [8]	ET$_A$	ASM relaxation, epithelium dependent Contraction of isolated lung parenchyma
	Bronchus	7.72 ± 0.12 [8]	ET$_B$	ASM contraction (atypical ET$_B$ receptor population)
Rat	*Trachea*	18 nM (pD$_2$ = 7.74) [46] (10–32 nM, 95% confidence limit)	ET$_A$ ET$_A$ ET$_B$	Increase in intracellular inositol phosphate and in intracellular Ca^{2+} ASM, involved in contraction induced by ET-1 ASM contraction
Mouse	*Trachea*	6.3 nM (pD$_2$ = 8.20) [31] (4.0–10, 95% confidence limit)	ET$_A$ ET$_B$	ASM, involved in contraction induced by ET-1 ASM contraction
Pig	*Trachea and bronchus*	22 nM (pD$_2$ = 7.66) [6] (9–55, 95% confidence limit)	ET$_A$ + ET$_B$	
Sheep	*Trachea*	8.36 ± 0.06 [60]	ET$_A$	ASM contraction
Rabbit	*Bronchus*	7.5 ± 0.1 [7]	ET$_B$	ASM contraction (heterogeneous ET$_B$ receptor population)

ASM: airway smooth muscle; EC$_{50}$: concentration giving 50% of maximal response; pD$_2$: −log EC$_{50}$; [§] for references: see text 2.2.

Table 2. Comparison of pD_2 values and E_{max} for some contractile agents on the human isolated bronchus [1, 47]

	pD_2		E_{max}	
Endothelin-1	7.90 ± 0.17	(n = 9)	82.5 ± 4.7	(n = 9)
Endothelin-2	7.23 ± 0.14	(n = 8)		
Endothelin-3	7.21 ± 0.08	(n = 7)		
Big Endothelin	7.53 ± 0.08	(n = 11)	78.5 ± 3.8 %	(n = 11)
Acetylcholine	4.96 ± 0.12	(n = 8)	100	
Histamine	6.02 ± 0.11	(n = 8)	92.4 ± 2.5	(n = 8)
LTD$_4$	7.67 ± 0.19	(n = 4)	80.5 ± 6.8	(n = 4)
Neurokinin A	8.48 ± 0.24	(n = 6)	77.6 ± 6.1	(n = 6)

E_{max} are expressed as % of ACh 3 mM; n = number of experiments.

Role of the airway epithelium: The airway epithelium not only acts as a physical barrier, but also contains enzymes which metabolize peptides, or releases mediators to modulate the responses of the underlying smooth muscle to a variety of contractile agents including ETs [22]. A potentiation of the contractile response to ET-1 after epithelium removal has already been reported in the guinea-pig trachea [12, 23–25] and bronchi [10] and the human bronchus [26]. In these studies, it was suggested that neutral endopeptidase EC 24.11 (NEP) was in part responsible for the potentiation of the ET-1 response after epithelium removal since inhibitors of this enzyme were able to partially mimic this [24, 26, 27]. However, the potentiation produced by these inhibitors was smaller than the effect of epithelium removal [26], suggesting the release of an epithelium-derived relaxant factor in response to ET-1 in human bronchi. In this preparation, a recent study has suggested the release of nitric oxide by airway epithelial cells and has shown a potentiation of the concentration-response curves to ET-1 in the presence of the NO synthase inhibitor, L-NAME, which was abolished by epithelium removal (Fig. 1) [28]. Furthermore, since this potentiation was not observed for the contractile response to the specific ET_B receptor agonist, IRL 1620, and was abolished by the ET_A receptor antagonist, BQ-123, it might be suggested that this phenomenon is related to ET_A receptor activation (Fig. 2) [28]. A release of prostaglandin E_2 from epithelial cells has also been suggested in guinea-pig [29, 30] or feline [31] airways.

Role of viral infection: In mice which developed a respiratory tract infection following influenza virus inoculation, ET-1 evoked concentration-dependent contraction of isolated tracheal smooth muscle similar to those obtained in tissues from naive animals, whereas the maximal contraction observed with the selective ET_B receptor agonist, sarafotoxin 6c, was reduced. In these animals, virus inoculation induced an increase in the density of ET_A sites and a decrease of ET_B sites [32, 33].

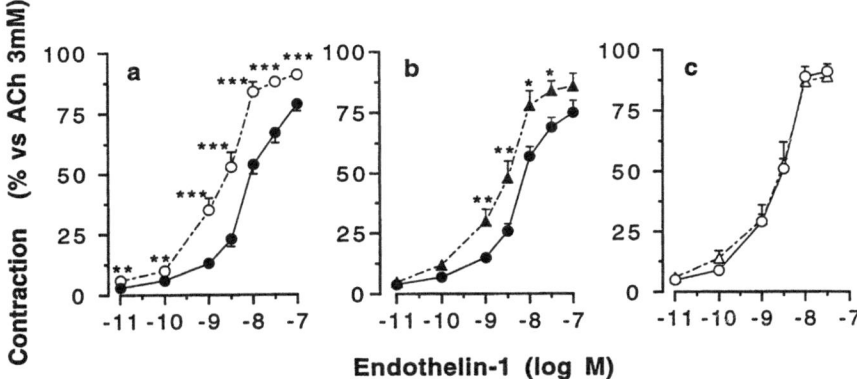

Figure 1. Influence of epithelium on the contractile effect of ET-1 on the human isolated bronchus: (a) Effects of ET-1 in preparation with (●) or without epithelium (○); (b) Influence of L-NAME (3 mM) on ET-1 concentration response curve in preparation with epithelium. Control (●) or with (▲) L-NAME; (c) Influence of L-NAME (3 mM) on ET-1 concentration response curve in preparation without epithelium. Control (○) or with (△) L-NAME. Significant differences from control are shown as: * $p < 0.05$; ** $p < 0.01$; *** $p < 0.001$. These results suggest that ET-1 induces a release of a relaxant factor from epithelium which might be NO [28].

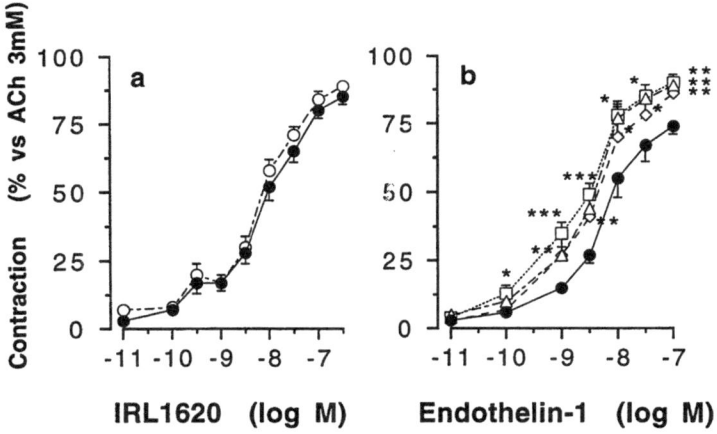

Figure 2. Analysis of receptors involved in the release of the relaxant factor (NO) by epithelial cells under the influence of ET-1: (a) Effect of epithelium removal on the contractile response of the specific ET_B receptor agonist IRL 1620. (●) control with epithelium, (○) without epithelium; (b) Effects of BQ-123 on the concentration response curve of ET-1 in preparation with epithelium. Control (●), BQ-123 10^{-7} M (△), 10^{-6} M (◇), 10^{-5} M (□). Significant differences from control are shown as: * $p < 0.05$; ** $p < 0.01$; *** $p < 0.001$. These results suggest that the release of the relaxant factor by epithelial cells under the influence of ET-1 is related to ET_A receptor stimulation [28].

2.1.2. Relaxant activity: In vitro studies have shown that ET-1 exerts a dual action on guinea-pig [29, 30, 34–36] or on rabbit [37] airway smooth muscle, an initial and transient relaxation followed by a sustained contraction. ET-1-induced-relaxation was attenuated by indomethacin and removal of the epithelium in rabbit airway [37] or guinea-pig trachea [36], suggesting that smooth muscle relaxant prostanoids (e.g., PGI_2, PGE_2) may mediate, at least in part, the relaxation. In the guinea-pig trachea, the relaxant activity of ET-1 was also found to be mediated by NO release from epithelial cells [38]. In addition, relaxation of the guinea-pig trachea by ET-1 or ET-3 was suggested to involve charybdotoxin-sensitive K^+ channels [35].

2.1.3. Indirect effects on neurotransmitter release: We will provide an overview of the organ bath studies looking at the modulation of electrically or capsaicin stimulated airways since this subject is discussed exensively in Chapter 11. Several groups have recently demonstrated that ET-1 and related peptides potentiate cholinergic nerve-induced contractions of preparations from rabbit, murine and human airways [39–41]. For instance, ET-1 markedly enhanced the contractile response elicited by electric field stimulation (EFS) of parasympathetic nerves in murine trachea [40]. Based on the findings that ET-1 enhanced the contractile responses to EFS but not those responses produced by exogenous administration of acetylcholine, it was proposed that the enhanced contractile response was the result of a prejunctional mechanism. This effect could be induced by ET-3 [39] as well suggesting the involvement of ET_B receptors located presynaptically on parasympathetic nerve endings. However, in the rat trachea, Knott et al., following studies using ET_A and ET_B selective agonists, have suggested the involvement of both receptors, and they have correlated the functional effect to the release of acetylcholine following the addition of ET-1 [42]. Recently, it has been demonstrated that ET-1 potentiated the release of calcitonin-gene related peptide and substance P induced by capsaicin (a C fiber activator) in guinea-pig airways [43]. The partial inhibition of this synergistic effect by ET_A or ET_B selective antagonists suggested the involvement of both receptor subtypes, located on the C fiber endings in the airways [43]. Taken together, these data demonstrated the importance of endothelin as a neuromodulator in the airways in addition to its direct effect on the smooth muscle.

2.1.4. Effect of big ET-1: ET-1 is derived from a 39 amino-acid precursor, named big ET-1, via a single cleavage between amino acids Trp-21 and Val-22 [44]. The enzyme responsible for this processing was called ET-converting enzyme (ECE) [45, 46]. Big endothelin 1 (Big ET-1) elicits a potent contraction of the human isolated bronchus ($pD_2 = 7.53 \pm 0.08$, $n = 11$) with a maximal effect of $78.5 \pm 3.8\%$ (% of acetylcholine 3 mM) [47]. Its potency was reduced by pretreatment of the preparation with phosphoramidon and to a lesser extent by thiorphan, in agreement with the concept that ECE is a metalloprotease different from the neutral endopeptidase (NEP, EC 2.24.11).

2.1.5. Signal transduction mechanisms

Intracellular calcium mobilization: Early studies have demonstrated that in the guinea-pig trachea, the contractile activity of ET-1 seemed to be dependent upon the influx of calcium ions from the extracellular space through dihydropyridine-sensitive calcium channels [48], or through receptor-opened channels [11, 21]. However, more extensive studies indicated that ET-1 induced contraction in airway is resistant to incubation in Ca^{2+}-free physiological buffer or Ca^{2+} channel inhibitors; this is particularly true for higher concentrations of ET-1 [10–12, 14, 34, 49, 50].

In human isolated bronchus, Advenier et al. [1] have described a two-step contraction: at low concentrations (below nM), the contraction was potentiated by Bay K 8644, reduced by nicardipine or in calcium free-medium. In contrast, at higher concentrations, the contractile effect was independent of voltage-dependent calcium channels. ET-induced contraction of human and animal isolated airways appears to be predominantly *via* mobilisation of intracellular Ca^{2+}, although the relative contribution of this signal may depend on the ET-receptor subtype and also the species [50, 51].

Phosphoinositide pathway: Although ET-1 increases the level of intracellular phosphatidylinositols under certain conditions, this mechanism seems to be only partially involved in the contraction of airway smooth muscle. In the guinea-pig trachea, ET-1 produced a concentration-dependent stimulation of phosphatidylinositol turnover, with an EC_{50} value of 45.9 nM, close to the EC_{50} of the contractile effect (30.9 nM) [11]. However, in bovine bronchial smooth muscle and in the human bronchus, Nally et al. [14, 15] have shown that ET-1 but not ET-3, at the same concentration, evoked a rise in the level of inositol (1,4,5) trisphosphate (IP_3). Similarly, in rat isolated tracheal smooth muscle, Henry has observed that ET-1 induced a seven-fold increase in intracellular [^{3}H] inositol phosphate accumulation over basal levels whereas sarafotoxin 6c increased it by only two-fold [50]. Although the effect of ET-1 was abolished by BQ-123, an antagonist of the ET_A receptors, this drug did not modify the contractile activity of ET-1, suggesting a minor role of the inositol phosphates in the airway contraction.

Role of protein kinase C: Phorbol (12,13) dibutyrate, which stimulates protein kinase C (PKC), enhanced both ET-1 and ET-3 evoked responses. This potentiation is abolished by Ro 31-8220, a selective inhibitor of PKC. However, staurosporine, another inhibitor of PKC, and Ro 31-8220 alone had no effect on ET-1 induced contraction, indicating that the role of PKC remains unclear [11, 14, 15].

2.2. Receptors Involved

In the context of this chapter, we will focus on the studies examining the effects of selective agonists and antagonists in the contraction and relaxation of isolated airways. Due to the quantitative and qualitative differences apparent in the pharmacological profile of the ET isoforms in airway isolated tissues, it was proposed that these effects may be mediated via multiple ET receptors. Binding and immunohistochemical studies have demonstrated the presence of both ET_A and ET_B receptors. Although the number of ET_B receptors is almost always superior to the number of ET_A receptors in the airways, it appears that the exact ratio is species-dependent [51] (Tab. 1). In recent years, the development of selective agonists and antagonists of ET_A and ET_B receptors has allowed careful pharmacological evaluation of the contraction and relaxation induced by ETs in the airways. First, sarafotoxin 6c and IRL 1620, the ET_B selective agonists [52–55], and peptide ET_A selective antagonists such as BQ-123 and FR 139317 [52, 53, 56] were characterized. Using these experimental tools, functional evidence was provided for a non-ET_A-mediated contraction (probably ET_B) in the guinea-pig trachea and bronchus [8, 49, 57] as well as in human bronchus [8]. In addition, regional differences in the relative distribution of ET_A and ET_B receptors occurs in guinea-pig airways. Thus, the contribution of ET_B receptors, assessed functionally using sarafotoxin 6c, predominated over ET_A receptors in guinea-pig bronchus, and increased from the upper trachea (little contribution) to the lower trachea (marked contribution) [8]. In a more recent study utilizing various ligands, including antagonists, which interact with ET_A and ET_B receptors, it was again concluded that ET_B receptors mediate contractions produced by ET-1 in guinea-pig trachea, bronchus and parenchyma, but ET_A receptor activation also contributes to the response in guinea-pig trachea and lung parenchyma [30]. However, the development of highly selective ET_B receptor antagonists such as BQ-788 [58] and IRL 2500 [59] allowed for further exploration of these responses. Indeed, although the selective ET_B agonist IRL1620 induced a contraction of human bronchi on its own, ET-1 demonstrated a higher intrinsic activity in this preparation [60], suggesting the involvement of non-ET_B receptors. Moreover, ET_A antagonist (BQ-123) or ET_B antagonist (IRL 2500) alone were unable to antagonize the contraction induced by ET-1 in the same preparation [60, 61]. However, the combination of both antagonists induced a leftward shift in the concentration-response curve to ET-1, suggesting the involvement of both ET_A and ET_B receptors in this preparation (Fig. 3). A recent study confirmed this data, using BQ-123, FR 139317 (endothelin ET_A receptor antagonist), BQ-788 or bosentan (a dual ET_A/ET_B receptor

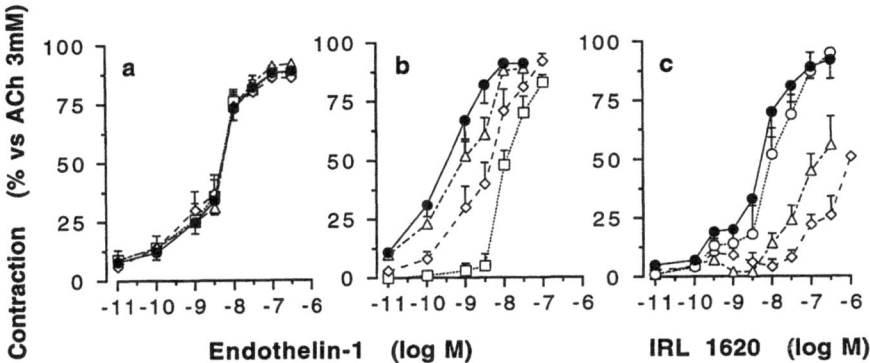

Figure 3. Analysis of the receptors involved in the contractile effect of endothelin-1 (a, b) and of the specific ET_B receptor agonist, IRL 1620 (c) on the human isolated bronchus without epithelium with the specific ET_B or ET_A receptor antagonists, IRL 2500 (a, b, c) and BQ-123 (10^{-6} M, b), respectively. Symbols are: Control (●), IRL 2500 10^{-8} M (○), 10^{-7} M (△), 10^{-6} M (◇), 10^{-5} M (□). This figure shows that IRL 2500 alone failed to modify the effects of ET-1 (a) but induced a significant shift to the right of the concentration response curve of IRL 1620 (c) ($pA_2 = 8.16 \pm 0.27$, slope: 1.27 ± 0.20). When combined with BQ-123 (10^{-6} M), IRL 2500 elicit a non-competitive antagonism vs ET-1 (b) ($pA'_2 = 7.32 \pm 0.16$, slope: 0.68 ± 0.06) [60].

antagonist), and suggested the necessity of dual blockade of ET_A and ET_B receptors for antagonism of ET-1-induced contraction in human bronchi [61, 62], as has been demonstrated in rat trachea [50]. The mechanism of the synergistic inhibition by ET_A and ET_B receptor antagonists of ET-1-induced contraction is not clear. One possible explanation in tissues where both receptors coexist is that a cross-talk mechanism exists in the signal transduction systems between ET_A and ET_B receptors.

3. Pulmonary Vascular Smooth Muscle

3.1. Functional Effects

3.1.1. Contractile activity: ET-1 is the most potent endogenous substance known to induce contraction of isolated blood vessels. Ever since the first observations with the peptidergic endothelium-derived contractile factor (EDCF) on porcine and bovine coronary arteries [63] and the bioassay of the isolated purified peptide on porcine coronary arteries [44], numerous studies confirmed the ability to induce potent contraction in arterial preparation of various anatomical origin isolated from a variety of animal species and humans [64]. The majority of studies also confirmed the original observation that ET-1 induced slowly developing and long lasting contractions. Although the majority of studies were performed on isolated large arteries, several studies compared the effect of ETs on large arteries and veins. In general, these studies found that ET-1 is three- to ten-fold more potent in veins than in arteries [64]. In contrast to these large blood vessels in the systemic circulation, arteries are more sensitive than veins in the pulmonary circulation. For instance, in the guinea-pig, pulmonary arteries were more sensitive to ET-1 than were pulmonary veins [65]. Overall ET-1 is a potent constrictor of mammalian pulmonary blood vessels, including human pulmonary artery and vein [8, 66, 67].

The response to ET-1 in human pulmonary artery appears to be mediated predominantly, if not solely, via ET_A receptor activation [8, 68–70]. Thus, ET-1 induced responses were potently antagonized by BQ-123, whereas sarafotoxin 6c, the ET_B selective agonist, was without effect on the level of tone on this tissue [8]. Recently, ET-3 was shown to contract human pulmonary arteries in a BQ-123-sensitive manner confirming the major role of ET_A receptor in the contraction by the endothelins in this preparation [71]. However, ET_B receptors have been proposed to mediate contraction in porcine pulmonary vein [72, 73] and artery [72] and rabbit pulmonary artery [74]. Recently, a complete pharmacological study was achieved in the rabbit pulmonary artery confirming the involvement of ET_B receptors [75]. However, in another study, the qualitative and quantitative differences in the relative profiles of the various structurally diverse peptide and non-peptide antagonists examined suggests that responses by ET-1 may not be mediated by a homogeneous ET_B receptor population [7]. Moreover, the involvement of both ET_A and ET_B receptors in the ET-1-induced contraction of pulmonary artery was suggested by the antagonist effect of BQ-123 in ET_B-desensitized preparations [76] or the complete blockade of the contraction using a combination of ET_A and ET_B receptor antagonists [77]. In guinea-pig pulmonary artery, evidence was provided that ET-1 and ET-2, on the one hand, and ET-3, on the other, produce contraction through two distinct receptors [78]. The same group demonstrated the involvement of ET_A receptors in this preparation using FR 139317, an ET_A receptor antagonist, as well as indirect evidence for ET_B receptor role since they were unable to antagonize ET-3-induced contraction with this ET_A receptor antagonist [49]. However, Matsuda et al. suggested that the contraction of this preparation by ET-1 was mainly mediated by ET_A receptors [79]. Furthermore, ET_A receptors were shown to be involved as well in the potentiation of EFS-induced contraction by ET-1 [79]. ET_A receptor blockade, with BQ-123 or FR 139317, was an effective means of attenuating ET-1 induced contraction of isolated rat pulmonary artery rings [80]. Furthermore, sarafotoxin 6c or IRL 1620 failed to induce any direct constriction in this preparation suggesting a major role for ET_A receptors [80]. In contrast, McCulloch et al. suggested the involvement of ET_B receptors since the order of potency of the agonists they were using, was sarafotoxin 6c = ET-3 > ET-1 and responses to ET-1 were unaffected by FR 139317 [70]. Other findings suggest that the contributions of ET_A and ET_B receptors to the contractile responses vary greatly depending on the vascular regions studied. For instance, in human coronary arteries, the contractile response of segment 8 to ET-1 was probably mediated by ET_A. However, in segments 5 and 6, it is likely that more than one receptor subtype is involved in the contractile response to ET-1 [81]. Recently, an interesting study indicated a role for ET_A receptors in the intralobar part of the rat pulmonary artery, whereas the contraction of the extralobar part by ET-1 was essentially ET_B mediated [82]. These differences could explain the discrepancies observed in previous studies.

3.1 2. Relaxant activity: Interaction of ETs with the vascular endothelium in human pulmonary vessels is not well documented. In various isolated vascular smooth muscles the ETs and sarafotoxins elicit an endothelium-dependent relaxation which is thought to be due to the release of nitric oxide from the endothelium [83, 84]. Endothelial ET_B receptors have been shown to mediate pulmonary vasodilatation in rats [85] and lambs [86], but whether this occurs in the human pulmonary circulation is not yet clear. In porcine pulmonary artery both ET-1 and ET-3 were potent relaxants via a mechanism which was not affected by an ET_A receptor antagonist, BQ-123 [72]. Numerous studies indicate that ETs stimulate ET_B receptors on the endothelium which release endothelium-derived vasodilator substances [73, 87, 88]. It has also been reported that responses to ET-1 in human large diameter intrapulmonary arteries were not affected by cyclo-oxygenase inhibition (indomethacin) or NO synthase inhibition (L-NOARG), suggesting that local endogenous release of endothelium-relaxant factors may not be important in regulating the contractile responses to ET-1 in human pulmonary arteries *in vitro* [89]. Therefore, by analogy to the contractile response of pulmonary vessels to ETs, it could be postulated that the endothelial ET_B-mediated vasodilatation varies depending on the pulmonary arterial region. Accordingly, the transient ET-3- or ET-1-induced vasodilatation in extrapulmonary rat arteries occurs at lower concentrations than in intrapulmonary arteries and it is mediated via ET_B receptors present on endothelial cells [90]. One possible explanation is that the population of functional ET_B receptors in the endothelium is greater in extrapulmonary than in intrapulmonary arteries.

3.2. Role of Hypoxia

Due to their potent vasoconstrictor properties and their proliferative effect on vascular smooth muscle cells, the endothelins have been implicated in many pathophysiological conditions, including pulmonary hypertension (PHT). Elevated circulating ET-1 levels have been observed in patients with PHT, both primary and secondary forms [91]. PTH is a major complication of chronic hypoxia that may result from chronic obstructive pulmonary disease, congestive heart failure, respiratory distress syndrome, cystic fibrosis and hypoventilation syndrome [92–94]. Several animal models of hypoxia-induced PTH have been developed, primarily in the rat, and these have been used to investigate the mechanisms underlying the vascular constriction and remodeling components of the disease [95, 96]. It has been shown that pulmonary arterial responses to ET-1 are increased in the chronic hypoxic rat model of PHT, and that hypoxia enhances expression of the ET-1 gene in the rat lung [80]. In PTH, it is the pulmonary resistance arteries which contribute most to elevations in pulmonary vascular resistance which, in humans, seems to express both ET_A and ET_B receptors [97]. In rat pulmonary resistance arteries ET_B receptors mediate most of the vasoconstriction induced by ET-1 [98]. This contrasts with rat and human large pulmonary arteries, where the ET_A receptor predominates [98, 99]. In a rat model, chronic hypoxia increased the maximum response to ET-1 in both the main pulmonary artery and resistance artery [80, 100]. This increase could be due to an increase in ET_A receptor number and/or activation since FR139317, an ET_A receptor antagonist, inhibited the response [100]. In contrast, in a guinea-pig model, hypoxia had little or no influence on the sensitivity of isolated pulmonary artery preparations to ET-1-induced contractions, but caused a marked reduction in the endothelium-dependent relaxation compared with that observed in tissues obtained from normoxic guinea-pigs [101]. Hence, not only do ET-receptor subtypes vary with the size or location of the pulmonary artery, but the effect of pulmonary hypertension, due to chronic hypoxic exposure, also varies with the vessel type studied. In conclusion, ET-1 has probably a significant role in the pathophysiology of pulmonary hypertension, not only by its increased production and release but also by the modulation of its receptor density.

4. Conclusion

Many pathologies of the airways such as asthma and chronic obstructive pulmonary diseases are complex diseases that involved bronchoconstriction to a significificant extent. In the present overview, we discussed the effects of endothelin on smooth muscle reactivity using *in vitro* models involving isolated airways or vessels in organ baths. These models have substantially contributed to the progress in our understanding of the mechanism of action of endothelin in the airways and pulmonary vessels. Furthermore, these studies have suggested the involvement of both ET_A and ET_B receptors in smooth muscle reactivity and the subtle changes in their distribution depending on localization and pathological conditions i.e., viral infections and hypoxia. The coupling of these receptors to the various effects produced by ETs is variable and encompasses multiple signal-transduction mechanisms. Although endothelin-induced contraction is strongly regulated by the airway epithelium and endothelium, this peptide is one of the most potent direct contractile agonists of the airways and pulmonary vessels. Therefore, these studies, together with the extensive literature suggesting the involvement of ETs in pulmonary diseases, would position ET_A and ET_B antagonists as a potential new therapeutic intervention at least as bronchodilators or pulmonary vasorelaxants.

References

1 Advenier C, Sarriá B, Naline E, Puybasset L, Lagente V (1990) Contractile activity of three endothelins (ET-1, ET-2 and ET-3) on the isolated human bronchus. *Br J Pharmacol* 100: 168–172

2 Hay DWP, Hubbard WC, Undem BJ (1993b) Endothelin-induced contraction and mediator release in human bronchus. *Br J Pharmacol* 110: 392–398

3 McKay KO, Black JL, Armour CL (1991) The mechanism of action of endothelin in human lung. *Br J Pharmacol* 102: 422–428

4 Hemsén A, Franco-Cereceda A, Matran A, Rudehill A, Lundberg JM (1990) Occurrence, specific binding sites and functional effects of endothelin in human cardiopulmonary tissues. *Eur J Pharmacol* 191: 319–328

5 Henry PJ, Rigby PJ, Self GJ, Preuss JM, Goldie RG (1990) Relationship between endothelin-1 binding sites densities and constrictor activities in human and animal airway smooth muscle. *Br J Pharmacol* 100: 786–792

6 Goldie RG, D'Aprile AC, Cvetkovski R, Rigby PJ, Henry PJ (1996) Influence of regional differences in ET(A) and ET(B) receptor subtype proportions on endothelin-1-induced contractions in porcine isolated trachea and bronchus. *Br J Pharmacol* 117: 736–742

7 Hay DWP, Luttman MA, Beck G, Ohlstein EH (1996) Comparison of endothelin$_B$ (ET$_B$) receptors in rabbit isolated pulmonary artery and bronchus. *Br J Pharmacol* 118: 1209–1217

8 Hay DWP, Luttmann MA, Hubbard WC, Undem BJ (1993) Endothelin receptor subtypes in human and guinea-pig pulmonary tissues. *Br J Pharmacol* 110: 1175–1183

9 Hay DWP, Hubbard WC, Undem BJ (1993) Relative contributions of direct and indirect mechanisms mediating endothelin-induced contraction of guinea-pig trachea. *Br J Pharmacol* 110: 955–962

10 Maggi CA, Patacchini R, Giulani S, Meli A (1989) Potent contractile effect of endothelin in isolated guinea-pig airways. *Eur J Pharmacol* 160: 189–182

11 Hay DWP (1990) Mechanism of endothelin-induced contraction of isolated guinea-pig trachea: comparison with rat aorta. *Br J Pharmacol* 100: 383–392

12 Sarriá B, Naline E, Morcillo E, Cortijo J, Esplugues J, Advenier C (1990) Calcium dependence of the contraction produced by endothelin (ET-1) in isolated guinea-pig trachea. *Eur J Pharmacol* 187: 445–453

13 Lee HK, Leikauf GD, Sperelakis N (1990) Electromechanical effects of endothelin on ferret bronchial and tracheal smooth muscle. *J Appl Physiol* 68: 417–420

14 Nally JE, McCall R, Young LC, Wakelam MJO, Thomson NC, McGrath JC (1994) Mechanical and biochemical responses to endothelin-1 and entothelin-3 in bovine bronchial smooth muscle. *Br J Pharmacol* 111: 1163–1169

15 Nally JE, McCall R, Young LC, Wakelam MJO, Thomson NC, McGrath JC (1994) Mechanical and biochemical responses to endothelin-1 and endothelin-3 in human bronchi. *Eur J Pharmacol* 288: 53–60

16 Nally JE, Bunton DC, Martin D, Thomson NC (1996) The role of cyclooxygenase and 5-lipoxygenase metabolites in potentiated endothelin-1-evoked contractions in bovine bronchi. *Pulm Pharmacol* 9: 211–217

17 McKay KO, Armour CL, Black JL (1996) Endothelin receptors and activity differ in human, dog, and rabbit lung. *Am J Physiol* 270: L37–L43

18 Battistini B, Sirois P, Braquet P, Filep JG (1990) Endothelin-induced constriction of guinea-pig airways: role of platelet-activating factor. *Eur J Pharmacol* 186: 307–310

19 Filep JG, Battistini B, Sirois P (1990) Endothelin induces thromboxane release and contraction of isolated guinea-pig airways. *Life Sci* 47: 1845–1850

20 Filep JG, Battistini B, Sirois P (1991) Pharmacological modulation of endothelin-induced contraction of guinea-pig isolated airways and thromboxane release. *Br J Pharmacol* 103: 1633–1640

21 Ninomiya H, Uchida Y, Saotome M, Nomura A, Ohse H, Matsumoto H, Hirata F, Hasegawa S (1992) Endothelins constrict guinea pig tracheas by multiple mechanisms. *J Pharmacol Exp Therapeutics* 262: 570–576

22 Bertrand C, Tschirhart E (1993) Epithelial factors: modulation of the smooth muscle tone. *Fund Clin Pharmacol* 7: 261–273

23 Hay DWP (1989) Guinea-pig tracheal epithelium and endothelin. *Eur J Pharmacol* 171: 241–245

24 Tschirhart EJ, Drijhout JW, Pelton JT, Miller RC, Jones CR (1991) Endothelins: functional and autoradiographic studies in guinea-pig trachea. *J Pharmacol Exp Ther* 258: 381–387

25 Tschirhart EJ, Miller R (1990) Airway epithelium actively metabolises endothelin and modulates endothelin-induced contraction in guinea-pig trachea. *Br J Pharmacol* 99: 57P

26 Candenas ML, Naline E, Sarriá B, Advenier C (1992) Effect of epithelium removal and enkephalin inhibition on the bronchoconstrictor response to three endothelins of the human isolated bronchus. *Eur J Pharmacol* 210: 291–297

27 Yamaguchi T, Kohrogi H, Kawano O, Ando M, Araka S (1992) Neutral endopeptidase inhibitor potentiates endothelin-1-induced airway smooth muscle contraction. *J Appl Physiol* 73: 1108–1113

28 Naline E, Bertrand C, Biyah K, Okada T, Fujitani Y, Sakaki J, Advenier C (1996) Characterization of endothelin-ET_A receptors in epithelium of human isolated airways responsible for a relaxant component of smooth muscle through NO release. *Am J Respir Crit Care Med* 153: A644

29 White SR, Hataway DP, Umans JG, Tallet J, Abrahams C, Leff AR (1991) Epithelial modulation of airway smooth muscle response to endothelin-1. *Am Rev Respir Dis* 144: 373–378

30 Battistini B, Warner TD, Fournier A, Vane JR (1994) Characterization of ETB receptors mediating contractions induced by endothelin-1 or IRL1620 in guinea pig isolated airways: effects of BQ-123, FR139317 or PD 145065. *Br J Pharmacol* 111: 1009–1016

31 Wu T, Rieves D, Laryvée P, Logun C, Lawrence MG, Shelhammer JH (1993) Production of eicosanoids in response to endothelin-1 binding sites in airway epithelial cells. *J Resp Cell Mol Biol* 8: 282–290

32 Henry PJ, Goldie RG (1994) ET_B but not ET_A receptor-mediated contractions to endothelin-1 attenuated by respiration tract viral infection in mouse airways. *Br J Pharmacol* 112: 1188–1194

33 Carr MJ, Goldie RG, Henry PJ (1996) Time course of changes in ET(B) receptor density and function in tracheal airway smooth muscle during respiratory tract viral infection in mice. *Br J Pharmacol* 117: 1222–1228

34 Turner NC, Dollery CT, Williams AJ (1989) Endothelin-1-induced contractions of vascular and tracheal smooth muscle: Effects of nicardipine and BRL 34915. *J Cardiovasc Pharmacol* 13: S180–S182

35 Hadj-Kaddour K, Michel A, Chevillard C (1996) Different mechanisms involved in relaxation of guinea-pig trachea by endothelin-1 and -3. *Eur J Pharmacol* 298: 145–148

36 El-Mowafy AM, Abou-Mohamed GA (1996) Endothelins-induce cyclic AMP formation in the guinea-pig trachea through and ET_A receptor- and cyclooxygenase-dependent mechanism. *Br J Pharmacol* 118: 531–536

37 Grunstein MM, Rosenberg SM, Shramm CM, Pawlowski NA (1991) Mechanisms of action of endothelin-1 in maturing rabbit airway smooth muscle. *Am J Physiol* 260: L434–L443

38 Filep JG, Battistini B, Sirois P (1993) Induction by endothelin-1 of epithelium-dependent relaxation of guinea pig trachea *in vitro*: role of nitric oxide. *Br J Pharmacol* 109: 637–644

39 McKay KO, Armour CL, Black JL (1993) Endothelin-3 increases transmission in the rabbit pulmonary parasymphathetic nervous system. *J Cardiovasc Pharmacol* 22 (Suppl. 8): S181–S184

40 Henry PJ, Goldie RG (1995) Potentiation by endothelin-1 of cholinergic nerve-mediated contractions in mouse trachea via activation of ET_B receptors. *Br J Pharmacol* 114: 563–569

41 Fernandes LB, Henry PJ, Rigby PJ, Goldie RG (1996) Endothelin$_B$ (ET$_B$) receptor-activated potentiation of cholinergic nerve-mediated contraction in human bronchus. *Br J Pharmacol* 118: 1873–1874

42 Knott PG, Fernandes LB, Henry PJ, Goldie RG (1996) Influence of endothelin-1 on cholinergic nerve-mediated contractions and acetylcholine release in rat isolated tracheal smooth muscle. *J Pharmacol Exp Ther* 279: 1142–1147

43 Bertrand C, Figini M, Javdan P, Emanueli C, Noguchi S, Geppetti P (1996) Synergistic effect of endothelin-1 on capsaicin-evoked substance P and CGRP release from sensory neurons of the guinea pig airways. *Am J Respir Crit Care Med* 153: A163

44 Yanagisawa M, Kurihara H, Kimura S, Tomobe S, Kobayashi M, Yazaki Y, Goto K, Masaki T (1988) A novel potent vasoconstrictor peptide produced by vascular endothelial cells. *Nature* 332: 411–415

45 Sawamura T, Kasuya Y, Matsuhita Y, Suzuki N, Shinmi O, Kishi N, Sugita Y, Yanagisawa M, Goto K, Masaki T et al (1991) Phosphoramidon inhibits the intracellular conversion of big endothelin-1 to endothelin-1 in cultured endothelial cells. *Biochem Biophys Res Commun* 174: 779–784

46 McMahon EG, Palomo MA, Moore WM, McDonald JF, Stern MK (1991) Phosphoramidon blocks the pressor activity of porcine big endothelin-1 (1-39) *in vivo* and conversion of big endothelin-1-(1-39) *in vitro*. *Proc Acad USA* 88: 703–707

47 Advenier C, Lagente V, Zhang Y, Naline E (1992) Contractile activity of big endothelin-1 on the human isolated bronchus. *Br J Pharmacol* 106: 883–887

48 Uchida Y, Ninomiya H, Saotome M, Nomura A, Ohtsuka M, Yanagisawa M, Goto K, Masaki T, Hasegawa S (1988) Endothelin, a novel vasoconstrictor peptide, as potent bronchoconstrictor. *Eur J Pharmacol* 154: 227–228

49 Cardell LO, Uddman R, Edvisson L (1993) A novel ET$_A$-receptor antagonist, FR 139317 inhibits endothelin-induced contractions of guinea-pig pulmonary arteries, but not trachea. *Br J Pharmacol* 108: 448–452

50 Henry PJ (1993) Endothelin-1 (ET-1)-induced contraction in rat isolated trachea: involvement of ET$_A$ receptors and multiple signal transduction systems. *Br J Pharmacol* 110 435–441

51 Hay DWP (1995) Endothelins. In: D Raeburn, MA Giembycz (eds). *Airways smooth muscle peptide receptors. Ion channels and signal transduction.* Birkhäuser Verlag, Basel 1–50

52 Ihara M, Ishikawa K, Fukuroda T, Saeki T, Funabashi K, Fukami T, Suda H, Yano M (1992 *In vitro* biological profile of a highly potent novel endothelin (ET) antagonis BQ-123 selective for the ET$_A$ receptor. *J Cardiovasc Pharmacol* 20: S11–S14

53 Ihara M, Noguchi K, Saeki T, Fukuroda T, Tsuchida S, Kimura S, Fukami T, Ishikawa K Nishikibe M, Yano M (1992) Biological profiles of highly potent novel endothelin antagonists selective for the ET$_A$ receptor. *Life Sci* 50: 247–255

54 Williams JDL, Jones KL, Pettibone DJ, Lis F V, B.V. C (1991) Sarafotoxin S6c, an agonis which distinguishes between endothelin receptor subtypes. *Biochem Biophys Res Commun* 175: 556–561

55 Takai M, Umemura I, Yamasaki K, Watakabe T, Fujitani Y, Oda K. Urade Y, Inui T, Yamamura T, Okada T (1992) A potent and specific agonist, Succinyl-[Glu^9, Ala$^{11,\,15}$]-endothelin-1(8-21), IRL 1620, for the ETB receptor. *Biochem Biophys Res Commun* 184 953–959

56 Sogabe K, Nirei H, Shoubo M, Nomoto A, Ao A, Notsu Y, Ono T (1993) Pharmacologica profile of FR 139317, a novel, potent endothelin ET$_A$ receptor antagonist. *J Pharmaco Exp Ther* 264: 1040–1046

57 Hay DWP (1992) Pharmacological evidence for distinct endothelin receptors in guinea-pig bronchus and aorta. *Br J Pharmacol* 106: 759–761

58 Ishikawa K, Ihara M, Noguchi K, Mase T, Mino N, Saeki T, Fukuroda T, Fukami T, Osak S, Nagase T et al (1994) Biochemical and pharmacological profile of a potent and selective endothelin B-receptor antagonist, BQ-788. *Proc Natl Acad Sci USA* 91: 4892–4896

59 Balwierczak JL, Bruseo CW, DelGrande D, Jeng AY, Savage P, Shetty SS (1995) Characterization of a potent and selective endothelin-B receptor antagonist, IRL 2500. *J Cardiovasc Pharmacol* 26: S393–S396

60 Bertrand C, Naline E, Biyah K, Fujitani Y, Okada T, Sakaki J, Advenier C (1996) Influenc of ET$_A$ receptor blockade on the ET$_3$ receptor antagonism following ET-1 and IRL 1620 induced contraction of the human bronchi. *Am J Respir Crit Care Med* 153: A465

61 Fukuroda T, Ozaki S, Ihara M, Ishikawa K, Yano M, Miyauchi T, Ishikawa S, Onizuka M, Goto K, Nishikibe M (1996) Necessity of dual blockade of endothelin ET(A) and ET(B) receptor subtypes for antagonism of endothelin-1-induced contraction in human bronchi. *Br J Pharmacol* 117: 995–999

62 Takahashi T, Barnes PJ, Kawikova I, Yacoub MH, Warner TD, Belvisi MG (1997) Contraction of human airway smooth muscle by endothelin-1 and IRL 1620: effect of bosentan. *Eur J Pharmacol* 324: 219–222

63 Hickey KA, Rubanyi GM, Paul RJ, Highsmith RF (1985) Characterization of a vasoconstrictor produced by cultured endothelial cells. *Am J Physiol* 248: C550–C556

64 Rubanyi GM, Polokoff MA (1994) Endothelins: molecular biology, biochemistry, pharmacology, physiology, and pathophysiology. *Pharmacol Rev* 46: 325–415

65 Cardell LO, Uddman R, Edvinsson L (1990) Analysis of endothelin-1-induced contractions of guinea-pig trachea, pulmonary veins and different types of pulmonary arteries. *Acta Physiol Scand* 139: 130–131

66 Brink C, Gillard V, Roubert P, Mencia-Huerta JM, Chabrier PE, Braquet P, Verley J (1991) Effects of specific binding sites of endothelin in lung preparations. *Pulmon Pharmacol* 4: 54–59

67 Hay DWP, Henry PJ, Goldie R (1993) Endothelin and the respiratory system. *Trends Pharmacol Sci* 14: 29–32

68 Buchan KW, Magnusson H, Rabe KF, Sumner MJ, Watts IS (1994) Characterisation of the endothelin receptor mediating contraction of human pulmonary artery using BQ-123 and Ro 46-2005. *Eur J Pharmacol* 260: 221–225

69 Russell FD, Davenport AP (1995) Characterization of endothelin receptors in the human pulmonary vasculature using bosentan, SB209670, and 97–139. *J Cardiovasc Pharmacol* Suppl 3:

70 McCulloch KM, MacLean MR (1995) Endothelin B receptor-mediated contraction of human and rat pulmonary resistance arteries and the effect of pulmonary hypertension on endothelin responses in the rat. *J Cardiovasc Pharmacol* 26: S169–S176

71 Holm P, Franco-Creceda A (1996) Tissue concentrations of endothelins and functional effects of endothelin-receptor activation in human arteries and veins. *J Thor Cardiovasc Surg* 112: 264–272

72 Fukuroda T, Nishikibe M, Ohta Y, Ihara M, Yano M, Ishikawa K, Fukami T, Ikemoto F (1992) Analysis of responses to endothelins in isolated porcine blood vessels by using a novel endothelin antagonist, BQ-153. *Life Sci* 50: PL107–PL112

73 Sudjarwo SA, Hori M, Takai M, Urade Y, Okada T, Karaki H (1993) A novel subtype of endothelin B receptor mediating contraction in swine pulmonary vein. *Life Sci* 53: 431–437

74 White DG, Cannon TR, Garratt H, Mundin JW, Sumner MJ, Watts IS (1993) Endothelin ETA and ETB receptors mediate vascular smooth muscle contraction. *J Cardiovasc Pharmacol* 22: S144–S148

75 Beck GR, Douglas SA, Elliott JD, Ohlstein EH (1995) Agonist-dependent inhibition by peptide and non-peptide endothelin receptor antagonists in the rabbit isolated pulmonary artery. *J Cardiovasc Pharmacol* 26: S385–S388

76 LaDouceur DM, Flynn MA, Keiser JA, Reynolds E, Haleen SJ (1993) ETA and ETB receptors coexist on rabbit pulmonary artery vascular smooth muscle mediating contraction. *Biochem Biophys Res Commun* 196: 209–215

77 Fukuroda T, Ozaki S, Ihara M, Ishikawa K, Yano M, Nishikibe M (1994) Synergistic inhibition by BQ-123 and BQ-788 of endothelin-1-induced contractions of the rabbit pulmonary artery. *Brit J Pharmacol* 113: 336–338

78 Cardell LO, Uddman R, Edvinsson L (1991) Two functional endothelin receptors in guinea-pig pulmonary arteries. *Neurochem Int* 18: 571–574

79 Matsuda H, Kawaguchi A, Uematsu M, Ohmori F, Nagata S, Miyatake K (1996) Endothelins contract guinea-pig pulmonary artery and enhance its adrenergic response via ETA receptors. *Clin Exp Pharmacol Physiol* 23: 379–385

80 MacLean MR, McCulloch KM, Baird M (1995) Pulmonary arterial ETA- and ETB-receptor-mediated vasoconstriction, tone and nitric oxide mediation in control and pulmonary hypertensive rats. *J Cardiovasc Pharmacol* 26: 822–830

81 Godfraind T (1993) Evidence for heterogeneity of endothelin receptor distribution in human coronary artery. *Brit J Pharmacol* 110: 1201–1205

82 Bialecki RA, Fisher CS, Murdoch WW, Barthlow HG, Bertelsen DL (1997) Functional comparison of endothelin receptors in human and rat pulmonary artery smooth muscle. *Am J Physiol* 16: L211–L218

83 Cocks TM, Faulkner NL, Sudhir K, Angus J (1989) Reactivity of endothelin-1 on human and canine large veins compared with large arteries *in vitro. Eur J Pharmacol* 171: 17–24

84 Karaki H, Sudjarwo SA, Hori M, Sakata K, Urade Y, Takai M, Okada T (1993) ETB receptor antagonist, IRL 1038, selectively inhibits the endothelin induced endothelium dependent vascular relaxation. *Eur J Pharmacol* 231: 371–374

85 Eddahibi S, Raffestin B, Braquet P, Chabrier PE, Adnot S (1991) Pulmonary vascular reactivity to endothelin-1 in normal and chronically hypertensive rats. *J Cardiovasc Pharmacol* 17: S358–S361

86 Wong J, Vanderford PA, Winters J, Soifer SJ, Fineman JR (1995) Endothelin-B receptor agonists produce pulmonary vasodilatation in intact newborn lambs with pulmonary hypertension. *J Cardiovascular Pharmacol* 25: 207–215

87 De Nucci G, Thomas R, D'Orleans-Juste P, Antunes E, Walder C, Warner TD, Vane JR (1988) Pressor effects of circulating endothelin are limited by its removal in the pulmonary circulation and by the release of prostacyclin and endothelium-derived relaxant factor. *Proc Natl Acad Sci USA* 85: 9797–9800

88 Warner TD, Mitchell JA, DeNucci G, Vane JR (1989) Endothelin-1 and endothelin-3 release EDRF from isolated perfused arterial vessels of the rat and rabbit. *J Cardiovasc Pharmacol* 13: S85–S88

89 Pussard G, Gascard JP, Gorenne I, Labat C, Norel X, Dulmet E, Brink C (1995) Endothelin-1 modulates cyclic GMP production and relaxation in human pulmonary vessels. *J Pharmacol Exp Ther* 274: 969–975

90 Higashi T, Ishizaki T, Shigemori K, Yamamura T, Nakai T (1997) Pharmacological characterization of endothelin-induced rat pulmonary arterial dilatation. *Brit J Pharmacol* 121: 782–786

91 Stewart DJ, Levy RD, Cernacek P, Langleben D (1991) Increased plasma endothelin-1 in pulmonary hypertension: marker or mediator of disease? *Ann Intern Med* 114: 464–469

92 Zapol WM, Snider MT (1977) Pulmonary hypertension in severe acute respiratory failure. *N Engl J Med* 296: 476–480

93 Cody RJ, Halas GJ, Binkley PF, Capers Q, Kelley R (1992) Plasma endothelin correlates with the extent of pulmonary hypertension in patients with chronic heart failure. *Circulation* 85: 504–509

94 MacNee W (1994) Pathophysiology of cor pulmonale in chronic obstructive pulmonary disease, state of the art: parts one and two. *Am J Respir Crit Care Med* 150: 833–852, 1158–1168

95 Rabinovitch M, Gamble W, Nadas AS, Miettinen OS, Reid L (1979) Rat pulmonary circulation after chronic hypoxia: hemodynamic and structural features. *Am J Physiol* 236: H818–H827

96 Thompson BT, Hassoun PM, Kradin RL, Hales CA (1989) Acute and chronic hypoxic pulmonary hypertension in guinea pigs. *Am J Physiol* 266: 920–928

97 McCulloch KM, Doherty CC, Morecroft I, MacLean MR (1996) Endothelin B receptors mediates contraction of human pulmonary resistance arteries. *Br J Pharmacol* 119: 917–930

98 MacLean MR, McCulloch KM, Baird M (1994) Endothelin ETA- and ETB-mediated vasoconstriction in rat pulmonary arteries and arterioles. *J Cardiovasc Pharmacol* 23: 838–845

99 Fukuroda T, Kobayashi M, Ozaki S, Yano M, Miyauchi T, Onizuka M, Sugishita Y, Goto K, Nishikibe M (1994) Endothelin receptor subtypes in human versus rabbit pulmonary arteries. *J Appl Physiol* 76: 1976–1982

100 MacLean MR, McCulloch KM (1998) Influence of applied tension and nitric oxide on responses to endothelins in rat pulmonary resistance arteries: effect of chronic hypoxia. *Br J Pharmacol* 123: 991–999

101 Underwood DC, Bochnowicz S, Osborn RR, Luttmann MA, Hay DWP (1997) Nonpeptide endothelin receptor antagonists. X. Inhibition of endothelin-1- and hypoxia-induced pulmonary pressor responses in the guinea pig by the endothelin receptor antagonist, SB 217242. *J Pharmacol Exp Ther* 283: 1130–1137

Pulmonary Actions of the Endothelins
ed. by R. G. Goldie and D. W. P. Hay
© 1999 Birkhäuser Verlag Basel/Switzerland

CHAPTER 7
In Vivo Effects of Endothelins in the Lung

William M. Abraham

*Division of Pulmonary and Critical Care Medicine, University of Miami at Mount Sinai
Medical Center, 4300 Alton Road, Miami Beach, Florida 33140*

1   Introduction
1.1 Bronchoconstrictor Effects of Parenteral Endothelins
1.2 Bronchoconstrictor Effects of Aerosolized Endothelins
1.3 Role of Endothelins in Allergic Responses
1.4 Lung Vascular Responses to Endothelins
2   Conclusion
     References

1. Introduction

The pathophysiologic role of the potent vasoconstrictor/bronchoconstrictor peptide endothelin-1 (ET-1) in lung disease is still unclear, but reports of increased expression of ET-1 and/or levels of ET-1 in the airways of patients with asthma [1–8], pulmonary hypertension [9–12] and adult respiratory distress syndrome (ARDS) [13, 14], would suggest a role in these conditions. Nevertheless, conclusive proof of the role of ETs in the pathogenesis of these diseases await clinical trials with appropriate antagonists and inhibitors demonstrating beneficial effects. However, to date much of our current knowledge about the role of ETs in airway disease has been derived from studies in experimental animals. This chapter will attempt to provide a mechanistic understanding of the *in vivo* effects of the ETs, in particular ET-1, in the lung.

1.1. Bronchoconstrictor Effects of Parenteral Endothelins

ET-1 induced effects in the lung are mediated through one of two receptors: ET_A receptor, which has a higher affinity for ET-1 and ET-2 than for ET-3; ET_B receptor which shows similar affinities for the three peptides [15]. The existence and role in mammalian lung of the putative ET_C receptor, which has a higher affinity for ET-3 than ET-1 [16], remains to be clarified. *In vivo*, ETs cause bronchoconstriction in guinea pigs, cats, pigs, sheep and asthmatic subjects [17 –24]. ET-1 is the most potent constrictor [21], with the mechanism(s) of ET-1 induced bronchoconstriction depending both on the species and the route of administration. In general, i.v. administration of ET-1 usually exerts its pulmonary effects, i.e., bronchoconstriction and pulmonary vasoconstriction, *via* the secondary generation of cyclooxygenase metabolites of arachidonic acid [19, 20, 25, 26]. There does not appear to be involvement of the 5-lipoxygenase pathway, although there is a report that an i.v. bolus of ET-1 in rats caused a significant increase in 15-hydroxyeicosatetraenoic acid in bronchoalveolar lavage fluid (BALF) and in lung homogenates [27].

The receptors involved in ET-1 induced bronchoconstriction differ in the various species. In guinea pigs, i.v. administration of ETs induces marked bronchoconstriction which is not inhibited by the ET_A receptor antagonists, BQ-123 and FR 139317, which suggests an ET_B receptor-mediated action [28, 29]. Consistent with these observations is a study showing that infusion of a selective ET_B receptor agonist, IRL 1620, produces a biphasic bronchoconstriction in guinea pigs, with the first phase observed within 2 min and the second phase following 5–10 min after injection [30]. The first phase of the response was found to be dependent on ET_B receptor-mediated thromboxane A_2 (TxA_2) production, as it was blocked with the TxA_2 inhibitor, OKY-046, and the ET_B receptor antagonist, BQ-788, but not the ET_A receptor antagonist, BQ-123. The second phase of the response was thought to result from non selective stimulation of the ET_A receptor, because BQ-123 blocked this late (5–10 min) constrictor response. This interplay between the two receptors appeared to be dose-dependent because the secondary response to IRL 1620, required a dose of at least 0.1 mg/kg; lower doses only induced a monophasic response. Thus, under normal conditions, significant ET_B receptor stimulation had to be achieved to elicit an ET_A receptor-mediated response. If, however, the animals were treated with phosphoramidon, an inhibitor of the enzyme neutral endopetidase (NEP), which metabolizes ET-1, then doses of IRL 1620, which previously produced only monophasic responses, elicited biphasic responses. Furthermore, the second phase of the bronchoconstriction in these phosphoramidon-treated animals was sensitive to BQ-123. These data suggest that primary activation of ET_B receptors leads to autocrine/paracrine ET-1 release that, subsequently, cause bronchoconstriction through both ET_A and ET_B receptor stimulation. This pathway becomes more sensitive if NEP activity of the airways is reduced. Since NEP is thought to be reduced in asthmatic airways [31], this could be one of the reasons for increased sensitivity to ET-1 seen in these patients (see Sect. 1.2 below).

The distribution of the different receptor subtypes in lung tissue will affect the various aspects of lung function upon ET-1 stimulation. In open-chest, mechanically ventilated guinea pigs, i.v. injection of ET-1 produced concentration-dependent increases in lung resistance, tissue resistance, airway resistance and lung elastance [32]. Treatment with the ET_A receptor antagonist, BQ-123, partially blocked the increases in lung and airway resistance, which are primarily determined by smooth muscle in the larger airways. BQ-123 did not affect the increase in tissue resistance or the lung elastance, which are influenced mostly by the status of the peripheral airways and alveoli. All parameters were significantly inhibited if the animals were treated with the ET_B receptor antagonist, BQ-788. The combination of both antagonists completely ablated the response to ET-1 [32]. These data are consistent with the distribution and densities of ET receptor subtypes in the guinea pig lung where the proportion of ET_B to ET_A receptors is increased as one goes from large airways to the alveoli [33]. Furthermore, it appears that it is important to inhibit the stimulation of both receptor subtypes to achieve complete blockade of bronchoconstrictor effects of ET-1.

Although ET-1 is a potent constrictor of airway smooth muscle, recent studies have demonstrated some pro-inflammatory actions as well. Bolus i.v. injection of ET-1 (0.1 – 1 nmol/kg) in anesthetized guinea pigs evoked increases in plasma albumin extravasation into the trachea and upper and lower bronchi, but not the lung parenchyma [34]. Qualitatively, similar changes were observed following i.v. injection of the ET_B receptor agonist, IRL 1620, although it was approximately three times less potent than ET-1. ET-1-induced responses were blocked by the ET_A receptor antagonist, FR 139317. The combined ET_A/ET_B receptor antagonist, bosentan, blocked the permeability effects of both ET-1 and IRL 1620. ET-1, but not IRL 1620, produced a dose-dependent neutropenia, but did not induce an influx of neutrophils into the lung. The ET-1-induced neutropenia was blocked by FR 139317 and bosentan. These studies indicate that the ET-1 induced pro-inflammatory changes occur primarily *via* activation of ET_A receptors. In spite of these results there is some controversy over the inflammatory actions of ET-1 in the lungs. For example, chronic infusion of ET-1 (2 nmol/day *via* jugular vein) in guinea pigs for 6 days did not induce histological changes in the lung or cause inflammation [35].

Although i.v. ET-1 may not elicit an inflammatory cell influx into the airways, airway inflammation is associated with increased ET-1 levels. Instillation of sephadex beads into the lungs of rats induced a prolonged inflammatory response that was associated with increased bronchomotor tone [36]. Using this model, instillation of sephadex beads caused a 3.5-fold increase in lung ET-1 content. The inflammatory response was blocked by treating the animals with budesonide and associated with this reduced inflammation was a lowering of the ET-1 levels by 72% (36). Interestingly, when the rats were treated with the combined ET_A/ET_B receptor antagonist, bosentan, there was no protection against the lung edema formation or leukocyte numbers in the BALF. These results suggest that the increase in ET-1 levels seen in this model is not responsible for the lung inflammation [37].

1.2. Bronchoconstrictor Effects of Aerosolized Endothelins

The results with aerosolized ET-1 differ from those results obtained with parenteral administration. In guinea pigs, parenteral administration of

ET-1 causes cyclooxygenase (primarily thromboxane mediated)-dependent bronchoconstriction, increases in pulmonary vascular resistance, and edema formation, whereas aerosol administration causes bronchoconstriction which is not cyclooxygenase-dependent and does not alter pulmonary vascular resistance and/or cause edema. Furthermore, in contrast to the additive effects of ET_B and ET_A receptor-mediated bronchoconstriction with i.v. ET-1 challenge, the effect of aerosolized ET-1 in guinea pigs is mediated by ET_B receptor stimulation, with the ET_A receptor acting to downregulate this response [38].

Aerosols of ET-1 or ET-3 cause bronchoconstriction in sheep [21]. In this species, ET-3 was about 400-fold less potent than ET-1. ET-1 was 500-fold more potent than LTD_4. ET-1-induced bronchoconstriction was blocked by BQ-123 or the ET-1 analog [diaminoproprionic acid[1]-Asp[15]] ET-1. Thus, in contrast to guinea pigs, the data indicate that in sheep airways ET-1 induced bronchoconstriction is primarily mediated via ET_A receptors. These findings are consistent with the receptor ligand binding data of Goldie et al. who found an abundance of ET_A receptors on ovine airway smooth muscle and lung [39]. The aerosolized ET-1-induced bronchoconstriction was not affected by the cyclooxygenase inhibitor, indomethacin, indicating that the effects of inhaled ET-1 were due to direct stimulation of the receptor and not the result of secondary prostanoid synthesis. Furthermore, the response does not involve thrombin because pretreatment of the sheep with a peptide that binds the thrombin receptor had no protective effect on the ET-1 induced response [21].

Inhaled ET-1 also caused a reduction in tracheal mucus velocity (TMV), a measure of mucociliary clearance, in sheep [40]. The response was most marked within the first 30 min of administration. A significant effect was seen with 10^{-8} M ET-1, which is comparable to the dose required to elicit the bronchoconstrictor response [21]. Like the airway effects, the TMV response appears to be mediated by ET_A receptors, because the reduction in mucus clearance was blocked by BQ-123 but not the ET_B receptor antagonist, BQ-788 (Fig. 1). Again, these data are consistent with autoradiographic studies of ET-1 receptors in sheep tracheal tissue, which indicate a high density of ET_A binding sites in cells in the submucosa and in submucosal glands [39]. The ET-1-induced fall in TMV was not due to the release of prostanoids and/or leukotrienes because neither indomethacin nor the LTD_4 antagonist, MK-571, blocked the response [40].

There have been two recent studies published documenting the response to ET-1 in humans. In the first study, ET-1 was given intranasally to symptomatic allergic and non-allergic individuals and indices of secretion and nasal symptoms were obtained [41]. ET-1 induced multiple nasal responses, including rhinorrhea, glandular secretion, sneezing and itching. Because there was no evidence of histamine release, which would be suggestive of mast cell degranulation, the ET-1-induced responses were thought to be a direct effect of ET receptor stimulation. However, the recep-

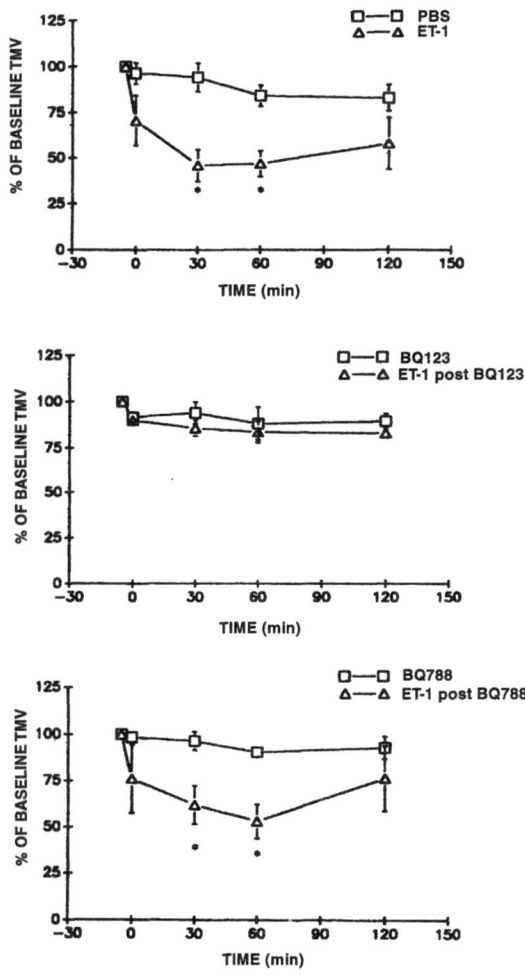

Figure 1. Effect of aerosolized ET-1 (50 breaths, 10^{-7} M) on tracheal mucous velocity (TMV) in sheep. ET-1-induced a fall in TMV (Top) which is blocked by pretreatment with the ET_A receptor antagonist, BQ-123 (1 mg/kg, aerosol – Middle) but not BQ-788 (1 mg/kg, aerosol – Bottom). Values are mean ± S.E. for 5–6 sheep (From ref. [40] with permission).

tor subtype was not identified since the effects of antagonists were not examined. In these studies, ET-1 was given by placing a small piece of filter paper in one side of the nose and then making measurements in both the ipsilateral and contralateral side to assess not only the direct effects, but also the reflex effects of ET-1. ET-1 induced responses in both nostrils in allergic and non-allergic subject groups, although the former had significantly greater responses to ET-1 in both nostrils relative to non-allergic individuals; furthermore, the responses were greater in the ipsilateral as

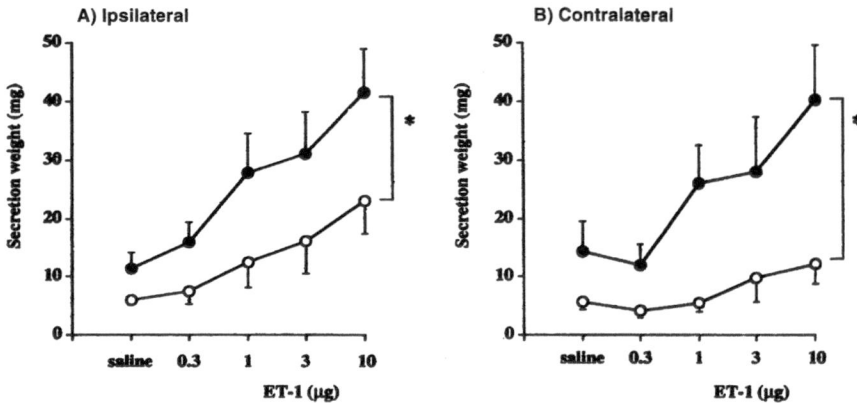

Figure 2. The effect of increasing doses of ET-1 on secretion weights of allergic (closed circles) and non-allergic (open circles) subjects in (A) the ipsilateral nostril and (B) the contralateral nostril. Data represent the mean and S.E.M. of nine allergic and nine non-allergic subjects. ET-1-induced significant increases in secretion weights for both groups in both nostrils. Allergic subjects had significantly higher secretion weights than non-allergic subjects (*p < 0.05). Ipsilateral secretion weights were significantly elevated in the ipsilateral compared with contralateral nostril of non-allergic subjects (p < 0.05). (From ref. [41] with permission).

compared to the contralateral nostril. Thus, not only did the allergic subjects show a heightened response to direct stimulation with ET-1, but there appeared to be an increased reflex response in the allergic patients (Fig. 2). These results support the concept that allergic subjects are hyperresponsive to ET-1 [41].

This conclusion was supported further in a recent study comparing the effects of aerosolized ET-1 in asthmatic patients and normal subjects [23]. In this study, aerosolized ET-1, across the dose range used (nebulized dose range 0.96 to 15.36 nmol) caused a rapid onset (< 5 min) dose-dependent bronchoconstriction in asthmatic subjects. The mean (range) ET-1 PC_{35SGaw} (the provocative concentration of ET-1 that caused a 35% decrease in specific airway conductance) in these asthmatic patients was 5.15 (1.4 to 13.9) nmol and 4.2 (1.2 to 8.3) nmol for two separate ET-1 challenges. In these same subjects, the PC_{35SGaw} value for methacholine was 0.42 (0.2 to 0.7) nmol (Fig. 3). The ET-1-induced bronchoconstriction lasted from 60 to 90 min but could be rapidly reversed by the β-adrenoceptor agonist, albuterol. These findings were in contrast to the results in normal subjects where the same doses of ET-1 failed to cause bronchoconstriction. Thus, in both the upper and lower airways, there appears to be a heightened response to ET-1 challenge in human subjects with nasal allergies or asthma.

The constrictor effects of ET-1 in human airways are thought to be mediated primarily by ET_B receptors. However, the heightened responses seen in patients is not related to changes in receptor number, because the number of binding sites in asthmatic and non-asthmatic airways are simi-

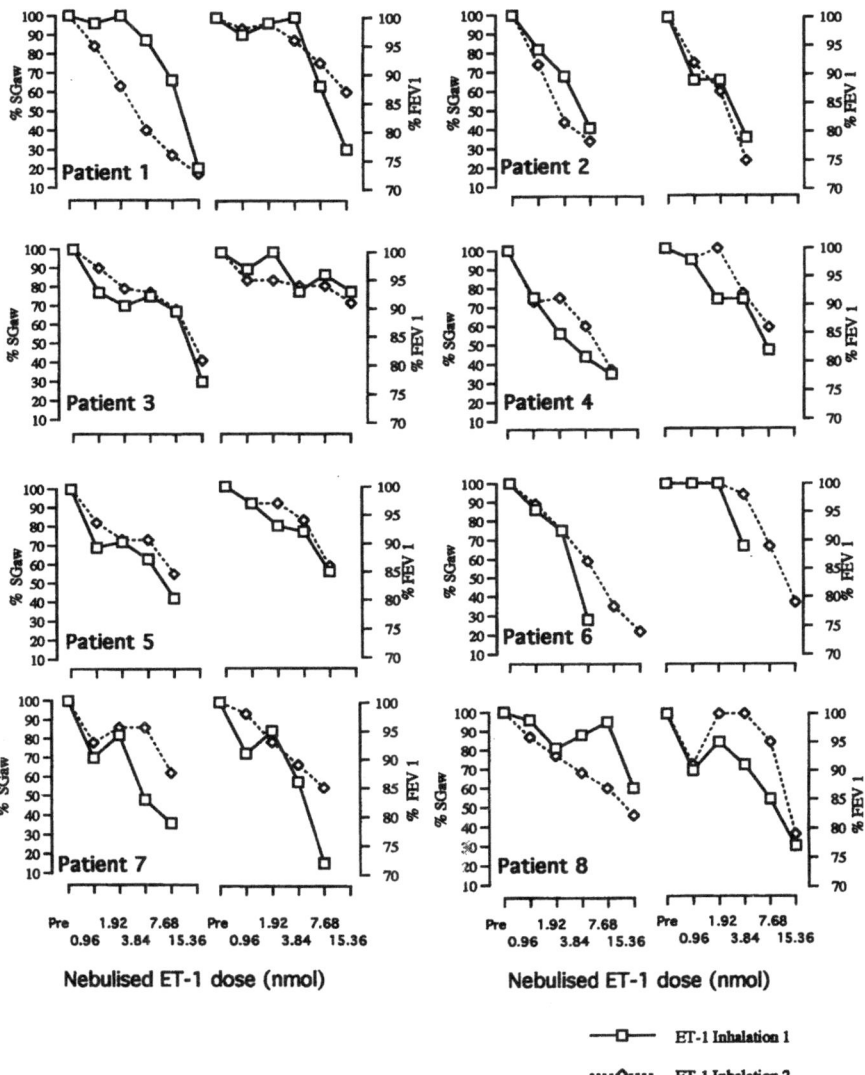

Figure 3. Individual percent change in specific airway conductance (S_{Gaw}) and FEV_1 during two endothelin-1 inhalations (dose range, 0.96 to 15.36 nmol) in eight asthmatic subjects. (From ref. [23] with permission).

lar [42]. The greatest density of ET receptors is located on the airway smooth muscle with the ET_B receptor subtype constituting 82% and 88% of the ET-1-sensitive receptors in the asthmatic and non-asthmatic airway smooth muscle. Given this receptor distribution and the heightened responsiveness to ET-1 *in vivo*, it is also difficult to explain why, *in vitro*, asthmatic airways, as compared to non-asthmatic airways, were less sensitive to the contractile effects of ET_B receptor stimulation induced by either the specific ET_B receptor agonist, sarafotoxin S6c, or ET-1 [42]. One proposed explanation was that the reduced sensitivity reflects desensitization of the ET_B receptor in response to increased production and release of ET-1 in asthmatic airways. This hypothesis has still to be tested directly.

1.3. Role of Endothelins in Allergic Responses

The production of asthma-like symptoms when ET-1 is inhaled or instilled into the airways suggests that ET-1 contributes to allergic responses. Several observations in human airways support this concept including the identification of increased expression and/or immunoreactive (ir-) ET-1, in tissues and/or BALF from asthma patients [2, 5–8]. In another study, asthmatic patients whose lung function returned to normal after treatment with inhaled β-agonists and oral corticosteroids had a reduction in ET levels [3]. Furthermore, a significant inverse correlation was found between the ET level in BALF and % predicted FEV_1 in non-steroid treated asthmatics [3]. In addition to these functional studies the ability of ETs to augment the proliferation of airway smooth muscle, vascular smooth muscle and fibroblasts and to stimulate collagen synthesis by fibroblasts suggests the participation of ET in lung tissue inflammation [1, 43]. ETs can also prime neutrophils to produce superoxide, stimulate neutrophil migration and induce monocytes to release interleukin 6, interleukin 8 and tumor necrosis factor α [44]. Because these events have been shown to contribute to the pathogenesis of asthma, it is reasonable to consider a role for ET in this disease. Although studies with ET receptor antagonists have not yet been completed in human subjects, the hypothesis of a pathophysiological role of ET has been tested in allergic sheep and sensitized guinea pigs using the ET_A receptor antagonist, BQ-123, and the ET_B receptor antagonist, BQ-788.

In allergic sheep, BQ-123 given either intravenously (100 µg/kg/min continuous infusion for 8 h) or by aerosol (1 mg/kg, administered 0.5 h before and 4, 8 and 24 h after antigen challenge) had no effect on the immediate bronchoconstriction to inhalation challenge with *Ascaris suum* antigen, but provided approximately 50% inhibition of the late antigen-induced response [22]. Perhaps more interesting was the fact that aerosolized BQ-123 blocked the increase in airway responsiveness measured 24 h after antigen provocation (Fig. 4). These results suggest a role for ET_A receptor stimulation in the late response and the post antigen-induced airway hyperresponsiveness in allergic sheep. In a second series of experiments, ET-1 itself was shown to produce airway hyperresponsiveness. In these studies, sheep were challenged with 20 µg/ml ET-1 by aerosol, and then 4 and 24 h later, airway responsiveness to inhaled carbachol was determined. ET-1

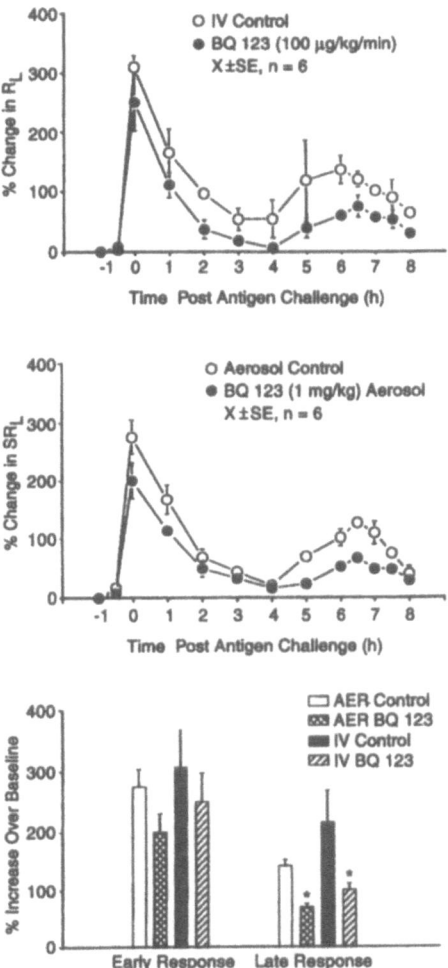

Figure 4. Top: Time-course of airway responses after antigen challenge in allergic sheep treated with continuous intravenous (IV) infusion of BQ-123 (100 µg/kg/min) or aerosolized (AER) BQ-123 (1 mg/kg in 3 ml buffer). Bottom: Comparative effects of BQ-123 on peak early and late responses. BQ-123 had no effect on early response but reduced the late response by 55 and 52% in IV and AER studies, respectively. *Significantly different compared with control values, $p < 0.05$. Values are mean ± S.E. for 6 sheep. (From ref. [22] with permission).

caused airway hyperresponsiveness, at both time points, which was blocked by pretreatment with BQ-123 (Fig. 5).

In cyclophosphamide-pretreated guinea pigs, a similar type of study was conducted with anti-ET-antiserum and selective receptor antagonists [45]. The anti-ET-1 antiserum reacts with guinea pig ET-1 and guinea pig ET-2 but not guinea pig ET-3 nor big ET-1. In these guinea pigs i.p. injection of

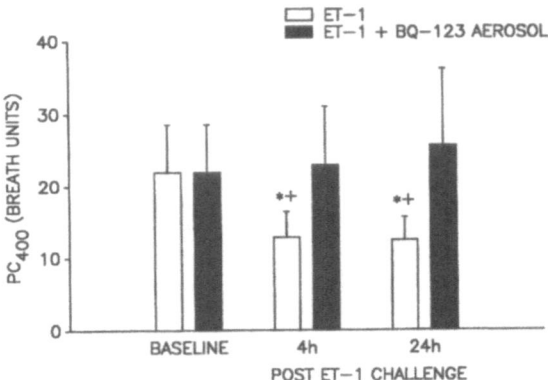

Figure 5. Production of airway hyperresponsiveness in sheep by ET-1. Inhaled ET-1 (25 breaths, 20 µg/ml) caused an increase in airway responsiveness to inhaled carbachol (i.e., decrease in PC_{400}) at 4 h, which persisted for the 24 h test period. BQ-123 (1 mg/kg AER) completely inhibited this ET-1-induced response. Significantly different ($p < 0.05$) compared with: * baseline values, and + BQ-123 + ET-1 values. Values are mean ± S.E. for 6 sheep. (From ref. [22] with permission).

anti-ET-1 antiserum 1 h before antigen challenge reduced the immediate bronchoconstriction (measured as a reduction in specific conductance [S_{gaw}]) by approximately 50% and slightly attenuated the late airway response (Fig. 6). These results suggest a contribution of ET-1 to the immediate response and possibly the late response to allergen in this model. To further explore the possible receptor subtypes involved, similar experiments were conducted using the ET_B receptor antagonist, BQ-788, and the ET_A receptor antagonist, BQ-123. Pretreatment with BQ-788 inhibited the maximum fall in S_{Gaw} during the immediate response by 60% but had no effect on the late response, whereas pretreatment with BQ-123 provided 25% protection of the late airway response, but had no effect on the immediate response. These findings are similar to those seen in allergic sheep [22] (Fig. 7). The late response in these guinea pigs was also blocked by a PAF antagonist and a thromboxane inhibitor. These latter observations are consistent with the known *in vivo* actions of ET in guinea pigs, where the effects are dependent on the secondary production of prostanoids. These investigators did not test the ability of these antagonists to modify antigen-induced bronchial hyperreactivity.

1.4. Lung Vascular Responses to Endothelins

ETs are potent constrictors of isolated pulmonary blood vessels from various species including humans. Thus, ETs have been speculated to contribute to a variety of diseases characterized by an increase in pulmonary

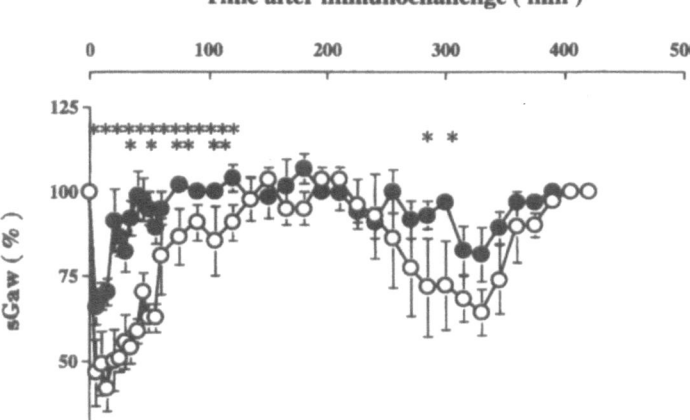

Figure 6. Effects of anti-ET antiserum and control serum on pulmonary mechanics in a guinea pig model of experimental asthma. Anti-ET antiserum (50 µg; closed circle) and control serum (50 µg; open circle) were injected i.p. 2 h before the immunochallenge and S_{Gaw} was measured. Each point represents the mean ± S.E. of 6 animals (*$p < 0.05$ vs. control by unpaired *t*-test). (From ref. [45] with permission).

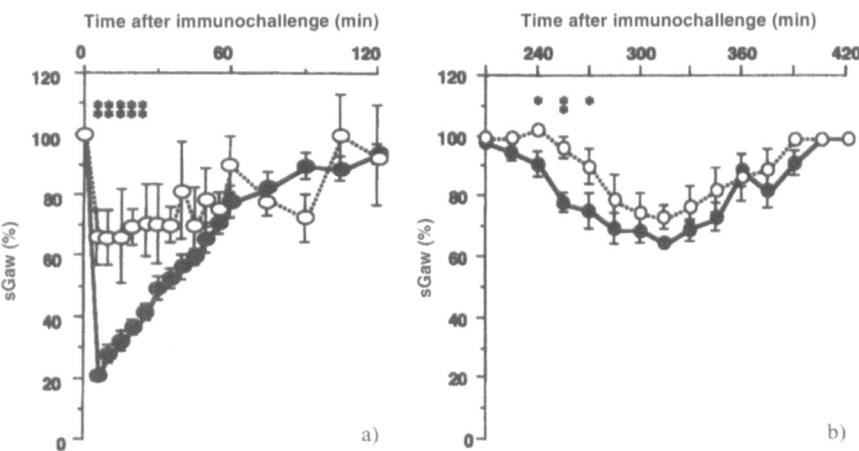

Figure 7. Effects of the ET_B receptor antagonist, BQ-788 (a), and the ET_A receptor antagonist, BQ-123 (b), on pulmonary mechanics in sensitized guinea pigs. Animals were pretreated with 10 mg/10 ml BQ-123 (b) open circle; n = 5) or 10 mg/10 ml BQ-788 (a; open circle; n = 5) 20 min before the immunochallenge. Control animals = closed circles; n = 6. (*$p < 0.05$ vs. control, by unpaired *t*-test). (From ref. [45] with permission).

vascular resistance. Increased levels of ET-1 have been found in the lungs of patients with pulmonary hypertension and adult respiratory distress syndrome (ARDS), suggesting a potential pathological role for ET-1 in these diseases [9–14]. However, it still remains to be determined if ET-1 contributes to the pathogenesis of these diseases or whether the elevated ET levels are surrogate markers of dysfunction.

The evidence supporting a role for ET-1 in the aforementioned diseases is gleaned from experimental studies in animals showing that infusion of ET-1 causes an increase in pulmonary vascular resistance and that ET receptor antagonists protect against this response, as well as the putative ET-1-mediated effects seen in these experimental models of disease. In humans, intravenous infusion of ET-1 (4 pmol/kg/min for 20 min) caused an 11-fold increase in ET-1-like immunoreactivity and resulted in a 67% increase in pulmonary vascular resistance and 25% increase in systemic vascular resistance [46]. Experimental studies in animals confirm that infusion of ET-1 causes increases in pulmonary vascular resistance. Under certain conditions, however, ET-1 can also induce vasodilation. Such diverse responses are dependent on factors such as the level of tone in the vasculature prior to infusion. Thus, in the setting of normal baseline pulmonary vascular resistance, ETs generally cause long lasting pulmonary vasoconstriction. In the presence of increased pulmonary vascular tone, however, ETs may produce concentration-dependent vasodilation [47]. For example, in new-born lambs ET-1 causes pulmonary vasodilation when tone is elevated [48]. ET-1 can also cause vasodilation during pulmonary hypertension induced by alveolar hypoxia [48]. The effect appears to involve nitric oxide (NO) and ATP-sensitive K^+ channels, but not prostaglandins [49]. In support of this, ET-1 elicited pulmonary vasodilation in cats was blocked by the ATP-sensitive K^+ channel antagonist, glybenclamide [50, 51].

The ET receptors responsible for vasoconstriction in the pulmonary circulation may change depending on vessel size. In rat large pulmonary arteries (2–3 mm in diameter) ET-1 produces constriction primarily through ET_A receptors, whereas in rat small intrapulmonary resistance vessels (100–150 µm in diameter) ET_B receptors contribute significantly to the ET-1-induced vasoconstriction. In spite of these data, the role of ET_B receptor stimulation in experimentally-induced pulmonary hypertension is still unclear. In rats, i.v., infusion of the ET_A receptor antagonist, BQ-123, 4 h before and 90 min during normobaric hypoxia (10% O_2) markedly attenuated the increase in mean pulmonary artery pressure [52]. Similarly, ET_A receptors appear to be involved in experimentally-induced pulmonary hypertension in the ovine fetus because the chronic pulmonary hypertension was attenuated by treatment with BQ-123 [53].

In addition to causing pulmonary vasoconstriction, infusing ET-1 into animals causes pulmonary edema. In isolated perfused rat lungs, ET-1 (1–10 nM) caused concentration-dependent pulmonary vasoconstriction and gross pulmonary edema at 10 nM. The combination of BQ-123 and BQ-788 inhibited the pulmonary vasoconstriction more effectively than BQ-123 alone. The combination of antagonists, but neither antagonist alone, prevented the increased hydrostatic pressure-induced pulmonary edema. Thus, ET-1-induced vasoconstriction and edema is mediated by both ET_A and ET_B receptors and blockade of both receptors is required for complete inhibition of these responses [54].

These results differ from studies in which ET receptor antagonists have been used to modify lung injury. For example, in an isolated perfused rat lung model of reperfusion injury (45 min of ischemia followed by 105 min of reperfusion), treatment with an ET_A receptor antagonist, BQ-610, and a combined ET_A/ET_B receptor antagonist (PD-156707-0015), given before the ischemic period, protected against the rise in lung filtration coefficient and the neutrophil accumulation in the lung. Administration of ET-1 alone, caused a dose-dependent increase in pulmonary artery pressure, but no measurable increase in microvascular permeability. These studies suggest that ET-1 is involved in ischemia reperfusion-induced lung endothelial injury through the stimulation of ET_A receptors. The inability of ET-1 alone to induce alterations in microvascular permeability suggest that the endothelial injury is dependent on ET interaction with other factors [55]. Thus, ET-1 has been shown to act synergistically with PAF to cause airway edema [55]. Similarly, increased protein extravasation was seen with a combination of ET-1 and an analog of L-arginine at doses where neither agonist alone produced an effect [56]. There is some evidence that ET-1-induced increases in microvascular permeability require the presence of leukocytes. In isolated rat lungs, ET-1 caused a rapid increase in pulmonary artery pressure, pulmonary microvascular pressure, and edema formation. Compared with Krebs albumin-perfused lungs, perfusion of lungs with blood accelerated the endemagenic effect of ET-1 [57]. Consistent with the latter results are findings that 2 h after ET-1 infusion in rats, histologic examination of the lungs showed adhesion of leukocytes to the vascular endothelium, in the pulmonary vessels and sequestration of leukocytes in the pulmonary capillaries [57].

Sepsis increases plasma ET-1 levels in mice, rats, pigs, sheep and dogs and, furthermore, the elevation in ET-1 levels correlates with the rise in pulmonary vascular resistance associated with this condition [58–61]. There is evidence to suggest that part of the ET-1-mediated effect in sepsis may be due to a change in both receptor number and receptor activity [62]. More importantly, it appears that the ability to survive an endotoxin insult is dependent on the levels of ET-1 generated. Thus, in sheep, the animals that survive endotoxin infusion had approximately three-fold less ET-1 in lung lymph compared to those animals that did not survive the challenge. The animals that failed to survive (i.e., those with higher ET-1 levels) had lower cardiac outputs and higher pulmonary vascular resistances. Thus, increased ET-1 levels appear to contribute to factors that determine mortality in septic shock [63].

2. Conclusion

It is evident that *in vivo* ETs cause bronchoconstriction, increases in pulmonary vascular resistance, edema formation and have pro-inflammatory actions in the lungs. These effects, combined with the reports of increased expression of ET and/or elevated ET levels in BALF, lung lymph and/or plasma in a variety of pulmonary disorders make this peptide a logical target for drug intervention. Future studies, using newly developed potent selective receptor antagonists and inhibitors should provide critical information as to the precise role of ET in pulmonary diseases.

References

1 Hay DWP, Goldie RG (1997) Endothelins. In: Barnes PJ, Grunstein MM (eds). *Asthma.* Lippincott-Raven, Philadelphia, 707–729
2 Springall DR, Howarth P, Counihan H, Djukanovic R, Holgate S, Polak JM (1991) Endothelin immunoreactivity in airway epithlium in asthmatic patients. *Lancet* 337: 697–701
3 Redington AE, Springall DR, Ghatei MA et al (1995) Endothelin in bronchoalveolar lavage fluid and its relation to airway obstruction in asthma. *Am J Respir Crit Care Med* 151: 1034–1039
4 Kraft M, Beam WR, Wenzel SE, Zamora MR, O'Brien RF, Martin RJ (1994) Blood and bronchoalveolar lavage endothelin-1 levels in nocturnal asthma. *Am J Respir Crit Care Med* 149: 947–952
5 Vittori E, Marini M, Fasoli A, De Franchis R, Mattoli S (1992) Increased expression of endothelin in bronchial epithelial cells of asthmatic patients and effect of corticosteroids. *Am Rev Respir Dis* 146: 1320–1325
6 Mattoli S, Mezzetti M, Riva G, Allegra L, Fasoli A (1990) Specific binding of endothelin on human bronchial smooth muscle cells in culture and secretion of endothelin-like material from bronchial epithelial cells. *Am J Respir Cell Mol Biol* 3: 145–151
7 Nomura A, Uchida Y, Kameyama M, Saotome M, Oki K, Hasegawa S (1989) Endothelin and bronchial asthma. *Lancet* 2: 747–748
8 Mattoli S, Soloperto M, Marini M, Fasoli A (1991) Levels of endothelin in the bronchoalveolar lavage fluid of patients with symptomatic asthma and reversible airflow obstruction. *J Allergy Clin Immunol* 88: 376–384
9 Cacoub P, Dorent R, Maistre G, Nataf P, Carayon A, Piette JC (1993) Endothelin-1 in primary pulmonary hypertension and the Eisenmenger Syndrome. *Am J Cardiol* 71: 448–450
10 Rosenberg A, Kennaugh J, Koppenhafer S, Loomis M, Chatfield BA, Abman SH (1993) Elevated immunoreactive endothelin-1 levels in newborn infants with persistent pulmonary hypertension. *J Pediatr* 123: 109–114
11 Cody RJ, Haas GJ, Binkley PF, Capers Q, Kelley R (1992) Plasma endothelin correlates with the extent of pulmonary hypertension in patients with chronic congestive heart failure. *Circulation* 85: 504–509
12 Giaid A, Yanagisawa M, Langleben D, Michel RP, Levy R, Shennib H (1993) Expression of endothelin-1 in the lungs of patients with pulmonary hypertension. *N Engl J Med* 328: 1732–1739
13 Druml W, Steltzer H, Waldhausl W et al (1993) Endothelin-1 in adult respiratory distress syndrome. *Am Rev Respir Dis* 148: 1169–1173
14 Langleben D, DeMarchie M, Laporta D, Spanier AH, Schlesinger RD, Stewart DJ (1993) Endothelin-1 in acute lung injury and the adult respiratory disress syndrome. *Am J Respir Dis* 148: 1646–1650
15 Masaki T, Yanagisawa M, Goto K (1992) Physiology and pharmacology of the endothelins. *Med Res Rev* 12: 391–421

16 Karne S, Jayawickreme C, Lerner M (1993) Cloning and characterization of an endothelin-3 specific receptor (ETC receptor) from *Xenopus laevis* dermal melanophores. *J Biol Chem* 268: 19126–19133

17 Kadowitz PJ, McMahon TJ, Hood JS, Feng C-J, Minkes RK, Dyson MC (1991) Pulmonary vascular and airway responses to endothelin-1 are mediated by different mechanisms in the cat. *J Cardiovasc Pharmacol* 17 (Suppl 7): S374–S377

18 Dyson MC, Kadowitz PJ (1991) Analysis of responses to endothelins 1, 2 and 3 and sarafotoxin 6b in airways of the cat. *J Appl Physiol* 71: 243–251

19 Macquin-Mauvier I, Levame M, Istin N, Harf A (1989) Mechanisms of endothelin-mediated bronchoconstriction in the guinea pig. *J Pharmacol Exp Ther* 250: 740–745

20 Payne AN, Whittle BJR (1988) Potent cyclooxygenase-mediated bronchoconstrictor effects of endothelin in the guinea pig *in vivo*. *Eur J Pharmacol* 158: 303–304

21 Abraham WM, Ahmed A, Cortes A, Spinella MJ, Malik A, Andersen TT (1993) A specific endothelin-1 antagonist blocks inhaled endothelin-1-induced bronchoconstriction in sheep. *J Appl Physiol* 74: 2537–2542

22 Noguchi K, Ishikawa K, Yano M, Ahmed A, Cortes A, Abraham WM (1995) Endothelin-1 contributes to antigen-induced airway hyperresponsiveness. *J Appl Physiol* 79: 700–705

23 Chalmers GW, Little S, Patel K, Thomson N (1997) Endothelin-1 induced bronchoconstriction in asthma. *Am J Respir Crit Care Med* 156: 382–388

24 Clement M, Dimori M, Albertini M (1993) Comparison of vascular and respiratory effects of endothelin-1 in the pig. *Med Inflamm* 2: 287–292

25 White SR, Hathaway DP, Umans JG, Tallet J, Abrahams C, Leff AR (1991) Epithelial modulation of airway smooth muscle response to endothelin-1. *Am Rev Respir Dis* 144: 373–378

26 Rae GA, Calixto JB, D'Orleans-Juste P (1995) Effects and mechanisms of action of endothelins on non-vascular smooth muscle of the respiratory, gastrointestinal and urogenical tracts. *Regul Pept* 55: 1–46

27 Nagase T, Fukuchi Y, Jo C et al (1990) Endothelin-1 stimulates arachidonate 15-lipoxygenase activity and oxygen radical formation in the rat distal lung. *Biochem Biophys Res Commun* 168: 485–489

28 Noguchi K, Noguchi Y, Hirose H et al (1993) Role of endothelin ET_B receptors in bronchoconstrictor and vasoconstrictor responses in guinea-pigs. *Eur J Pharmacol* 233: 47–51

29 Sogabe K, Nirei H, Shoubo M et al (1993) Pharmacological profile of FR 139317, a novel, potent endothelin ET_A receptor antagonist. *J Pharmacol Exp Ther* 264: 1040–1046

30 Noguchi S, Kashihara Y, Bertrand C (1996) The induction of a biphasic bronchospasm by the ET_B agonist, IRL 1620, due to thromboxane A_2 generation and endothelin-1 release in guinea-pig. *Br J Pharmacol* 118: 1397–1402

31 Borson DB (1991) Roles of neutral endopeptidase in airways. *Am J Physiol Lung Cell Mol Physiol* 260: L212–L225

32 Nagase T, Fukuchi Y, Matsui H, Aoki T, Matsuse T, Orimo H (1995) *In vivo* effects of endothelin A- and B-receptor antagonists in guinea pigs. *Am J Physiol Lung Cell Mol Physiol* 268: L846–L850

33 Goldie RG, D'Aprile AC, Self GJ, Rigby PJ, Henry PJ (1996) The distribution and density of receptor subtypes for endothelin-1 in peripheral lung of the rat, guinea-pig and pig. *Br J Pharmacol* 117: 729–735

34 Filep JG, Fournier A, F'Ides-Filep E (1995) Acute pro-inflammatory actions of endothelin-1 in the guinea-pig lung: involvement of ET_A and ET_B receptors. *Br J Pharmacol* 115: 227–236

35 Pons F, Boichot E, Lagente V, Touvay C, Menci-Huerta JM, Braquet P (1992) Role of endothelin in pulmonary function. *Pulm Pharmacol* 5: 213–219

36 Andersson SE, Zackrisson C, Hems NA, Lundberg JM (1992) Regulation of lung endothelin content by the glucocorticosteroid budesonide. *Biochem Biophys Res Commun* 188: 1116–1121

37 Andersson SE, Hems NA, Lundberg JM (1996) The effect of endothelin receptor blockade on the development of the Sephadex-induced inflammation in the rat lung. *Acta Physiol Scand* 158: 189–193

38 Polakowski J, Opgenorth TJ, Pollack D (1996) ET_A receptor blockade potentiates the bronchoconstrictor response to ET-1 in the guinea pig airway. *Biochem Biophys Res Comm* 225: 225–231

39 Goldie RG, Grayson PS, Knott PG, Self GJ, Henry PJ (1994) Predominance of endothelin A (ETA) receptors in ovine airway smooth muscle and their mediation of ET-1-induced
   contraction. *Br J Pharmacol* 112: 749–756
40 Sabater JR, Otero R, Abraham WM, Wanner A, O'Riordan TG (1996) Endothelin-1 depresses tracheal mucus velocity in ovine airways via ET-A receptors. *Am J Respir Crit Care Med*
   154: 341–345
41 Riccio MM, Reynolds CJ, Hay DWP, Proud D (1995) Effects of intranasal administration of
   endothelin-1 to allergic and nonallergic individuals. *Am J Respir Crit Care Med* 152:
   1757–1764
42 Goldie RG, Henry PJ, Knott PG, Self GJ, Luttmann MA, Hay DWP (1995) Endothelin-1
   receptor density, distribution, and function in human isolated asthmatic airways. *Am
   J Respir Crit Care Med* 152: 1653–1658
43 Takuwa N, Takuwa Y, Yanagisawa M, Yamashita K, Masaki T (1989) A novel vasoactive
   peptide endothelin stimulates mitogenesis through inotisol lipid turnover in Swiss 3T3
   fibroblasts. *J Biol Chem* 264:7856–7861
44 Helset E, Sildness T, Seljelid R, Konoski ZS (1993) Endothelin-1 stimulates human monocytes *in vitro* to release TNF-α, IL-β and IL-6. *Mediators Inflamm* 2: 417–422
45 Uchida Y, Jun T, Ninomiya H, Ohse H, Hasegawa S, Nomura A et al (1996) Involvement of
   endothelins in immediate and late asthmatic responses of guinea pigs. *J Pharm Exp Ther*
   277: 1622–1629
46 Weitzberg E, Ahlborg G, Lundberg JM (1993) Differences in vascular effects and removal
   of endothelin-1 in human lung, brain, and skeletal muscle. *Clin Physiol* 13: 653–662
47 Perreault T, De Marte J (1991) Endothelin-1 has a dilator effect on neonatal pig pulmonary
   vasculature. *J Cardiovasc Pharm* 18: 43–50
48 Wong J, Vanderford PA, Fineman JRCR, Soifer SJ (1993) Endothelin-1 produces pulmonary vasodilation in the intact newborn lamb. *Am J Physiol* 265: H1318–H1325
49 Lippton HL, Cohen GA, McMurtry IF, Hyman AI (1991) Pulmonary vasodilation to endothelin isopeptides *in vivo* is mediated by potassium channel activation. *J Appl Physiol* 70:
   947–952
50 Minkes RK, Bellan JA, Saroyan RM, Kerstein MD, Coy DH, Murphy WA (1990) Analysis
   of cardiovascular and pulmonary responses to endothelin-1 and endothelin-3 in the anesthetized cat. *J Pharm Exp Ther* 253: 1118–1125
51 Lippton HL, Cohen GA, Knight M, McMurtry IF, Gillot D, Arena F et al (1991) Evidence
   for distinct endothelin receptors in the pulmonary vascular bed *in vivo*. *J Cardiovasc Pharm*
   17: S370–S373
52 Oparil S, Chen S, Meng Q, Elton TS, Yano M, Chen Y (1995) Endothelin-A receptor antagonist prevents acute hypoxia-induced pulmonary hypertension in the rat. *Am J Phys* 268:
   L95–L100
53 Ivy DD, Parker TA, Ziegler JW, Galan HL, Kinsella JP, Tuder RN et al (1997) Prolonged
   endothelin A receptor blockade attention chronic pulmonary hypertension in the ovine
   fetus. *J Clin Invest* 99: 1179–1186
54 Sato K, Oka M, Hasunuma K, Ohnishi M, Kira S (1995) Effects of separate and combined
   ET$_A$ and ET$_B$ blockade on ET-1-induced constriction in perfused rat lungs. *Am J Physiol*
   269: L668–L672
55 Khimenko P, Moore TM, Taylor AE (1996) Blocked ET$_A$ receptors prevent ischemia and
   reperfusion injury in rat lungs. *J Appl Physiol* 80: 203–207
56 Filep JG, Földes-Filep E, Rousseau A, Sirois P, Fournier A (1993) Vascular responses to
   endothelin-1 following inhibition of nitric oxide synthesis in the conscious rat. *Br J Pharmacol* 110: 1213–1221
57 Helset E, Ytrehus K, Tveita T, Kjaeve J, Jorgensen L (1994) Endothelin-1 causes accumulation of leukocytes in the pulmonary circulation. *Circ Shock* 44: 201–209
58 Morel DR, Pittet JF, Gunning K, Hemsen A, Lacroix JS (1991) Time course of plasma and
   pulmonary lymph endothelin-like immunoreactivity during sustained endotoxaemia in
   chronically instrumented sheep. *Clin Sci* 81: 357–365
59 Takahashi K, Silva A, Cohen J, Lam H-C, Ghatei MA, Bloom SR (1990) Endothelin immu-
   noreacitivity in mice with gram-negative bacteraemia: relationship to tumour necrosis fac-
   tor-α. *Clin Sci* 79: 619–623
60 Lundblad R, Giercksky K-E (1995) Endothelin concentrations in experimental sepsis: profiles of big endothelin and endothelin 1-21 in lethal peritonitis in rats. *Eur J Surg* 161: 9–16

61 Nakamura T, Kasai K, Sekiguchi Y, Banba N, Takahashi K, Emoto T et al (1991) Elevation of plasma endothelin concentrations during endotoxin shock in dogs. *Eur J Pharm* 205: 227–282

62 Curzen NP, Griffiths MJD, Evans TW (1995) Contraction to endothelin-1 in pulmonary arteries from endotoxin-treated rats is modulated by endothelium. *Am J Physiol* 268: H2260-H2266

63 Armstead V, Perkowski S, Woolkalis M, Spath JJ, Gee MH (1995) An association between lung lymph endothelin concentration and survival during endotoxemia in awake sheep. *Shock* 4: 361–367

Pulmonary Actions of the Endothelins
ed. by R. G. Goldie and D. W. P. Hay
© 1999 Birkhäuser Verlag Basel/Switzerland

CHAPTER 8
Endothelin in Lung Development and Tissue Growth

Vera P. Krymskaya and Reynold A. Panettieri, Jr.

Pulmonary and Critical Care Division, Department of Medicine, University of Pennsylvania, School of Medicine, Philadelphia, PA 19104

1   Introduction
2   Endothelins and Lung Development
2.1 The Role of ET-1 in Cardiovascular and Pulmonary Development
2.2 ET-3 and ET_B Receptor Expression and Development
2.3 Endothelins in Fetal and Postnatal Growth
3   Endothelin and Airway and Pulmonary Vascular Remodeling
3.1 Effects of ETs *in vivo*
3.2 Endothelin-activated Receptors Coupled to G-Proteins and Phospholipase C Activation
3.3 ET-1 Induces Synthesis of Prostaglandins (PG) and TXA_2
3.4 ETs and Smooth Muscle Cell Proliferation
4   Summary
     References

1. Introduction

Endothelins (ETs), a family of polypeptides with a wide range of activities in the airways and pulmonary vascular system, are synthesized by respiratory epithelium and by pulmonary vascular endothelium [1–4].

The synthesis of the ETs is comprehensively described in Chapter 4. *In situ* hybridization studies revealed the presence of large quantities of ET mRNA in human lung [5] and in bronchial epithelial cells of rat fetal lung [6]. Mattoli et al. demonstrated that ET-1, secreted by human bronchial epithelial cells, specifically binds to receptors on human bronchial smooth muscle cells in culture [7]. Interestingly, some growth factors and inflammatory cytokines stimulate synthesis and release of ETs from epithelium. For example, IL-8, $TNF\alpha$ and $TGF\beta$ stimulate synthesis of ET-1 in guinea pig airway epithelial cells [8]. Endo et al. also demonstrated that IL-1 induced the synthesis of the ET-1 precursor big ET-1, whereas IL-2, IL-6 and IGF-1 stimulated the synthesis of big ET-1 and proliferation of guinea pig airway epithelial cells. Thrombin induced release of ET from isolated rabbit tracheal epithelial cells [9]. Taken together, these data suggest that proinflammatory cytokines and growth factors may play a role in ET synthesis which, in turn, may modulate airway smooth muscle contractility and proliferation. As a component of a complex homeostatic mechanism,

ETs may play a role in the pathogenesis of pulmonary hypertension and bronchial asthma. Developmental studies have also elucidated the importance of ET-1 for cardiopulmonary development.

Pharmacological studies have suggested the presence of specific ET receptors (see Chapter 2). Briefly, two subtypes of ET receptors have been cloned [10, 11]. The ET_A receptor subtype exhibits specific and high-affinity binding in rank order toward endothelin isopeptide ligands: ET1 ≥ ET-2 ≫ ET-3 [12]. The ET_B receptor subtype equally associates with all ET isotypes. Specific high-affinity binding sites for ET-1 have been demonstrated in human, guinea pig, pig, rat and mouse airway smooth muscle, peripheral lung alveolar wall tissue and pulmonary arteries [13–16].

2. Endothelins and Lung Development

2.1. The Role ET-1 in Cardiovascular and Pulmonary Development

Mutation and gene knockout studies have demonstrated the important role of ET-1 in cardiovascular and pulmonary development. Complete knockout of the ET-1 gene in mouse (ET-1–/– homozygous mice) induced morphological abnormalities of the pharyngeal-arch-derived craniofacial tissues and newborn death due to respiratory failure [17]. ET-1+/– heterozygous mice developed high blood pressure, suggesting the existence of an ET-1-mediated mechanism that may alter hemodynamic homeostasis. Studies with ET-1–/– homozygous mice revealed multiple developmental cardiovascular malformations that include interrupted aortic arch, tubular hypoplasia of the aortic arch, aberrant right subclavian artery, and ventricular septal defect with abnormalities of the outflow tract [18]. In parallel studies, a selective ET_A receptor antagonist or neutralizing antibody to ET enhanced the frequency and the extent of these morphological abnormalities. These data suggest a possible role of the ET_A receptor in the physiological and pathophysiological effects of ET-1 in cardiac and pulmonary development.

Analysis of microvascular endothelial cells from ET-1 knockout mice, however, revealed that lack of ET-1 expression did not alter endothelial cell morphology and growth [19]. Maemura et al. studied gene expression of ET-2, ET-3, ET_A and ET_B receptors in ET-1+/+ wild-type, ET-1+/– heterozygous, and ET-1–/– homozygous mice [20]. ET-1 gene knockout had little effect on endothelial cell mRNA levels of ET receptors. These data demonstrate that ET-1 expression is not essential for growth of endothelial cells. Experiments with transgenic ET-1 knockout mice, however, suggest that ET-1 plays a role in the normal development of pulmonary and cardiovascular systems.

Hocher et al. studied the effects of human ET-1 gene under the control of its natural promoter transferred into the germ line of mice [21]. This gene was predominantly expressed in brain, lung and kidney and was associated with development of glomerulosclerosis, interstitial fibrosis, and renal cysts. ET-1 gene expression had little effect on plasma and tissue levels of ET-1. Despite significant pathophysiological changes, ET-1 overexpression did not induce hypertension. These data suggest the existence of complex ET-mediated mechanisms involved in the development of such pathophysiological conditions.

2.2. ET-3 and ET_B Receptor Expression and Development

Recent studies have demonstrated an important role of ET-3 and ET_B receptor genes in normal development of epidermal melanocytes and enteric ganglion neurons in humans and in mice. In humans, mutations in ET_B receptor gene mapped to chromosome 13 led to the development of Hirschsprung disease (aganglionic megacolon) [22–24]. Heterozygous ET_B receptor gene mutations accounted for a frequent congenital malformation related to multigenic neurocristopathy. Homozygous ET_B receptor gene mutations in animals result in Hirschsprung disease-Waardenburg-Shah syndrome, aganglionic megacolon associated with pigment abnormalities [12, 24].

Hosoda et al. also studied the effects of targeted mutations of ET_B receptor gene [23]. Knockout of the murine ET_B receptor gene induced aganglionic megacolon and coat color spotting. These developmental defects were identical to those of piebald-lethal mice, and were the result of complete deletion of ET_B receptor gene. Mutations in the ET_B receptor locus resulted in lower levels of ET_B receptor mRNA and receptor expression, which was manifest as white coat spotting, but not aganglionic megacolon. These data suggest that disruption of ET_B receptor gene may alter normal colonic development that appears to mimic hereditary Hirschsprung disease.

Other studies of ET_B receptor and ET-3 gene expression in animals with hereditary defects in the development of epidermal melanocytes and ganglion neurons also revealed the presence of null mutation in these genes. Analysis of the ET_B receptor gene of the spotting lethal rat, which exhibit aganglionic megacolon and pigment abnormalities, demonstrated a 301 base pair deletion in the exon 1-intron 1 junction [12]. This deletion produced an aberrant ET_B receptor lacking the first and second putative transmembrane domains of the G-protein coupled receptor, which completely abrogated functional activity of the ET_B receptor. Baynash et al. found that interaction of ET-3, a ligand for ET_B receptor, was also necessary for development of epidermal melanocytes and enteric neurons [22]. Other investigators have shown that the lethal spotting mice carry a missense mutation in ET-3 gene. This mutation produces an inactive ET-3 intermediate with a substituted arginine residue with tryptophan in the C-terminal domain of big ET-3. This substitution makes it impossible to cleave big ET-3 by

ECE-1 and produce mature ET-3. These data demonstrate the essential role of ET-3, the ET_B receptor subtype and their interactions in normal mouse, rat and human organogenesis.

Enteric neurons and epidermal melanocytes are derived from neural crest cells. During embryonic development, neural crest cells are formed from the dorsal neural tube. After neural crest development, pluripotent cells migrate to form diverse tissues. Nataf et al. studied spatiotemporal expression pattern of the ET_B receptor gene in neural crest cells of normal avian embryo [25]. *In vivo* and *in vitro* experiments revealed that ET_B receptor mRNA was abundantly expressed in neural crest cells as well as in migratory neural crest cells. ET-3 and activation of the ET_B receptor were necessary for growth and differentiation of neural crest cells *in vitro* [26]. ET-3 also induced neural crest cell growth and differentiation into melanocytes. These data clearly suggest involvement of ET-3 and ET_B receptors in tissue development derived from neural crest cells.

2.3. Endothelins in Fetal and Postnatal Growth

Multiple studies have demonstrated differences in ET localization and expression in fetal, postnatal and adult lung [4, 27, 28]. Giaid et al., utilizing immunohistochemistry and *in situ* hybridization techniques, examined the localization of endothelin-producing cells in developing and adult lung [4]. Endothelin-like immunoreactivity and ET-1 mRNA were found predominantly in pulmonary endocrine and endothelial cells of fetal lung. The level of ET-1 mRNA expression and ET-1 immunoreactivity were highest in fetal lung, decreased before birth, and lowest in adult. Chan et al. studied preproET-1 expression in different tissues from embryonic mouse [27]. PreproET-1 expression was detected in the bronchial epithelium and the endothelial cells of dorsal aorta and aortic arches. In the late stages of embryonic development, the level of preproET-1 expression decreased in the bronchial epithelium. Interestingly, in lung vascular endothelium, high levels of preproET-1 expression were found at the late phase of development. Apparently, high levels of ET-1 in human and mouse fetal lung are necessary for normal lung development. Quantitative autoradiographic studies also revealed the presence of both ET_A and ET_B receptor subtypes in alveolar wall tissue of fetal pig lung, indicating their possible role in fetal lung development [16].

Using rat pulmonary artery and vein, the distribution of ET-1 immunoreactivity appeared to be localized to pulmonary vascular endothelium in postnatal and adult animals [29]. These studies also suggest that ET-1 may play a role in modulating pulmonary vascular function.

ET-1 effects on tracheal smooth muscle contractility also revealed age-dependent differences [30]. The potency of ET-1-induced smooth muscle segment contraction of the two-week-old rabbit was significantly higher as compared with the airway smooth muscle of adult animal. Nifedipine, a Ca^{2+} channel antagonist, decreased ET-1-induced contraction of tracheal smooth muscle segment from the two-week-old rabbit. The ET-1-induced formation of inositol 1,4,5-trisphosphate was also significantly higher in mature smooth muscle. H-7, a protein kinase C (PKC) inhibitor, antagonized ET-1-induced contraction in both smooth muscle segments, although to a greater extent than observed in adult tissue. Interestingly, at 10 nM, ET-1 induced relaxation of tracheal smooth muscle, although this effect was significantly greater in the adult. The relaxation of adult smooth muscle was associated with ET-1-induced release of prostaglandins (PG) E_2 and I_2. These data demonstrate significant differences in ET-1-induced cell signaling, which may depend on the stage of maturation of tracheal smooth muscle.

3. Endothelin and Airway and Pulmonary Vascular Remodeling

3.1. Effects of ETs in vivo

ETs, first discovered as vasoconstrictors, have been reported in numerous studies to act as potent bronchoconstrictor *in vitro* and *in vivo* [31–34]. The schematic presentation of vasodilation and vasoconstriction induced by ET-1 or ET-3 is shown in Figure 1. This ability to produce bronchospasm has been the most widely studied effect of ET-1 in the lung. Studies of ET-1 effects on airway smooth muscle in guinea pig revealed that intravenous administration of ET-1 induced immediate bronchoconstriction and an increase in airway resistance. A biphasic response was also observed using determinations of tracheal smooth muscle active tension: a short initial relaxation response was followed by a prolonged increase in force generation [9, 35].

3.2. Endothelin-activated Receptors Coupled to G-Proteins and Phospholipase C Activation

ETs signal by binding to receptors, which belong to the G-protein-coupled seven-transmembrane domain receptor superfamily. In most vascular smooth muscle preparations, ET-1 binds to the ET_A receptor and in airways, ET-1 preferentially interacts with the ET_B receptor [34]. Stimulated receptors associate with G-proteins and activate phospholipase C (PLC) with hydrolysis of phosphatidylinositol-4,5-diphosphate (PIP_2) to inositol-1,4,5-trisphosphate (IP_3) and diacylglycerol (DAG) (see Chapter 5). ET-1 stimulated adenosine 3',3'-cyclic monophosphate accumulation in canine tracheal epithelial cells [36] and in cultured rat vascular smooth muscle cells [37]. Thus, ETs can induce PLC and adenylate cyclase activation, which is cell specific and mediated through multiple G-proteins.

The second messengers of ET-induced PLC activation, IP_3 and DAG, also play an important role in intracellular signaling. IP_3, through binding to its receptors on sarcoplasmic reticulum, induces release of Ca^{2+} [38, 39]. ET-induced activation of PLC and PI hydrolysis results in accumulation of DAG in vascular smooth muscle [40, 41], a second messenger involved in activation of PKC. The specific PKC isoenzyme stimulated by ET remains unknown, and further studies are needed to define the role of PKC activation in mediating ET effects on smooth muscle cells. In addition, PLC-induced PI hydrolysis of DAG may also be a product of phosphatidylcholine hydrolysis by activated PLC or phosphatidic acid degradation by phospholipase D [3]. Resink et al. have shown in vascular smooth muscle cells that ET-1 stimulation induced hydrolysis of phosphatidylcholine and activation of phospholipase D [42]. These signaling events may also be important in regulation of smooth muscle cell growth and matrix production.

3.3. ET-1 Induces Synthesis of Prostaglandins (PG) and TXA_2

De Nucci et al. have determined that ET-1 induced release of PGI_2 and TXA_2, products of arachidonic acid metabolism and phospholipase A_2 activation, in perfused isolated rat and guinea pig lung preparations [43]. ET-1 induced synthesis and release of PGI_2 and PGE_2 in rabbit tracheal epithelium and this effect was attenuated by pretreatment of cells with the cyclooxygenase inhibitor indomethacin [30, 44]. In vascular smooth muscle, ET-1 induced activation of phospholipase A_2 [45]. Schumacher et al. also reported that release of TXA_2 and activation of TXA_2 receptors are involved in ET-1-induced bronchoconstriction [46]. The consequence of ET-induced effects on arachidonic acid metabolism remains unknown. Clearly, blockade of cyclooxygenase abrogates the transient relaxation phase induced by ET in a variety of smooth muscle preparations, but why a spasmogen induces relaxation and then sustained contraction remains unknown. The role of arachidonic acid metabolites in airway remodeling and tissue growth is also unknown.

3.4. ETs and Smooth Muscle Cell Proliferation

ET-1 is a weak mitogen for cultured airway and vascular smooth muscle cells. Although the mitogenic effects of ET-1 are relatively minor in comparison with the effects of EGF and fetal calf serum, ET-1-induced mitogenesis is dependent on both pertussis toxin-insensitive and -sensitive pathways in specific cell types [47, 48]. ET-1 stimulates DNA and protein synthesis in pulmonary artery smooth muscle [49]; these data suggest that ET-1 may play a role in modulating the growth of pulmonary vascular smooth muscle.

ET-1 alone, however, does not induce significant human airway smooth muscle cell proliferation [50]. Interestingly, ET-1 potentiated mitogenesis induced by EGF, apparently *via* an ET_A receptor-mediated mechanism [50]. ET_A receptor activation is involved in ET-1-induced mitogenic signaling in other mesangial cells [51]. Studies on lung organ cultures from dogs revealed that ET-1 increases positive staining for proliferation cell nuclear antigen in lung parenchyma [52]. BQ-123, an ET_A receptor antagonist, did not abolish this effect suggesting that in dog lung ET-1 signals through other ET receptor isotypes.

Takimoto et al. studied the involvement of ET_A and ET_B receptors in ET-1-induced DNA synthesis in a human lung cell line, CCD-18Lu, which expresses both receptor isotypes [48]. FR139317, a selective antagonist of ET_A receptor, completely inhibited ET-1-induced DNA synthesis. BQ-788 and IRL 2500, selective antagonists of ET_B receptor, partly inhibited ET-1-induced DNA synthesis, suggesting that the ET_B receptor does not play a dominant role in ET-1-induced mitogenesis. In contrast to the mitogenic effect of ET-1, ET-3 did not stimulate DNA synthesis [48].

Although ET-1-induced mitogenesis may require 12 to 18 h to manifest, rapid desensitization of ET-1-stimulated ET_A and ET_B receptors occurs [53]. Receptor-specific or homologous desensitization was observed within 4 min in A10 cells transfected with human ET_A receptor, and in 293 cells expressing either ET_A or ET_B receptor. Receptor desensitization coincided with ET-1-induced phosphorylation of each receptor. Freedman et al. demonstrated that inhibition of G-protein-coupled receptor kinases (GRKs) or dominant negative GRK2 mutant reduced ET-1-induced ET_A and ET_B receptor phosphorylation and abrogated receptor desensitization [53]. These data demonstrate that ET-1-induced signaling involves an activation of ET_A and ET_B receptors followed by rapid desensitization.

The activation of ET_A and ET_B receptors by ETs stimulates the mitogen-activated protein kinase (MAPK) cascade [54–57]. A potential mechanism of MAPK kinase activation by ET-1 is shown in Figure 2. MAPK activation has been studied to determine the mechanisms underlying ET-induced mitogenesis. ET-1-stimulated MAPK activation was markedly inhibited by PKC downregulation [55, 56]. Investigators using PKC inhibitors, however, showed that PKC may be necessary but not sufficient for ET-1-induced mitogenesis, suggesting that alternative signaling pathways may be involved in ET-1-induced MAPK activation. Stimulation of MAPK by ET-1 is pertussis toxin-sensitive [57]. Daub et al. demonstrated that transphosphorylation of EGF receptor may also occur in response to ET-1 stimulation and is dependent on G-protein-coupled receptors [58].

ET-1-induced signaling involves other members of the MAPK family. Aquilla et al. demonstrated that ET-1 activates the ET_B receptor and stimulates the extracellular regulated kinase (ERK) 2, c-Jun N-terminal kinase 1 (JNK) and p38 kinase [59]. Using a mutant of the ET_B receptor, they found

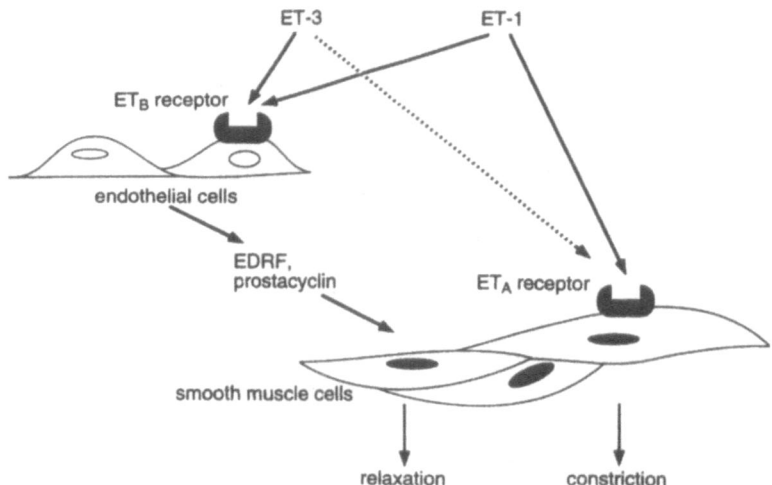

Figure 1. Endothelin-1 and endothelin-3 mediate vasodilation and vasoconstriction. ET_A receptor expression on smooth muscle manifests high affinity for ET-1 but low affinity for ET-3. Activation of ET_A receptor induces vasoconstriction. Expression of ET_B receptor on endothelium, which has equal affinity for both peptides, can release endothelium-derived relaxing factor (EDRF) or prostacyclin. The physiologic consequences of activation of a cell depend on the balance between expression of the ET_A and ET_B receptors (reproduced with permission [60]).

that the C-terminal tail of the receptor is required for activation of these multiple MAPK signal transduction pathways.

ET-induced mitogenesis likely involves an activation of signaling pathways that alters gene transcription. In cultured rat vascular smooth muscle cells, ET-1 induced the expression of the immediate early response genes, *c-fos* and *c-myc* [60]. ET-1 stimulated tyrosine phosphorylation of the nonreceptor protein tyrosine kinase, pp60c-src, which suggests a possible contribution of this tyrosine kinase in ET-1-induced mitogenic signaling [51, 61]. Simonson et al. also demonstrated that expression of dominant negative c-Src mutant and the COOH-terminal Src kinase (Csk), which both inhibit c-Src kinase activity, abrogated ET-1-stimulated *c-fos* transcription [62]. These data suggest that non-receptor tyrosine kinases play a role in ET-1-induced mitogenic signaling.

4. Summary

Current studies suggest a critical biological role of ETs in modulating the development of the lung and vasculature. Pathophysiologically, ETs may also play a role in Hirschsprung disease, asthma and pulmonary hypertension. Mutation and gene knockout studies have also demonstrated an important influence of ET-1 in cardiovascular and pulmonary function and

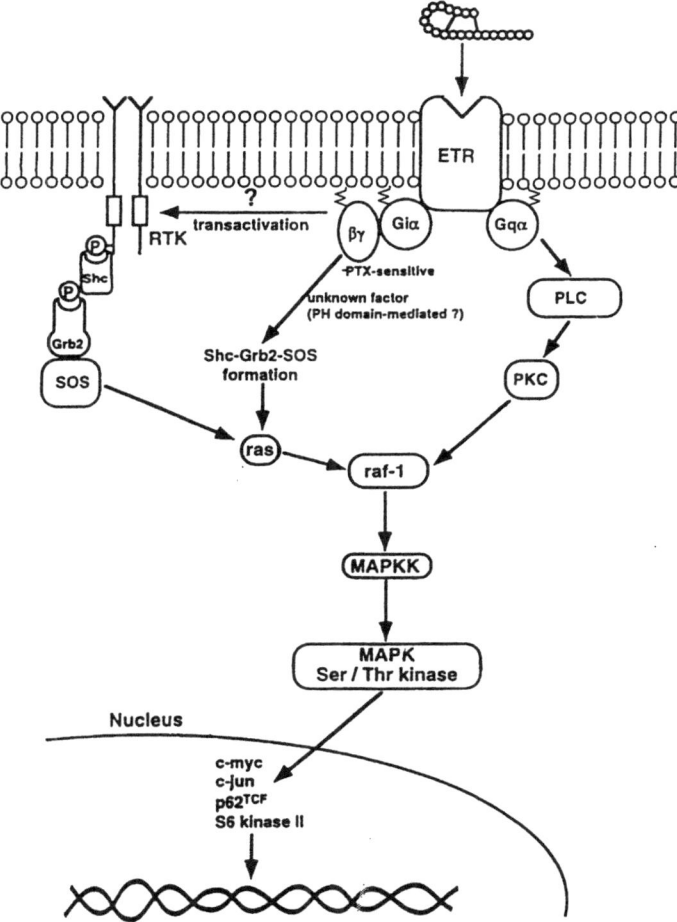

Figure 2. Endothelin-1 receptor activation which is coupled to G-proteins may induce activation of phospholipase C or possibly activate ras. The consequences of PKC and/or ras activation may lead to raf-dependent MAP kinase activation. Recent studies have suggested that ET receptor activation may alter responses generated from receptor tyrosine kinase (RTK) coupled pathways. ETR: endothelin receptor; RTK: receptor tyrosine kinase; G: GTP-binding protein; PLC: phospholipase C; PKC: protein kinase C; PTX: pertussis toxin; Shc: SH2-containing protein; Grb2: growth factor receptor bound 2; SOS: son of sevenless; MAPK: mitogen-activated protein kinase; MAPKK: MAPK kinase (reproduced with permission from [60]).

development. Knockout experiments of the ET-1 gene reveal that lack of ETs induces abnormalities of the pharyngeal-arch-derived craniofacial tissues and death by respiratory failure at birth. ET-3 and ET_B receptor genes may also play an important role in normal development of epidermal melanocytes and enteric ganglion neurons in humans and mice. In humans, mutations in the ET_B receptor gene mapped to chromosome 13 resulted in development of Hirschsprung disease. ETs, first discovered as vasoconstrictors, were also reported to act as a potent bronchoconstrictor and mitogen. Taken together, compelling evidence suggests that ET-1 may contribute to smooth muscle growth, vasculature and airways remodeling as well as modulating cardiopulmonary development.

References

1 Black PN, Ghatei MA, Takahashi K et al (1989) Formation of endothelin by cultured airway epithelial cells. *FEBS Lett* 255: 129–132
2 Rozengurt N, Springall DR, Polak JM (1990) Localization of endothelin-like immunoreactivity in airway epithelium of rats and mice. *J Pathol* 160: 5–8
3 Rubanyi GM, Polokoff MA (1994) Endothelins: molecular biology, biochemistry, pharmacology, physiology, and pathophysiology. *Pharmacological Reviews* 46: 325–415
4 Giaid G, Polak JM, Gaitonde V et al (1991) Distribution of endothelin-like immunoreactivity and mRNA in the developing and adult lung. *Am J Respir Cell Mol Biol* 4: 50–58
5 Nunez DJR, Brown MJ, Davenport AP, Neylon CB, Schofield JP, Wyse RK (1990) Endothelin-1 mRNA is widely expressed in porcine and human tissues. *J Clin Invest* 85: 1537–1541
6 MacCumber MW, Ross CA, Glaser BM, Snyder SH (1989) Endothelin: visualization of mRNAs by *in situ* hybridization provides evidence for local action. *Proc Natl Acad Sci USA* 86: 7285–7289
7 Mattoli S, Mezzetti M, Riva G, Allegra L, Fasoli A (1990) Specific binding of endothelin on human bronchial smooth muscle cells in culture and secretion of endothelin like material from bronchial epithelial cells. *Am J Respir Cell Mol Biol* 3: 145–151
8 Endo T, Uchida Y, Matsumoto H et al (1992) Regulation of endothelin-1 synthesis in cultured guinea pig airway epithelial cells by various cytokines. *Biochem Biophys Res Com* 186: 1594–1599
9 White SR, Hathaway DP, Umans JG, Leff AR (1992) Direct effects on airway smooth muscle contractile response caused by endothelin pig trachealis. *Am Rev of Respir Dis* 145: 491–493
10 Arai H, Hori S, Aramori I, Ohkubo H. Nakanishi S (1990) Cloning and expression of a cDNA encoding an endothelin receptor. *Nature* 348: 730–732
11 Sakurai M, Yanagisawa M, Takuwa Y et al (1990) Cloning of a cDNA encoding a non-isopeptide-selective subtype of the endothelin receptor. *Nature* 348: 732–735
12 Gariepy CE, Cass DT, Yanagisawa M (1996) Null mutation of endothelin receptor type B gene in spotting lethal rats causes aganglionic megacolon and white coat color. *Proc Natl Acad Sci USA* 93: 867–872
13 Power RF, Wharton J, Zhao Y, Bloom SR, Polak JM (1989) Autoradiographic localization of endothelin-1 binding sites in the cardiovascular and respiratory systems. *J Cardiovasc Pharmacol* 13 (Suppl 5): S50–S56
14 Henry PJ, Rigby PJ, Self GJ, Preuss JM, Goldie RG (1990) Relationship between endothelin-1 binding site densities and constrictor activities in human and animal airway smooth muscle. *Brit J Pharmacol* 100: 786–792
15 Hemsen A, Franco-Cereceda A, Matran R, Rudehill A, Lundberg JM (1990) Occurrence, specific binding sites and functional effects of endothelin in human cardiopulmonary tissue. *Eur J Pharmacol* 191: 319–328
16 Goldie RG, D'Aprile AC, Self GJ, Rigby PJ, Henry PJ (1996) The distribution and density of receptor subtypes for endothelin-1 in peripheral lung of the rat, guinea-pig and pig. *Brit J Pharmacol* 100: 729–735
17 Kurihara Y, Kurihara H, Suzuki H et al (1994) Elevated blood pressure and craniofacial abnormalities in mice deficient in endothelin-1. *Nature* 368: 703–710
18 Kurihara Y, Kurihara H, Oda H et al (1995) Aortic arch malformations and ventricular septal defect in mice deficient in endothelin-1. *J Clin Invest* 96: 293–300

19 Maemura K, Kurihara H, Kurihara Y, Nagai R, Yazaki Y (1994) Isolation and characterization of vascular endothelial cells derived from mice lacking endothelin-1. *Biochem Biophys Res Com* 201: 538–545
20 Maemura K, Kurihara H, Kurihara Y, Kuwaki T, Kumada M, Yazaki Y (1995) Gene expression of endothelin isoforms and receptors in endothelin-1 knockout mice. *J Cardiovasc Pharmacol* 26 (Suppl 3): S17–S21
21 Hocher B, Thone-Reineke C, Rohmeiss P et al (1997) Endothelin-1 transgenic mice develop glomerulosclerosis, interstitial fibrosis, and renal cysts but not hypertension. *J Clin Invest* 99: 1380–1389
22 Baynash AG, Hosoda K, Giaid A et al (1994) Interaction of endothelin-3 with endothelin-B receptor is essential for development of epidermal melanocytes and enteric neurons. *Cell* 79: 1277–1285
23 Hosoda K, Hammer RE, Richardson JA et al (1994) Targeted and natural (piebald-lethal) mutations of endothelin-B receptor gene produce megacolon associated with spotted coat color in mice. *Cell* 79: 1267–1276
24 Amiel J, Attie T, Jan D et al (1996) Heterogeneous endothelin receptor B (EDNRB) mutations in isolated Hirschsprung disease. *Human Molecular Genetics* 5: 355–357
25 Nataf V, Lecoin L, Eichmann A, Le Douarin NM (1996) Endothelin-B receptor is expressed by neural crest cells in the avian embryo. *Proc Natl Acad Sci USA* 93: 9645–9650
26 Lahav R, Ziller C, Dupin E, Le Douarin NM (1996) Endothelin 3 promotes neural chest cell proliferation and mediates a vast increase in melanocyte number in culture. *Proc Natl Acad Sci USA* 93: 3892–3897
27 Chan TS, Lin CX, Chan SS, Chung SK (1995) Mouse preproendothelin-1 gene. cDNA cloning, sequence analysis and determination of sites of expression during embryonic development. *Eur J Biochem* 234: 819–826
28 Guembe L, Villaro AC, Bodegas ME (1996) Immunocytochemical detection of endothelin during the development of murine lung. *Internatl J Dev Biol* (Suppl 1): 257S–258S
29 Loesch A, Burnstock G (1996) Ultrastructural localization of nitric oxide synthase and endothelin in rat pulmonary artery and vein during postnatal development and aging. *Cell and Tissue Research* 283: 355–365
30 Grunstein MM, Rosenberg SM, Schramm CM, Pawlowski NA (1991) Mechanisms of action of endothelin 1 in maturing rabbit smooth muscle. *Am J Physiol* 260: L434–L443
31 Mavquin-Mavier I, Levame M, Istin N, Harf A (1989) Mechanisms of endothelin mediated bronchoconstriction in the guinea pig. *J Pharmacol Exp Ther* 250: 740–745
32 Uchida Y, Ninomiya H, Saotome M et al (1988) Endothelin, a novel vasoconstrictor peptide as potent bronchoconstrictor. *Eur J Pharmacol* 154: 227–228
33 Maggi CA, Patacchini R, Giuliani S, Meli A (1989) Potent contractile effect of endothelin in isolated guinea-pig airways. *Eur J Pharmacol* 160: 179–182
34 Hay DWP, Henry PJ, Goldie RG (1993) Endothelin and the respiratory system. *Trends Pharmacol Sci* 14: 29–32
35 White SR, Hathaway DP, Umans JG, Tallet J, Abrahams C, Leff AR (1991) Epithelial modulation of airway smooth muscle response to endothelin-1. *Am Rev of Respir Dis* 144: 373–378
36 Plews PI, Abdel-Malek ZA, Doupnik CA, Leikauf GD (1991) Endothelin stimulates chloride secretion across canine tracheal epithelium. *Am J Physiol* 261: L188–194
37 Eguchi S, Hirata Y, Ihara M, Yano M, Marumo F (1992) A novel ETA antagonist (BQ 123) inhibits endothelin 1 induced phosphoinositide breakdown and DNA synthesis in rat vascular smooth muscle cells. *FEBS Lett* 302: 243–246
38 Bialecki RA, Izzo NJ jr, Colucci WS (1989) Endothelin 1 increases intracellular calcium mobilization but not calcium uptake in rabbit vascular smooth muscle cells. *Biochem Biophys Res Com* 164: 474–479
39 Kai H, Kanaide H, Nakamura M (1989) Endothelin sensitive intracellular Ca^{2+} store overlaps with caffeine sensitive one in rat aortic smooth muscle cells in primary cultures. *Biochem Biophys Res Com* 158: 235–243
40 Lee TS, Chao T, Hu KQ, King GL (1989) Endothelin stimulates a sustained 1,2 diacylglycerol increase and protein kinase C activation in bovine aortic smooth muscle cells. *Biochem Biophys Res Com* 162: 381–386
41 Griendling KK, Tsuda T, Alexander RW (1989) Endothelin stimulates diacylglycerol accumulation and activates protein kinase C in cultured vascular smooth muscle cells. *J Biol Chem* 264: 8237–8240

42 Resink TJ, Scott-Burden T, Buhler FR (1990) Activation of multiple signal transduction pathways by endothelin in cultured human vascular smooth muscle cells. *Eur J Biochem* 189: 415–421

43 De Nucci G, Thomas R, D'Orleans Juste P et al (1988) Pressor effects of circulating endothelin are limited by its removal in the pulmonary circulation and by the release of prostacyclin and endothelium derived relaxing factor. *Proc Natl Acad Sci USA* 85: 9797–9800

44 Grunstein MM, Chuang ST, Schramm CM, Pawlowski NA (1991) Role of endothelin 1 in regulating rabbit airway contractility. *Am J Physiol* 260: L75–L82

45 Resink TJ, Scott-Burden T, Buhler FR (1989) Activation of phospholipase A_2 by endothelin in cultured vascular smooth muscle cells. *Biochem Biophys Res Com* 158: 279–286

46 Schumacher WA, Steinbacher TE, Allen GT, Ogletree ML (1990) Role of thromboxane receptor activation in the bronchospasmic response to endothelin. *Prostaglandins* 40: 7!–9

47 Bobik A, Grooms A, Millar JA, Mitchell A, Grinpukel S (1990) Growth factor activity of endothelin on vascular smooth muscle. *Am J Physiol* 258: C408–C415

48 Takimoto M, Oda K, Fruh T, Takai M, Okada T, Sasaki Y (1996) ETA and ETB receptors cooperate in DNA synthesis *via* opposing regulations of cAMP in human lung cell line. *Am J Physiol* 271 (3, Pt 1): L366–L373

49 Janakidevi K, Fisher MA, Del Veccio PJ, Tirappathi C, Figge J, Malik AB (1992) Endothelin 1 stimulates DNA synthesis and proliferation of pulmonary artery smooth muscle cells. *Am J Physiol* 263: C1295–C1301

50 Panettieri RA, Goldie RG, Rigby PJ, Eszterhas AJ, Hay DWP (1996) Endothelin-1-induced potentiation of human airway smooth muscle proliferation: an ET_A receptor-mediated phenomenon. *Brit J Pharmacol* 118: 191–7

51 Simonson MS, Herman WH (1993) Protein kinase C and protein tyrosine kinase activity contribute to mitogenic signaling by endothelin-1. *J Biol Chem* 268: 9347–9357

52 Ricagna F, Miller VM, Tazelaar HD, McGregor CG (1996) Endothelin-1 and cell proliferation in lung organ cultures. Implications for lung allografts. *Transplantation* 62: 1492–1498

53 Freedman NJ, Ament AS, Oppermann M, Stoffel RH, Exum ST, Lefkowitz RJ (1997) Phosphorylation and desensitization of human endothelin A and B receptors. Evidence for G protein-coupled receptor kinase specificity. *J Biol Chem* 272: 17734–17743

54 Wang Y, Rose PM, Webb ML, Dunn MY (1994) Endothelins stimulate mitogen-activated protein kinase cascade through either ET_A or ET_B. *Am J Physiol* 267: C1130–C1135

55 Cazaubon S, Parker PJ, Strosberg AD, Couraud P-O (1993) Endothelins stimulate tyrosine phosphorylation and activity of p42/mitogen-activated protein-kinase in astrocytes. *Biochem J* 293: 381–386

56 Wang Y, Simonson MS, Pouyssegur J, Dunn MJ (1992) Endothelin rapidly stimulates mitogen-activated protein kinase activity in rat mesangial cells. *Biochem J* 287: 589–594

57 Kasuya Y, Abe Y, Hama H et al (1994) Endothelin-1 activates mitogen-activated protein kinases through two independent signaling pathways in rat astrocytes. *Biochem Biophys Res Com* 204:1325–1333

58 Daub H, Weiss FU, Wallasch C, Ullrich A (1996) Role of transactivation of the EGF receptor in signaling by G-protein-coupled receptors. *Nature* 379: 557–560

59 Aquilla E, Whelchel A, Knott HJ, Nelson M, Posada J (1996) Activation of multiple mitogen-activated protein kinase signal transduction pathways by the endothelin B receptor requires the cytoplasmic tail. *J Biol Chem* 271: 31572–31579

60 Goto K, Hama H, Kasuya Y (1996) Molecular pharmacology and pathophysiological significance of endothelin. *Jpn J Pharmacol* 72: 261–290

61 Stewart AG, Grigoriadis G, Harris T (1994) Mitogenic actions of endothelin-1 and epidermal growth factor in cultured airway smooth muscle. *Clin Exp Pharmacol Physiol* 21: 277–285

62 Simonson MS, Wang Y, Herman WH (1996) Nuclear signaling by endothelin-1 requires Src protein-tyrosine kinases. *J Biol Chem* 271: 77–82

Pulmonary Actions of the Endothelins
ed. by R. G. Goldie and D. W. P. Hay
© 1999 Birkhäuser Verlag Basel/Switzerland

CHAPTER 9
Endothelin and the Airway Epithelium

Joaquim Mullol[1], James N. Baraniuk[2], Cesar Picado[3]
and James H. Shelhamer[4]

[1] *Fundació Clínic per a la Recerca Biomèdica, Institut d'Investigacions Biomèdiques ":August Pi i Sunyer", Hospital Clínic i Universitari. Barcelona, Catalonia, Spain*
[2] *Division of Reumathology, Al·lergy and Immunology, Department of Medicine, Georgetown University Medical Center, Washington DC, USA*
[3] *Servei de Pneumologia i Al·lèrgia Respiratòria, Hospital Clínic i Universitari, Departament de Medicina, Universitat de Barcelona. Barcelona, Catalonia, Spain*
[4] *Critical Care Medicine Department, Clinical Center, National Institutes of Health. Bethesda, Maryland, USA*

1   Introduction
2   Endothelin Synthesis and Release
2.1 Lower Airway Epithelium
2.2 Upper Airway Epithelium
3   Endothelin Metabolism and Proteolytic Regulation
4   Endothelin Receptors In Airway Epithelium
4.1 Lower Airway Epithelium
4.2 Upper Airway Epithelium
4.3 Receptor Antagonists
5   Endothelin in Airway Inflammation
5.1 Proinflammatory Mediators
5.2 Arachidonic Acid Metabolites
5.3 Glucocorticoids
6   Epithelium-Derived Endothelin in Airway Physiology
6.1 Effects in the Lower Airways
6.2 Effects in the Upper Airways
7   Epithelium-Derived Endothelin in Airway Pathology
7.1 Bronchial Asthma
7.2 Rhinitis
7.3 Other Respiratory Diseases
8   Concluding Remarks
     References

1. Introduction

The airway epithelium plays an important role in host defense as a barrier against physical, pathological and chemical stimuli. The epithelium is also an active metabolic and biosynthetic site for production and release of mediators, such as arachidonic acid metabolites, cytokines and a putative epithelium-derived refaxing factor, associated with the regulation of the bronchomotor tone and airway inflammation.

The role of endothelins (ETs) in the respiratory tract has been previously reviewed elsewhere [1–12]. ETs constitute a family of 21-amino acid peptides originally isolated from the culture supernatant of porcine aortic endothelial cells [13]. ET has potent vasoconstrictor [14–15], broncho-constrictor [15] and glandular secretory effects [16–18]. ET has at least three genomic isoforms, ET-1, ET-2 and ET-3 [19], which act through to at least two different receptors, ET_A and ET_B [20–22]. ET-1 and ET-2 bind to ET_A receptors more avidly than ET-3, whereas all three bind to ET_B receptors with equal affinity.

2. Endothelin Synthesis and Release

Endothelin may be also produced by non-endothelial cells such as epithelial cells. ET production by airway epithelial cells likely exerts both paracrine and autocrine effects.

2.1. Lower Airway Epithelium

2.1.1. Animal Studies: Immunoreactivity to endothelin (irET-1 and irET-3) is detected in conditioned culture medium from both canine and porcine tracheal epithelial cells [23]. IrET is present in ciliated and secretory bronchiolar epithelial cells in Wistar rats and mice [24]. In the cultured and intact epithelium from rabbit trachea, irET-1 is found, together with SP and arginine-vasopressin [25]. In piglet lung, focal irET-1 is seen over epithelial cells of bronchi, bronchioles and terminal bronchioles [26], raising the possibility that ET-1 could play a role in the regulation of bronchial, as well as vascular, tone and development. In the guinea-pig, the non-ciliated epithelial (Clara) cells are also a source of ET-1, but not ET-2 and ET-3 [27, 28]. In guinea-pig tracheal epithelial cells, ET-1 content of the submucosal side is over 30 times higher than that of the apical side, suggesting the release of ET-1 from airway epithelial cells toward the submucosal side [29]. Bronchiolar epithelial cells from fetal and adult rat lungs express ET mRNA [30].

Several stimuli may induce ET production from epithelial cells. Isolated epithelial cells from rabbit trachea selectively released ET in response to thrombin in a mechanism dependent on protein synthesis and connected with activation of phospholipase C [31]. However, in piglet lung, no changes in the distribution of ET-1 in airway epithelium were observed after exposure to several stimuli such as hypoxia and α-thrombin [26].

Bacterial lipopolysaccharide (LPS) endotoxin is known to produce airway epithelial damage and airway hyperreactivity in guinea pigs. The increase in ET-1 [32] and ET-1 mRNA [33] induced by bacterial endotoxin in cultured epithelial cells may be related to this hyperreactivity. In guinea pig cultured tracheal epithelial cells, cytokines involved in damage, inflammation and repair may regulate ET-1 synthesis. ET-1 production is increased by IL-1, IL-8, TNF-α and TGFβ while big ET-1 is enhanced by IL-2, IL-6 and IGF-1 enhanced big ET-1 release. Several other cytokines (IL-2, IL-6, GM-CSF, EGF, PDGF, IGF-1) also enhance epithelial cell mitogenesis [34]. Conversely, steroids attenuate the release of ET-1 by guinea-pig Clara cells [27].

2.1.2. Human Studies: ET-immunoreactive material (irET) is released from human tracheal [35] and bronchial [36] epithelial cells. Although ET mRNA has been demonstrated in epithelial cells by McCumber et al. [30], other authors have found that only 50% of human adults showed irET and mRNA transcripts in airway epithelial cells [37]. Immunoreactive pro-ET-1 and pro-ET-3 are located along with mature ET in the airway epithelia of the adult human lung [38]. Takimoto et al. have also recently shown that human bronchial squamous cells, but not basal cells, secrete ET-1 [39].

Early *in situ* hybridation studies showed expression of ppET-1 mRNA in human bronchial biopsies [40]. More recently, Sun and coworkers have reported that human bronchial epithelial cells express two different forms (2.3 and 2.5 kb) of ppET-1 mRNA [41].

Proinflammatory mediators such as endotoxin and cytokines (TNF-α, IL-1β), increased ET release and ET mRNA expression from human epithelial cells [42, 43]. Histamine and IL-1β also modulated ET-1 expression in bronchial epithelial cells from asthmatic and non-asthmatic patients [44].

2.1.3. Cell Line Studies: Since several animal and human airway cell lines express and release ET, they may be used to investigate the role of ET in airway pathophysiology. L2 cells, a cloned rat alveolar epithelial cell line, secrete ET-1 and express ET-1 mRNA [45]. IL-2 and fetal bovine serum increase both ET-1 release and mRNA ET-1 expression in A549 cells, a human pulmonary epithelial cell line [46]. Another human bronchial epithelial cell line, BEAS-2B, expressed mRNA for ET-1 but not for ET-2 and ET-3 [47].

2.2. Upper Airway Epithelium

Few ET studies have been done in either human or animal nasal mucosa. MacCumber and coworkers have reported the expression of ET mRNA by fetal rat nasal mucosa [30]. In human nasal mucosa, the epithelium shows moderate irET-1 but not expression of ET-1 mRNA, suggesting that ET-1 synthesized by other cells could be bound and released by epithelial cells [18]. Other authors have also reported the presence of irET in epithelial cells as well as endothelial cells and the connective tissue [48]. Epithelial cells from allergic nasal mucosa and predominantly from nasal polyps also release detectable amounts of ET-1 into culture supernatants [49]. A recent study reports that epithelial cells from human nasal mucosa expressed both irET-1 and ET-1 mRNA [50].

3. Endothelin Metabolism and Proteolytic Regulation

Neutral endopeptidase (NEP), a metallo-endopeptidase (EC3.4.24.11, enkephalinase) is released by human airway epithelial cells [51]. NEP induces the degradation by hydrolysis of ETs. This effect is inhibited by protease inhibitors such as phosphoramidon [52, 53]. Guinea-pig [33], human nasal [50] and bronchial [43] airway epithelial cells co-express both ET-1 and endothelin converting enzyme (ECE), an enzyme which converts big ET in ET-1.

Phosphoramidon, an inhibitor of ECE (metalloproteinase), and NEP, increases the contractile potencies of ET-1 and ET-2 on guinea-pig trachea, suggesting a degradation of ETs by the latter protease [54]. Big-ET induces potent contractile activity in isolated human bronchus due to a conversion of big-ET to ET-1 by a phosphoramidon-sensitive metalloprotease similar to ECE [55]. Conversely, phosphoramidon had no effect on ET-1 and ET-3-induced bovine bronchial smooth muscle contraction [56], while it decreased baseline growth and ET-1 production of porcine cultured tracheal epithelial cells [57]. Moreover, the generation of ET-1 by guinea-pig cultured Clara cells was also inhibited by phosphoramidon, whereas thiorphan, an inhibitor of neutral endopeptidase 24.11, increased the levels of ET-1 by reducing ET degradation [27, 28].

Several studies [58, 59] have demonstrated that the action of ET-1 on guinea-pig airways is strengthened when the epithelium is removed. Hay and coworkers reported that the increased responsiveness of guinea pig trachea to the contractile effect of ET induced by epithelium denudation may be due in part to the removal of NEP or another phosphoramidon-sensitive peptidase which degrades ET [58]. Other studies also suggest that the inhibitory effect exerted by the epithelium on the contraction of isolated human bronchi induced by ETs (ET-1, ET-2, ET-3) may be attributed in part to an enkephalinase activity [60].

However, in addition to the degradation by NEP and other proteases, other events such as the release of epithelium-derived-relaxant agents, may contribute to the modulation of endothelin-induced responses by airway epithelium.

4. Endothelin Receptors in Airway Epithelium

4.1. Lower Airway Epithelium

The presence and number of ET receptors, as well as the predominant receptor subtypes, in the airway epithelium differs among species.

4.1.1. Animal Studies: Early studies performed on rat lungs reported that bronchiolar epithelial cells from fetal and adult animals expressed ET specific binding sites [30], involving two different ET receptors [61]. Using ET specific receptor antagonists, two distinct ET receptors, ET_A and ET_B, have been found in porcine pulmonary tissues, with the ET_A subtype preferentially localized in the bronchial epithelium and vasculature [62]. In the porcine lung, little epithelial binding was found using [^{125}I]ET-1 (non-selective for ET_A and ET_B receptors) and [^{125}I]BQ3020 (highly selective for ET_B receptors), suggesting low levels of ET receptors in the epithelium [63]. The bronchiolar and bronchial epithelium of rat and rabbit lungs express abundant receptors of the ET_B subtype [64]. Two saturable binding sites for ET-1 (K_d = 35.3 and 205.9 pM) are present in cultured epithelial cells from feline trachea [65], while canine tracheal epithelial cells express a high-affinity specific-binding site to ET-1 (K_d = 0.2 nM) which shows the characteristics of the ET_A receptor subtype [66].

4.1.2. Human Studies: Although early autoradiographic studies in human trachea showed that ET-1 receptors were present in smooth muscle and parasympathetic nerves but not in the epithelium [35], specific binding sites for ET-1, ET-2 and ET-3 have been found recently in airway epithelium and submucosa of the adult human lung [38]. Goldie and coworkers demonstrated low levels of the ET_A, but not ET_B, receptor subtype in the bronchial epithelium of both asthmatic and non-asthmatic lungs [67]. Similarly, Takimoto et al. have found binding sites to ET-1 in both squamous and basal human bronchial epithelial cells, demonstrating the existence of only ET_A receptors [39].

4.1.3. Cell Line Studies: L2 cells, a cloned rat alveolar epithelial cell line, show a single binding site for ET-1 (ET_A, K_d = 4.03 nM) [45]. Similarly, BEAS-2B cells, a human bronchial epithelial cell line, also expresses mRNA for ET_A but not for ET_B [47] (Fig. 1).

4.2. Upper Airway Epithelium

While fetal rat nasal mucosa expresses binding sites to ET [30], binding sites for [^{125}I]ET-1 were found in submucosal glands, arterioles and venous sinusoids in human nasal mucosa, but no constitutive ET receptors were seen in the epithelium [18].

4.3. Receptor Antagonists

Evidence for a role of ET in airway pathophysiology includes studies using ET-receptor antagonists [68]. Nakamichi and coworkers, using ET specific

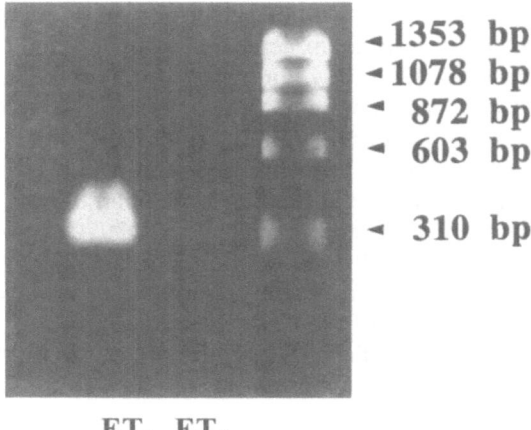

◄ 1353 bp
◄ 1078 bp
◄ 872 bp
◄ 603 bp

◄ 310 bp

ET$_A$ ET$_B$

Figure 1. Human airway epithelial cells (BEAS-2B) express mRNA (300 bp) for ET$_A$ but not ET$_B$ receptor subtype.

receptor antagonists, found two distinct ET receptors, ET$_A$ and ET$_B$, in porcine pulmonary tissues, ET$_A$ being mainly localized in the bronchial epithelum and vasculature [62]. Recently, Markewitz et al. reported that BQ-123, an ET$_A$ selective receptor antagonist, blocked ET-1 binding and the inhibition of ET-1-induced PGE$_2$ and cAMP production by L2 cells, a rat alveolar epithelial cell line [45]. BQ-123 also inhibited ET-1 binding in canine tracheal epithelial cells [66]. Both BQ-123 and PD-145065 (mixed ET$_A$-ET$_B$ receptor antagonist) also decreased ET-1-induced growth of cultured porcine tracheal epithelial cells, suggesting that ET-1 acts as an autocrine growth factor for airway epithelial cells [57].

Recent reports also suggest the possibility that ET-1 receptor antagonists directed at the appropriate receptor subtype may be useful in the treatment of asthma. For example, aerosolized BQ-123 blocked the late, but not the immediate, antigen-induced airway hyperresponsiveness to carbachol in allergic sheep [69]. Further, pretreatment with BQ-123, but with the ET$_B$ selective receptor antagonist BQ-788, blocked the decrease in sheep trachea mucociliary clearance induced by aerosolized ET-1 [70].

5. Endothelin in Airway Inflammation

5.1. Proinflammatory Mediators

Recent observations suggest that, in airway inflammation, proinflammatory mediators such as allergens, bacterial endotoxin and cytokines stimulate airway epithelial cells to produce ET, which may be related to acute exacerbation of asthma.

5.1.1. Animal Studies: Cytokines involved in damage, inflammation and repair regulate ET-1 synthesis in cultured guinea pig tracheal epithelial cells. IL-1, IL-8, TNF-α and TGF-β increase ET-1 synthesis while IL-2, IL-6 and IGF-1 enhance the synthesis of big ET-1 [34].

5.1.2. Human Studies: In human cultured bronchial epithelial cells, bacterial LPS endotoxin and proinflammatory cytokines such as IL-1 and TNF-α increased ET gene expression (ppET-1 mRNA) and ET-1 release [42, 43], while IL-1β and histamine, probably released by inflammatory cells during acute asthma exacerbation, upregulated both ET-1 gene expression and ET protein secretion from bronchial epithelial cells of asthmatic patients [44].

5.1.3. Cell Line Studies: IL-2 and fetal bovine serum increased, while steroids inhibit, both ET-1 release and ET-1 mRNA expression in A549 cells, a human pulmonary epithelial cell line [46]. Furthermore, LPS, IL-1, TNF-α and TGF-β, increase, while IFNγ and PGDF decrease, ET-1 secretion from L2 cells, a cloned rat alveolar epithelial cell line, raising the possibility that in patients with sepsis, pulmonary infection or inflammation increase alveolar ET-1 production which may contribute to the worsening of these processes [45]. More recently, studies done in BEAS-2B cells showed that ET-1 induced the release of IL-8, IL-6 and GM-CSF, but not G-CSF, without affecting cell proliferation [71]. Since these cells express ET$_A$ receptors and may also produce ET [47], it is possible that ET of epithelial origin may act in an autocrine-paracrine fashion to stimulate the production of IL-8, IL-6 and GM-CSF, and other substances which may contribute to airway inflammation.

5.2. Arachidonic Acid Metabolism

5.2.1. Animal Studies: In the rat lung, ET-1 stimulates the release of 15-HETE into bronchoalveolar lavage (BALF) [72]. Since it is a potent chemoattractant for eosinophils, 15-HETE could contribute to airway inflammation. Lower doses of ET-1 decrease the acetylcholine-induced rabbit tracheal smooth muscle contraction by increasing the release of bronchodilatory PGs, including PGI$_2$ and PGE$_2$, while at higher concentrations ET-1 elicits a Ca^{2+}-dependent airway constrictor response [73]. In canine tracheal epithelial cells, secondary synthesis of cyclooxygenase (COX) metabolites contributed to ET-1 – enhanced chloride secretion [74, 75]. In the guinea-pig trachea, ET-1 causes an increased release of prostaglandins such as PGE$_2$, PGD$_2$, PGI$_2$, PGF$_{2\alpha}$, TXA$_2$ and TXB$_2$ [76, 77].

Moreover, this effect was not affected by the removal of the epithelium and did not have a major influence on ET-1-induced smooth muscle contractions [77]. In cultured epithelial cells from feline trachea, Wu et al. have reported that ET-1 increased the release of PGE_2, $PGF_{2\alpha}$, 5-HETE and 15-HETE, by acting *via* specific receptors [65] (Fig. 2).

5.2.2. Human Studies: ET-1 does not augment the spontaneous release of histamine or immunoreactive leukotrienes from human bronchi, but it caused a concentration-dependent increase in PG_s (PGD_2, PGE_2, $PGF_{2\alpha}$, prostacyclin and TXB_2) release from human bronchi via ET_A receptors [77]. Despite this, COX inhibitors, in addition to atropine and receptor antagonists to PAF, TXA_2, leukotrienes or H_1-histamine, have no effect on ET-1-induced contraction in human isolated bronchi. In human bronchial epithelial cells, it has been recently reported that ET-1 stimulates the release of prostanoids (PGE_2, TXB_2) from squamous but not from basal cells [39]. ET-1 also stimulates the release of arachidonic acid products from human nasal mucosa [78], which may modulate the airway inflammatory response. In nasal mucosal explants, ET-1 induced the release of PGE_2, PGD_2, $PGF_{2\alpha}$, TXB_2 and 15-HETE suggesting a predominant effect of ET on the release of COX products [78].

5.2.3. Cell Line Studies: In L2 cells, a cloned rat alveolar epithelial cell line, exogenous ET-1 increased cAMP and PGE_2 production, whereas blockade of endogenous ET-1 decreased PGE_2 production [45]. These findings also suggest that ET may act as an autocrine factor in alveolar epithelial cells.

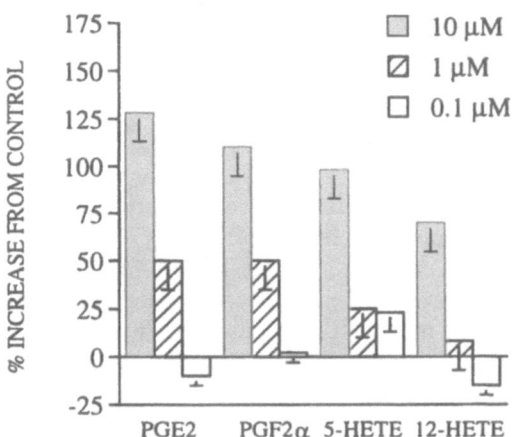

Figure 2. ET-1 causes a dose-related release of eicosanoids (PGE_2, $PGF_{2\alpha}$, 5-HETE, 12-HETE) by feline tracheal epithelial cells (n = 4) after 1 h of incubation. (Reprinted with permission from [65]).

5.3. Glucocorticoids

Several studies in different animal species, as well as in humans, show the antiinflammatory effect of steroids on ET production and ET-induced effects.

5.3.1. Animal Studies: Dexamethasone attenuated the release of ET-1 by guinea-pig cultured Clara cells [27]. Acute treatment with dexamethasone in guinea-pig isolated airways does not prevent ET-1-induced contractions, but does produce relaxation when airways are precontracted with ET-1 in the presence or absence of epithelium [79]. In the rat lung, ET levels are elevated in the setting of inflammation. This increase is sensitive to aerosolized budesonide and to surgical adrenalectomy [80].

5.3.2. Human Studies: Corticosteroids decreased the release of irET material from bronchial epithelial cells of patients with symptomatic asthma, without modifying the expression of ppET transcripts [40]. In contrast, a separate study suggests that steroids do not reduce the BAL concentration of ET-1 in steroid-treated asthmatic patients [81].

5.3.3. Cell Line Studies: In A549 cells, a human pulmonary epithelial cell line, corticosteroids inhibited both baseline and IL-2-induced ET-1 release and ET-1 mRNA expression, this effect being reversed by mifepristone, a competitive steroid receptor antagonist [46].

6. Epithelium-Derived Endothelin in Airway Physiology

6.1. Effects in the Lower Airways

6.1.1. Animal Studies ET effects: Early studies done in the guinea-pig showed that ET is a potent bronchoconstrictor [15]. In allergic sheep, ET-1 is released in the airways after antigen challenge, while aerosolized ET-1 induces bronchoconstriction by stimulating ET_A receptors, suggesting a contribution of ET to the severity of allergic reaction by increasing airway smooth muscle responsiveness [69]. In addition, aerosolized ET-1 decreased mucociliary clearance of sheep trachea [70], but it failed to induce the release of oxygen radicals from guinea pig tracheal epithelial cells [82].

ET, formed by canine tracheal epithelial cells, is also a potent agonist of epithelial chloride secretion. The effect of ET-1 on chloride and cAMP production is probably due to the formation of COX products. In the same study, ET also increased intracellular Ca^{2+} [83]. ET-1, but not ET-2 or ET-3, applied to the luminal surface of canine tracheal epithelial cells, increased chloride secretion and the transepithelial potential difference across the epithelium [74, 75]. While ET applied to the submucosal side had no effect on chloride movement, secondary synthesis of COX metabolites appears to contribute to ET-1 enhanced chloride secretion [74, 75].

In the ferret trachea, ET-1 contracted airway smooth muscle and increased transepithelial potential differences [84]. ET also inhibited glandular secretion and epithelial albumin transport induced by muscarinic and α-adrenoceptor stimulation [84, 85]. In feline trachea, ET-1, in contrast to ET-2 and ET-3, produced secretion of a mucus glycoprotein from isolated submucosal glands via Ca^{2+} influx, whereas it may induce an inhibitory factor for glandular secretion from epithelial cells [17]. ET-1, through a possible extracellular Ca^{2+}-dependent mechanism, also enhanced ciliary beat frequency (32%) in canine tracheal epithelial cells [75], and in preparations of rabbit nasal and tracheal mucosa [86]. Finally, ET-1, in contrast to ET-3, is mitogenic for porcine tracheal epithelial cells since it enhanced their growth *via* ET_A receptor subtype stimulation [57].

Modulation of ET effects by the Epithelium: Epithelium denudation increases the responsiveness of guinea pig trachea and bronchi to the contractile effect of ET-1 and ET-2 [54, 58, 59], in part due to removal of metabolic effect of NEP. ET-1 induced an early relaxation of the guinea-pig sensitized tracheal smooth muscle, which was somewhat enhanced by epithelium removal [87], suggesting the release of a potential smooth muscle relaxing factor by ET [88]. In a report from Filep et al., ET-1 induced a dual action, evoking contractions at low resting tone and relaxations at higher resting tone in guinea-pig tracheal strips. This ET-1 relaxant effect may be mediated by NO released from epithelial cells [76, 89]. In dogs, ET-1-induced bronchial contraction was inhibited by epithelium removal [90]. In contrast, epithelium removal did not modify the ET-1 and ET-3-induced contraction of bovine bronchial smooth muscle [56].

Ninomiya and coworkers found that epithelium removal enhanced ET-induced guinea-pig tracheal contractions by allowing a direct effect in smooth muscle cells as well as inducing mediator release from inflammatory cells [91]. Other reports suggest that ET-1 may induce an early relaxation of guinea pig airway smooth muscle and a late contractile effect requiring the synthesis of PGs and the presence of an intact epithelium, since the contractile effect is inhibited by COX inhibitors [92]. The early relaxation was not observed when ET is topically administered, suggesting that epithelial modulation of ET effects depend upon the route of administration [93]. The addition of corticosteroids to guinea pig isolated airways produced relaxation of airways precontracted with ET-1, in the presence or absence of epithelium, but it did not prevent ET-1-induced contraction [79].

6.1.2. Human Studies
Effects: ET is a potent bronchoconstrictor of human isolated bronchus [35, 55, 56, 67, 94]. The release of ET from epithelial cells may directly induce bronchoconstriction by binding to specific receptors on bronchial smooth

muscle in a way that has been mimicked at least partially, in culture [36, 95]. ET-1, ET-2 and ET-3 potently contract human isolated airway smooth muscle [55, 67]. A direct effect on airway smooth muscle through ET_B receptors, an indirect effect secondary to the release of mediators such as PAF and PGs from inflammatory cells that infiltrate asthmatic airways, and a neuromodulator effect that potentiated nerve-mediated contractions [96] may be the potential mechanisms involved. In human bronchial explants ET-1 also induced the release of respiratory glycoproteins from sub-mucosal glands [16]. In human bronchial epithelial cells ET-1 also upregulated the gene expression and release of fibronectin, an important component of the extracellular matrix [97]. Recently, ET-1 has shown the ability to induce DNA synthesis in human bronchial basal epithelial cells, suggesting a co-mitogenic role for ET [39] as has been reported in human airway smooth muscle cells in culture [95].

Modulation of ET effects by the Epithelium: Removing the epithelium from human bronchi significantly increased the constriction caused by all three ETs [60], suggesting an important role of the epithelium in modulating ET-induced contraction, either by serving as a barrier, degrading the ETs, or releasing an epithelial-dependent relaxing factor [60] (Fig. 3). The enhanced response after epithelium removal has obvious implications in asthma and acute lung injury, where inflammation frequently damages the airway epithelium [98]. In contrast to findings from animal studies [76,

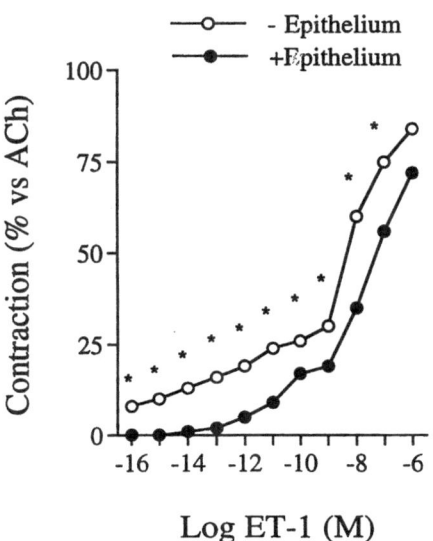

Figure 2. The response of human isolated bronchi to ET-1 is higher in the absence of epithelium than in its presence (* = p < 0.05, n = 4–9). (Reprinted with permission from [60]).

89], ET-1 failed to stimulate iNOS expression in human bronchial epithelial cells [39].

6.1.3. Cell Line Studies: Takuwa and coworkers reported that ET acts as a potent mitogen in Swiss 3T3 fibroblasts [99]. Since asthmatic airways show collagen deposition in the basement membrane, this mitogenic effect of ET-1 on fibroblasts may also be of relevance to asthma. However, no effect of ET-1 in the proliferation of bronchial epithelial cell line, BEAS-2B, could be demonstrated [47].

6.2. Effects in the Upper Airways

Locally released, ET-1 may participate in the regulation of nasal blood flow and secretory cell exocytosis *via* paracrine mechanisms [18, 100].

6.2.1. Animal Studies: Studies in rabbits have shown that ET-1 induced a potent and prolonged vasoconstriction [48, 100]. These findings suggest that locally released ET-1, in association with other neurotransmitters, could participate in the control of nasal blood flow.

6.2.2. Human Studies: The fact that epithelium is an ET-producing tissue and that ET specific binding sites as well as ET receptor mRNA are expressed in epithelial cells, suggest that ETs are locally acting peptides (autocrine-paracrine) rather than as circulating hormones. Circulating plasma concentrations of endothelin are normally below one picomolar, far below pharmacological levels [22]. In addition to its vasoconstrictor properties, ET stimulates cell proliferation, neurotransmitter release and smooth muscle contraction [6, 7]. ET-1 and ET-2, in contrast to ET-3, increase both serous (lactoferrin) and mucous (glycoproteins) gland product release in normal human nasal mucosa suggesting, in addition to regulating the nasal vascular tone, a role for this peptide in stimulating nasal glandular secretion [18]. Intranasal administration of ET-1 also induces an increase in nasal symptoms and the release of glandular products [101].

7. Epithelium-Derived Endothelin in Airway Pathology

7.1. Bronchial Asthma

Epithelial injury is one of the characteristic features of bronchial asthma [98] and in this situation ET might be released from airway epithelial cells and have proinflammatory effects in target tissues. Evidence that ET may have a role in airway pathophysiology was first obtained by Nomura et al., who reported that ET-1 levels rise in BALF from 0.05 to 0.3 pg/ml during acute asthmatic episode [102]. ET-1 has been localized to vascular endo-thelium, connective tissue cells and bronchial epithelium in asthma [103]; the percentage of cells staining position for expression was higher (70%) in biopsy samples from asthmatic than in controls patients (10%). Mattoli and coworkers also reported higher concentrations of irET (ET-1 and ET-3) in the BAL of patients with symptomatic asthma (250 pg/L) compared to patients with chronic bronchitis (80 pg/L) and normal volunteers (60 pg/L), while the plasma ET-1 levels were similar in all three groups [104]. In that study, after treatment with β_2-agonists and steroids, ET-1 content significantly decreased to levels found in the BALF of normals. Furthermore, adults with asthma managed without steroids have increas-ed amounts of ET-1 and ET-3 in their BALF compared with normal con-trol subjects [81, 102, 104], with concentrations of $0.1-1$ pM that corre-late with the FEV_1. The cellular sources of the ET in the BALF are unknown, but irET in the airway is predominately in endothelial, glandu-lar and epithelial cells.

In situ hybridization studies have also shown that human bronchial epithelial cells from patients with symptomatic asthma express increased levels of ppET-1 and are able to release irET-1 material compared to healthy subjects and patients with chronic bronchitis and comparable airflow obstruction. The addition of steroids decreased irET release but not the expression of ppET transcripts [40].

More recent findings also suggest that ET-1 may act as a proinflamma-tory mediator with a role in the pathogenesis of asthma. Several cytokines stimulate airway epithelial cell production of ET-1 and, during asthma treatment, glucocorticoids may downregulate ET-1 production from these cells, preventing ET-1-induced effects on its cellular targets and helping to restore pulmonary homeostasis. ET-1 expression is upregulated in bron-chial epithelial cells of patients with symptomatic asthma, and factors released by inflammatory cells (IL-1β, histamine) during acute asthma exacerbation, increased both gene expression (preproET-1 mRNA) and ET-1 secretion from bronchial epithelial cells of patients with asymptom-atic asthma [44]. ET-1 content in the BAL of asthmatic patients was in-creased compared with levels in healthy volunteers, being correlated with the extent of airflow obstruction. This study did not find a reduction of ET-1 concentration in the BAL of these patients after steroid-treatment [81]. ET, probably from cytokine-stimulated endothelium or airway epi-thelium, increased in urine as well as during acute asthma [105].

Other proinflammatory effects induced by ET may also be of importance in the pathogenesis of asthma. Bronchial epithelial cells from asthmatics release higher quantities of ET, 15-HETE and PGE_2 than do cells from healthy subjects [106] ·Since the impairment of mucociliary clearance cor-relates with the severity of asthma, the increased production of ET-1 in asthma may contribute also to the mucociliary dysfunction of the disease [70, 86].

Finally, recent reports have shown that anti-IgE stimulation induced ET release from bronchial epithelial cells of asthmatic patients which express the low affinity receptor for IgE (FcεRII) [107]. Moreover, Kurokawa and coworkers have found that ET-1 increased in plasma and in sputum of asthmatic patients during the immediate and late allergic reaction following allergen provocation [108]. In contrast, Shokeir and coworkers have recently demonstrated that cigarette smoke does not induce ET production in bronchiolar epithelial cells, and the airway hyperresponsiveness observed in patients with lung disease does not appear to be related to an increased production of ET by epithelial cells [109].

Moreover, the production of fibronectin, an important component of the extracellular matrix and a potent chemotactic factor for fibroblasts, which is deposited in excess in asthmatic bronchial mucosa, increased in response to ET-1 [97, 99].

7.2. Rhinitis

Intranasal ET-1 stimulated secretions and triggered rhinorrea, itching, and sneezing in both allergic and nonallergic individuals [101]. ET-1 induced an increase in nasal symptoms in both allergic and nonallergic subjects. ET induces nociceptive nerve activation, sneezing and parasympathetic glandular secretion [101], this effect being similar to that of bradykinin [110]. ET does not induce vascular permeability. These *in vivo* observations suggest that ET activates a nociceptive nerve-parasympathetic nervous system reflex that is upregulated in allergic rhinitis, providing evidences for a role of allergic inflammation in the nasal mucosa hyperresponsiveness to ET-1 [101]. In addition, a recent study from Furukawa et al. reports an increased co-expression of ET-1 and ECE-1 in human nasal mucosa from patients with chronic inflammation [50].

7.3. Other Respiratory Diseases

Marciniak et al. reported the presence of mature irET in epithelia of human airways (main bronchi, small bronchi and bronchioles) in patients with pulmonary hypertension, cystic fibrosis and cryptogenic fibrosing alveolitis [38]. In the same study, specific binding sites for ET-1, ET-2 and ET3 were localized in airway epithelium and submucosa.

Giaid et al. reported immunostaining for ET-1 and big ET-1 and expression of ET-1 mRNA (*in situ* hybridation) in airway epithelium and type II pneumocytes of patients with cryptogenic fibrosing alveolitis, suggesting a pathogenic role for this peptide in the morphological changes and activity of the disease [111]. Other studies report enhanced levels of ET-1 in the lungs of patients with systemic sclerosis in addition to a role for

ET-1 in BALF-induced fibroblast proliferation, suggesting that ET-1 may play a role in the early pathogenesis of this disease and that an increased presence of ET-1 in the epithelial lining fluid may be a marker of early lung involvement [112]. A recent study by Saleh and coworkers demonstrated [43] that epithelial and other cells from patients with idiopathic pulmonary fibrosis showed an increase in ECE-1 and ET-1 production which may contibute to the disease abnormalities by promoting epithelial cells and fibroblast growth and by increasing collagen production.

8. Concluding Remarks

Epithelial cells line the upper and lower airways and play a barrier role to protect the airways against inhaled pollutants. The airway epithelium may play an active role, through the synthesis and release of mediators which modulate inflammatory cell recruitment, airway smooth muscle tone, epithelium permeability and vascular tone. The presence of specific binding sites for ET in the airways and the demonstration of ET production by epithelial cells strengthen the hypothesis that this peptide plays an important role in respiratory diseases.

It seems likely that in human airways ET functions as a local mediator, being mainly elaborated by epithelial cells, in addition to glands and endothelial cells, and acting on adjacent smooth muscle and other target tissues. Since shedding of the epithelium is a feature of the asthmatic airways, ET clearly might be released by airway epithelial cells and have proinflammatory effects in bronchial and vascular smooth muscle, airway epithelium or proinflammatory cells. It seems reasonable to speculate that ET may participate in modulating not only bronchoconstriction in patients with asthma, but also epithelial function and, hence, mucociliary transport in the airways.

Extrapolation from experiments done in animals, human beings and epithelial cell lines clearly suggests that proinflammatory mediators such as TNF-α, IL-1β, allergens or bacterial endotoxin may increase ET production in conditions characterized by airway inflammation such as asthma and rhinitis (Fig. 4). In consequence ET would cause: a) an autocrine effect on epithelial cells by increasing the release of more ET and proinflammatory mediators that could contribute to a relatively rapid inflammatory response to the airway injury, b) a paracrine effect on several airway target cells inducing glandular secretion, arachidonic acid metabolite release, vascular and bronchial smooth muscle contraction, proinflammatory cell secretion and fibroblast chemotaxis and hyperplasia, and leading to airflow limitation.

Once initiated, it is possible that cytokines and ET could establish an autocrine and/or a paracrine proinflammatory loop that could became independent of the original stimulus, and contribute to long-term inflammatory

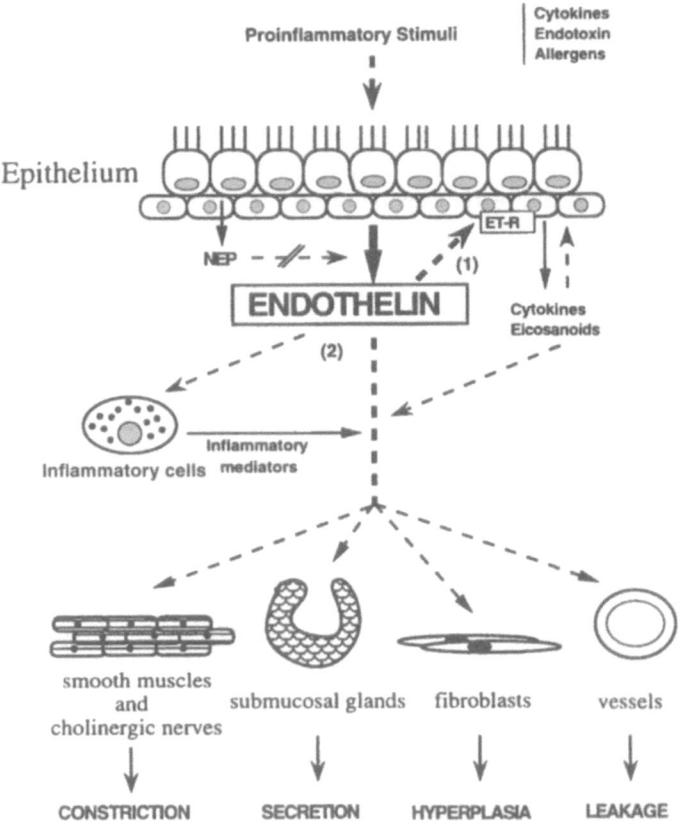

Figure 4. Hypothetical role of epithelial cell-derived endothelin in airway inflammation. In response to proinflammatory stimuli, epithelial cells release ET, which may be degraded by neutral endopeptidase (NEP). ET may act, through an autocrine pathway (1), on epithelial cell ET_A receptors to produce inflammatory mediators such as cytokines and eicosanoids which may create a proinflammatory loop. Furthermore, and through a paracrine pathway (2), ET may act on proinflammatory cells infiltrating the airway mucosa and on target tissues to induce several proinflammatory effects such as bronchoconstriction, mucus hypersecretion, fibrosis or vascular edema. [*plain arrows* = secretion, *dashed arrows* = effects].

changes in the respiratory mucosa. ET-induced cytokine release could contribute to chemoattraction of proinflammatory cells (eosinophils, neutrophils, T and B lymphocytes) that could in turn contribute to the perpetuation of this hypothetical inflammatory loop. The importance of this endothelin cascade will be determined hopefully in the future with the clinical evaluation of specific endothelin receptor antagonists in pulmonary diseases.

Acknowledgements

Figure 2 is reprinted from *Am J Respir Cell Mol Biol*, Volume 8, Wu et al., "Production of eicosanoids in response to endothelin-1 and identification of specific endothelin binding sites in airway epithelial cells", 282–290, 1993, with kind permission of the American Thoracic Society, 1740 Broadway, New York, NY 10019, USA.
Figure 3 is reprinted from *Eur J Pharmacol*, Volume 210, Candenas et al., "Effect of epithelium removal and of enkephalin inhibition on the bronchoconstrictor response to three endothelins of the human isolated bronchus", 291–297, 1992, with kind permission of Elsevier Science – NL, Sara Burgerhartstraat 25, 1055 KV Amsterdam, The Netherlands.

References

1 Barnes PJ (1994) Endothelins and pulmonary diseases. *J Appl Physiol* 77: 1051–1059
2 Bax WA, Saxena PR (1994) The current endothelin receptor classification: time for reconsideration? *TiPS* 15: 379–386
3 Douglas SA, Meek TD, Ohlstein EH (1994) Novel receptor antagonists welcome a new era in endothelin biology. *TiPS* 15: 313–316
4 Filep JG (1993) Endothelin peptides: biological actions and pathophysiological significance in the lung. *Life Sci* 52: 119–133
5 Hay DW, Henry PJ, Goldie RG (1993) Endothelin and the respiratory system. *TiPS* 14: 29–32
6 Hay DWP, Henry PJ, Goldie RG (1996) Is endothelin-1 a mediator in asthma? *Am J Respir Crit Care Med* 154: 1594–1597
7 Michael JR, Markewitz BA (1996) Endothelins and the lung. *Am J Respir Crit Care Med* 154: 555–581
8 Miller RC, Pelton JT, Huggins JP (1993) Endothelins – from receptors to medicine. *TiPS* 14: 54–60
9 Opgenorth TJ, Wu-Wong JR, Shiosaki S (1992) Endothelin-converting enzymes. *FASEB J* 6: 2653–2659
10 Pons F, Boichot E, Lagente V, Touvay C, Mencia-Huerta JM, Braquet P (1992) Role of endothelin in pulmonary function. *Pulm Pharmacol* 5: 213–219
11 Sakurai T, Yanagisawa M, Masaki T (1992) Molecular characterization of endothelin receptors. *TiPS* 13:103–107
12 Webb DJ (1991) Endothelin receptors cloned, endothelin converting enzyme characterized and pathophysiological roles for endothelin proposed. *TiPS* 12:43–46
13 Yanagisawa M, Kurihara H, Kimura S, Tomobe Y, Kobayashi M, Mitsui Y, Yazaki Y, Goto K, Masaki T (1988) A novel potent vasoconstrictor peptide produced by vascular endothelial cells. *Nature* 332: 411–415
14 De Nucci G, Thomas R, d'Orleans-Juste P, Antunes E, Walder C, Warner TD, Vane JR (1988) Pressor effects of circulating endothelin are limited by its removal in the pulmonary circulation and by the release of prostacyclin and endothelin-derived relaxing factor. *Proc Natl Acad Sci USA* 85: 9797–9800
15 Uchida Y, Ninomiya H, Saotome M, Nomura A, Ohtsuka M, Yanagisawa M, Goto K, Masaki T, Hasegawa S (1988) Endothelin a novel vasoconstrictor peptide, as potent bronchoconstrictor. *Eur J Pharmacol* 154: 227–228
16 Johnson CW, Rieves RD, Logun C et al (1991) Endothelin-1 stimulates secretion of respiratory glycoprotein from human airways *in vitro*. *Am Rev Respir Dis* 143: A138
17 Shimura S, Ishihara H, Satoh M, Masuda T, Nagaki N, Sasaki H, Takishima T (1992) Endothelin regulation of mucus glycoprotein secretion from feline tracheal submucosal glands. *Am J Physiol* 262: L208–L213
18 Mullol J, Chowdhury BA, White MV, Ohkubo K, Rieves RD, Baraniuk J, Hausfeld JN, Shelhamer JH, Kaliner MA (1993) Endothelin in human nasal mucosa. *Am J Respir Cell Mol Biol* 8: 393–402
19 Inoue A, Yanagisawa M, Kimura S, Kasuya Y, Miyauchi T, Goto K, Masaki T (1989) The human endothelin family: Three structurally and pharmacologically distinct isopeptides predicted by three separate genes. *Proc Natl Acad Sci USA* 86: 2863–2867

20 Hosoda K, Nakoa K, Arai H, Suga S, Ogawa Y, Mukoyama M, Shirakami G, Saito Y, Nakanishi S, Imura H (1991) Cloning and expression of human endothelin-1 receptor cDNA. *FEBS* 287: 23–26

21 Sakamoto A, Yanagisawa M, Sakurai T, Takuwa Y, Yanagisawa H, Makai T (1991) Cloning and functional expression of human cDNA for the ET$_B$ endothelin receptor. *Biochem Biophys Res Commun* 178: 656–663

22 Sakurai T, Yanagisawa M, Masaki T (1992) Molecular characterization of endothelin receptors. *TiPS* 13: 103–107

23 Black PN, Ghatei MA, Takahashi K, Bretherton-Watt D, Krausz T, Dollery CT, Bloom SR (1989) Formation of endothelin by cultured airway epithelial cells. *FEBS Lett* 255: 129–132

24 Rozengurt N, Springall DR, Polak JM (1990) Localization of endothelin-like immunoreactivity in airway epithelium of rats and mice. *J Pathol* 160: 5–8

25 Rennick RE, Loesch A, Burnstock G (1992) Endothelin, vasopressin, and substance P like immunoreactivity in cultured and intact epithelium from rabbit trachea. *Thorax* 47: 1044–1049

26 Perreault T, Stewart DJ, Cernacek P, Wu X, Ni F, De Marte J, Giaid A (1993) Newborn pliget lungs release endothelin-1: effect of alpha-thrombin and hypoxia. *Can J Physiol Pharmacol* 71: 227–233

27 Laporte J, D'Orleans-Juste P, Singh G, Sirois P (1995) Dexamethasone and phosphoramidon inhibit endothelin release by cultured nonciliated bronchiolar epithelial (Clara) cells. *J Cardiovasc Pharmacol* 26 (Suppl 3): S53–S55

28 Laporte J, D'Orléans-Juste P, Sirois P (1996) Guinea pig Clara cells secrete endothelin-1 through a phosphoramidon-sensitive pathway. *Am J Respir Cell Mol Biol* 14: 356–362

29 Noguchi Y, Uchida Y, Endo T, Ninomiya H, Nomura A, Sakamoto T, Goto Y, Haraoka S, Shimokama T, Watanabe T et al (1995) The induction of cell differentation and polarity of tracheal epithelium cultured on the amniotic membrane. *Biochem Biophys Res Commun* 210: 302–309

30 MacCumber MW, Ross CA, Snyder SH (1989) Endothelin: visualization of mRNA by *in situ* hybridization provides evidence for local action. *Proc Natl Acad Sci USA* 86: 7285–7289

31 Rennick RE, Milner P, Burnstock G (1993) Thrombin stimulates release of endothelin and vasopressin, but not substance P, from isolated rabbit tracheal epithelial cells. *Eur J Pharmacol* 230: 367–370

32 Ninomiya H, Uchida Y, Ishii Y, Nomura A, Kameyama M, Saotome M, Endo T, Hasegawa S (1991) Endotoxin stimulates endothelin release from cultured epithelial cells of guinea-pig trachea. *Eur J Pharmacol* 203: 299–302

33 Shima H, Yamanouchi M, Omori K, Sugiura M, Kawashima K, Sato T (1995) Endothelin-1 production and endothelin converting enzyme expression by guinea pig airway epithelial cells. *Biochem Mol Biol Int* 37: 1001–1010

34 Endo T, Uchida Y, Matsumoto H, Suzuki M, Nomura A, Hirata F, Hasegawa S (1992) Regulation of endothelin-1 synthesis in cultured guinea pig airway epithelial cells by various cytokines. *Biochem Biophys Res Commun* 186: 1594–1599

35 Armour CL, MacKay KO, Diment LM, Black JM (1990) Endothelin-functional and autoradiographic studies in human airways. *Eur J Pharmacol* 183: 183

36 Mattoli S, Mezzetti M, Riva G, Allegra L, Fasoli A (1990) Specific binding of endothelin on human bronchial smooth muscle cells in culture and secretion of endothelin-like material from bronchial epithelial cells. *Am J Respir Cell Mol Biol* 3: 145–151

37 Giaid A, Polak JM, Gaitonde V, Hamid QA, Moscoso G, Legon S, Uwanog-ho D, Roncalli M, Shinmi O, Sawamura T, Kimura S, Yanagisawa M, Masaki T, Springall DR (1991) Distribution of endothelin-like immunoreactivity and mRNA in the developing and adult human lung. *Am J Respir Cell Mol Biol* 4: 50–58

38 Marciniak SJ, Plumpton C, Barker PJ, Huskisson NS, Davenport AP (1992) Localization of immunoreactive endothelin and proendothelin in the human lung. *Pulm Pharmacol* 5: 175–182

39 Takimoto M, Oda K, Sasaki Y, Okada T (1996) Endothelin-A receptor-mediated prostanoid secretion via autocrine and deoxyribonucleic acid synthesis via paracrine signaling in human bronchial epithelial cells. *Endocrinology* 137: 4542–4550

40 Vittori E, Marini M, Fasoli A, De Franchis R, Mattoli S (1992) Increased expression of endothelin in bronchial epithelial cells of asthmatic patients and effect of corticosteroids. *Am Rev Respir Dis* 146: 1320–1325

41 Sun G, de Angelis G, Nucci F, Ackerman V, Bellina A, Mattoli S (1996) Functional analysis of the preproendothelin-1 gene promoter in pulmonary epithelial cells and monocytes. *Biochem Biophys Res Commun* 221: 647–652

42 Nakano J, Takizawa H, Ohtoshi T, Shoji S, Yamaguchi M, Ishii A, Yanagisawa M, Ito K (1994) Endotoxin and pro-inflammatory cytokines stimulate endothelin-1 expression and release by airway epithelial cells. *Clin Exp Allergy* 24: 330–336

43 Saleh D, Furukawa K, Tsao M-S, Maghazachi A, Corrin B, Yanagisawa M, Barnes PJ, Giaid A (1997) Elevated expression of endothelin-1 and endothelin-converting enzyme-1 in idiopathic pulmonary fibrosis: Possible involvement of proinflammatory cytokines. *Am J Respir Cell Mol Biol* 16: 187–193

44 Ackerman V, Carpi S, Bellini A, Vassalli G, Marini M, Mattoli S (1995) Constitutive expression of endothelin in bronchial epithelial cells of patients with symptomatic and asymptomatic asthma and modulation by histamine and interleukin-1. *J Allergy Clin Immunol* 96: 618–627

45 Markewitz BA, Kohan DE, Michael JR (1995) Endothelin-1 synthesis, receptors, and signal transduction in alveolar epithelium: evidence for an autocrine role. *Am J Physiol* 268: L190–L200

46 Calderón E, Gómez-Sánchez CE, Cozza EN, Zhou M, Coffey RG, Lockey RF, Prockop LD, Szentivanyi A (1994) Modulation of endothelin-1 production by a pulmonary epithelial cell line. I. regulation by glucocorticoids. *Biochem Pharmacol* 48: 2065–2071

47 Mullol J, Baraniuk JN, Logun C, Picado C, Shelhamer JH (1995) Endothelin-1 (ET-1) induces GM-CSF, IL-6 and IL-8 but not G-CSF release from a human bronchial epithelial cell line. *Am J Respir Crit Care Med* 151: A367

48 Casasco A, Benazzo M, Casasco M, Cornaglia AI, Springall DR, Calligaro A, Mira E, Polak JM (1993) Occurrence, distribution and possible role of the regulatory peptide endothelin in the nasal mucosa. *Cell Tissue Res* 274: 241–247

49 Ohkubo K, Ohnishi M, Yokoshima K, Takizawa R, Okuda M, Kaliner MA (1994) Study of endothelin: distribution in the airway and release from nasal epithelial cells. *Arerugi* 43: 448–457

50 Furukawa K, Saleh D, Bayan F, Emoto N, Kaw S, Yanagisawa M, Giaid A (1996) Co-expression of endothelin-1 and endothelin-converting enzyme-1 in patients with chronic rhinitis. *Am J Respir Cell Mol Biol* 14: 248–253

51 Ohkubo K, Baraniuk JN, Hohman RJ, Kaulbach HC, Hausfeld JN, Mérida M, Kaliner MA (1993) Human nasal mucosal neutral endopeptidase (NEP): location, quantitation, and secretion. *Am J Respir Cell Mol Biol* 9: 557–567

52 Sokolovsky M, Galron R, Kloog Y, Bdolah A, Indig FE, Blumberg S (1990) Endothelins are more sensitive than serafotoxins to neutral endopeptidase: possible physiological significance. *Proc Natl Acad Sci USA* 87: 4702–4706

53 Vijayaraghavan J, Scicli AG, Carretero OA, Slaughter C, Moomaw C, Hersh LB (1990) The hydrolysis of endothelins by neutral endopeptidase 24.11 (enkephalinase). *J Biol Chem* 265: 14150–14155

54 Tschirhart EJ, Drijfhout JW, Pelton JT, Miller RC, Jones CR (1991) Endothelins: functional and autoradiographic studies in guinea pig trachea. *J Pharmacol Exp Ther* 258: 381–387

55 Advenier C, Lagente V, Zhang Y, Naline E (1992) Contractile activity of big endothelin on the human isolated bronchus. *Br J Pharmacol* 106: 883–887

56 Nally JE, McCall R, Young LC, Wakelam MJ, Thomson NC, McGrath JC (1994) Mechanical and biochemical responses to endothelin-1 and endothelin-3 in bovine bronchial smooth muscle. *Br J Pharmacol* 111: 1163–1169

57 Murlas CG, Gulati A, Singh G, Najmabadi F (1995) Endothelin-1 stimulates proliferation of normal airway epithelial cells. *Biochem Biophys Res Commun* 212: 953–959

58 Hay DWP (1989) Guinea-pig tracheal epithelium and endothelin. *Eur J Pharmacol* 171: 241–245

59 Maggi CA, Patacchini R, Giuliani S, Meli A (1989) Potent contractile effect of endothelin in isolated guinea-pig airways. *Eur J Pharmacol* 160: 179–182

60 Candenas ML, Naline E, Sarria B, Advenier C (1992) Effect of epithelium removal and of enkephalin inhibition on the bronchoconstrictor response to three endothelins of the human isolated bronchus. *Eur J Pharmacol* 210: 291–297

61 Masuda Y, Miyazaki H, Kondoh M, Watanabe H, Yanagisawa M, Masaki T, Murakami K (1989) Two different forms of endothelin receptors in rat lung. *FEBS Lett* 257: 208–210

62 Nakamichi K, Ihara M, Kobayashi M, Saeki T, Ishikawa K, Yano M (1992) Different distribution of endothelin receptor subtypes in pulmonary tissues revealed by the novel selective ligands BQ-123 and [Ala1,3,11,15]ET-1. *Biochem Biophys Res Commun* 182: 144–150

63 Kobayashi M, Ihara M, Sato N, Saeki T, Ozaki S, Ikemoto F, Yano M (1993) A novel ligand, [125]BQ-3020, reveals the localization of endothelin ETB receptors. *Eur J Pharmacol* 235: 95–100

64 Durham SK, Goller NL, Lynch JS, Fisher SM, Rose PM (1993) Endothelin receptor B expression in the rat and rabbit lung as determined by *in situ* hybridization using non-isotopic probes. *J Cardiovasc Pharmacol* 22 (Suppl 8): S1–S3

65 Wu T, Rieves RD, Larivee P, Logun C, Lawrence MG, Shelhamer JH (1993) Production of eicosanoids in response to endothelin-1 and identification of specific endothelin binding sites in airway epithelial cells. *Am J Respir Cell Mol Biol* 8: 282–290

66 Ninomiya H, Yu XY, Uchida Y, Hasegawa S, Spannhake EW (1995) Specific binding of endothelin-1 to canine tracheal epithelial cells in culture. *Am J Physiol* 268: L424–L431

67 Goldie RG, Henry PJ, Knott PG, Self GJ, Luttmann MA, Hay DWP (1995) Endothelin-1 receptor density, distribution, and function in human isolated asthmatic airways. *Am J Respir Crit Care Med* 152: 1653–1658

68 Clozel M, Breu V, Burri K, Cassal JM, Fischli W, Gray GA, Hirth G, Löffler BM, Müller M, Neidhart W, Ramuz H (1993) Pathophysiological role of endothelin revealed by the first orally active endothelin receptor antagonist. *Nature* 365: 759–761

69 Noguchi K, Ishikawa K, Yano M, Ahmed A, Cortes A, Abraham WM (1995) Endothelin-1 contributes to antigen-induced airway hyperresponsiveness. *J Appl Physiol* 79: 700–705

70 Sabater JR, Otero R, Abraham WM, Wanner A, O'Riordan TG (1996) Endothelin-1 depresses tracheal mucus velocity in ovine airways via ET-A receptors. *Am J Respir Cell Mol Biol* 154: 341–345

71 Mullol J, Baraniuk JN, Logun C, Picado C, Shelhamer JH (1996) Endothelin-1 (ET-1) induces GM-CSF, IL-6 and IL-8 but not G-CSF release from a human bronchial epithelial cell line (BEAS-2B). *Neuropeptides* 30: 551–556

72 Nagase T, Fukuchi Y, Jo C et al (1990) Endothelin-1 stimulates arachidonate 15-lipoxygenase activity and oxygen radical formation in the rat distal lung. *Biochem Biophys Res Commun* 168: 485–489

73 Grunstein MM, Chuang ST, Schramm CM, Pawlowski NA (1991) Role of endothelin-1 in regulating rabbit airway contractility. *Am J Physiol* 260: L75–L82

74 Satoh M, Shimura S, Ishihara H, Nagaki M, Saski H, Takishima T (1992) Endothelin-1 stimulates chloride secretion across canine tracheal epithelium. *Respiration* 59: 145–150

75 Tamaoki J, Kanemura T, Sakai N, Isono K, Kobayashi K, Takizawa T (1991) Endothelin stimulates ciliary beat frequency and chloride secretion in canine cultured tracheal epithelium. *Am J Respir Cell Mol Biol* 4: 426–431

76 Filep JG, Battistini B, Sirois P (1993) Induction by endothelin-1 of epithelium-dependent relaxation of guinea-pig trachea *in vitro*: role for nitric oxide. *Br J Pharmacol* 109: 637–644

77 Hay DWP, Hubbard WC, Undem BJ (1993) Relative contributions of direct and indirect mechanisms mediating endothelin-induced contraction of guinea-pig trachea. *Br J Pharmacol* 110: 955–962

78 Wu T, Mullol J, Rieves RD, Logun C, Hausfield J, Kaliner MA, Shelhamer JH (1992) Endothelin-1 stimulates eicosanoid production in cultured human nasal mucosa. *Am J Respir Cell Mol Biol* 6: 168–174

79 Filep JG, Battistini B, Fournier A, Sirois P (1993) Relaxation by dexamethasone isolated of guinea-pig airways precontracted with endothelin-1. *Eur J Pharmacol* 240: 315–318

80 Andersson SE, Eirefelt S, Zackrisson C, Hemsen A, Dahlback M, Lundberg JM (1995) Regulatory effects of aerosolized budesonide and adrenalectomy on the lung content of endothelin-1 in the rat. *Respiration* 62: 34–39

81 Redington AE, Springall DR, Ghatei MA, Lau LCK, Bloom SR, Holgate ST, Polak JM, Howarth PH (1995) Endothelin in bronchoalveolar lavage fluid and its relation to airflow obstruction in asthma. *Am J Respir Crit Care Med* 151: 1034–1039

82 Adler KB, Kinula VL, Akley N, Lee J, Cohn LA, Crapo JD (1992) Inflammatory mediators and the generation and release of reactive oxygen species by airway epithelium *in vitro. Chest* 101: 53S–54S

83 Plews PI, Abdel-Malek ZA, Doupnik CA, Leikauf GD (1991) Endothelin stimulates chloride secretion across canine tracheal epithelium. *Am J Physiol* 261: L188–L194

84 Webber SE, Yurdakos E, Woods AJ, Widdicombe JG (1992) Effects of endothelin-1 on tracheal submucosal gland secretion and epithelial function in the ferret. *Chest* 101 (Suppl 3): 63S–67S

85 Yurdakos E, Webber SE (1991) Endothelin-1 inhibits prestimulated tracheal submucosal gland secretion and epithelial albumin transport. *Br J Pharmacol* 104: 1050–1056

86 Amber FR, Lindberg SO, McCaffrey TV, Runer T (1993) Mucociliary function and endothelin 1, 2 and 3. *Otolaryngol Head Neck Surg* 109: 634–645

87 Saotome M, Ninomiya H, Nomura A, Ohse H, Endoh T, Hasegawa S, Uchida Y (1991) Endothelin-1 induced relaxation of guinea pig trachea after an anaphylactic reaction. *Arerugi* 40: 1377–1383

88 Uchida Y, Saotome M, Normura A et al (1990) Endothelin-1 releases an epithelial-derived relaxing factor(s) from the guinea-pig trachea. *J Vasc Med Biol* 2: P3–P20

89 Howarth PH, Redington AE, Springall DR, Martin U, Bloom SR, Polak JM, Holgate ST (1995) Epithelially derived endothelin and nitric oxide in asthma. *Int Arch Allergy Immunol* 107 (1–3): 228–230

90 McLarty AJ, Miller VM, Tazelaar HD, McGregor CGA (1993) Bronchial contractions in transplanted lungs. Influence of denervation, acute rejection, and the bronchial epithelium. *J Thorac Cardiovasc Surg* 106: 797–804

91 Ninomiya H, Uchida Y, Saotome M, Nomura A, Ohse H, Matsumuro H, Hirata F, Hasegawa S (1992) Endothelins constrict guinea pig tracheas by multiple mechanisms. *J Pharmacol Exp Ther* 262: 570–576

92 White SR, Hathaway DP, Umans JG, Tallet J, Abrahams C, Leff AR (1991) Epithelial modulation of airway smooth muscle response to endothelin-1. *Am Rev Respir Dis* 144: 373–378

93 White SR, Hathaway DP, Umans JG, Leff AR (1992) Directs effects on airway smooth muscle contractile response caused by endothelin in guinea pig trachealis. *Am Rev Respir Dis* 145: 491–493

94 Hay DWP, Hubbard WC, Luttmann MA, Undem BJ (1993) Endothelin receptor subtypes in human and guinea-pig pulmonary tissues. *Br J Pharmacol* 110: 1175–1183

95 Panettieri RH, Goldie RG, Rigby PJ, Eszterhas AJ, Hay DWP (1996) Endothelin-induced potentiation of human airway smooth muscle proliferation: An ET_A receptor-mediated phenomenon. *Br J Pharmacol* 118:191–197

96 Fernandes LB, Henry PJ, Rigby PJ, Goldie RG (1996) Endothelin B (ET_B) receptor-activated potentiation of cholinergic nerve-mediated contraction in human bronchus. *Br J Pharmacol* 118: 1873–1874

97 Marini M, Carpi S, Bellini A, Patalano F, Mattoli S (1996) Endothelin-1 induces increased fibronectin expression in human bronchial epithelial cells. *Biochem Biophys Res Commun* 220: 896–899

98 Laitinen LA, Heino M, Laitinen A, Kava T, Haahtela T (1985) Damage of the airway epithelium and bronchial reactivity in patients with asthma. *Am Rev Respir Dis* 131: 599–606

99 Takuwa N, Takuwa Y, Yanagisawa M, Yamashita K, Masaki T (1989) A novel vasoactive peptide endothelin stimulates mitogenesis through inositol lipid turnover in Swiss 3T3 fibroblasts. *J Biol Chem* 264: 7856–7861

100 Benazzo M, Casasco A, Lovotti P, Cornaglia AI, Casasco M, Polak JM (1994) Endothelin-induced vasoconstriction in rabbit nasal mucosa. *Acta Otolaryngol* (Stockh) 114: 544–546

101 Riccio MM, Reynolds CJ, Hay DWP, Proud D (1995) Effect of intranasal administrations of endothelin-1 to allergic and nonallergic individuals. *Am J Respir Crit Care Med* 152: 1757–1764

102 Nomura A, Uchida Y, Kameyama M, Saotome M, Oki K, Hasegawa S (1989) Endothelin and bronchial asthma. *Lancet* 2: 747–748

103 Springall DR, Howarth PH, Counihan H, Djukanovic R, Holgate ST, Polak JM (1991) Endothelin immunoreactivity of airway epithelium in asthmatic patients. *Lancet* 337: 697–701
104 Mattoli S, Soloperto M, Marini M, Fasoli A (1991) Levels of endothelin in the broncho-alveolar lavage fluid of patients with symptomatic asthma and reversible airflow obstruction. *J Allergy Clin Immunol* 88: 376–384
105 Baraniuk JN, Sabol MB, Milzman D, Becker KL (1995) Endothelin is increased in urine during acute asthma. *Am J Respir Crit Care Med* 151: A388
106 Vignola AM, Chanez P, Campbell AM, Fournier M, Godard P (1994) Mechanisms of bronchial hyperreactivity. Role of the epithelium. *Rev Mal Respir* 11: 141–147
107 Campbell AM, Vignola AM, Chanez P, Godard P, Bousquet J (1994) Low-affinity receptor for IgE on human bronchial epithelial in asthma. *Immunology* 82: 506–508
108 Kurokawa M, Konno S, Gonokami Y, Kouno Y, Kawazu K, Okamoto M, Adachi M (1996) Endothelin-1 levels in sputum and plasma of asthmatic patients after allergic provocation. *Arerugi* 45: 386–392
109 Shokeir MO, Paré P, Wright JL (1994) Relation of smoking to immuno-reactive endothelin in the bronchiolar epithelial cells. *Thorax* 49: 786–788
110 Baraniuk JN, Silver PB, Kaliner MA, Barnes PJ (1994) Perennial rhinitis subjects have altered vascular, glandular, and neural responses to bradykinin nasal provocation. *In Arch Allergy Immunol* 103: 202–208
111 Giaid A, Michel RP, Stewart DJ, Shepperd M, Corrin B, Hamid Q (1993) Expression of endothelin-1 in lungs of patients with cryptogenic fibrosing alveolitis. *Lancet* 341: 1550–1554
112 Cambrey AD, Harrison NK, Dawes KE, Southcott AM, Black CM, du Bois RM, Laurent GJ, McAnulty RJ (1994) Increased levels of endothelin-1 in bronchoalveolar lavage fluid from patients with systemic sclerosis contribute to fibroblast mitogenic activity *in vitro*. *Am J Respir Cell Mol Biol* 11: 439–445

Pulmonary Actions of the Endothelins
ed. by R. G. Goldie and D. W. P. Hay
© 1999 Birkhäuser Verlag Basel/Switzerland

CHAPTER 10
Endothelin as a Proinflammatory Mediator

János G. Filep[1] and Douglas W. P. Hay[2]

[1] *Research Center, Maisonneuve-Rosemont Hospital, University of Montréal, Montréal, Québec, Canada H1T 2M4*
[2] *Department of Pulmonary Pharmacology, SmithKline Beecham Pharmaceuticals, King of Prussia, Pennsylvania 19406-0939, USA*

1   Introduction
2   Proinflammatory Actions of Endothelins
2.1   Effects on Inflammatory Cells
2.1.1 Neutrophil granulocytes
2.1.2 Monocytes/macrophages
2.1.3 Cells obtained from bronchoalveolar lavage
2.1.4 Platelets
2.1.5 Mast cells
2.1.6 Eosinophils
2.1.7 Lymphocytes
2.2   Effects on Vascular Endothelial Cells
2.3   Effects on Vascular Permeability
2.4   Mucus Secretion
2.5   Airway Epithelial Cells
2.6   Tissue Remodeling
2.7   Cytotoxicity
3   Potential Proinflammatory Roles of Endothelins in Pulmonary Diseases
4   Conclusions
      References

1. Introduction

Pulmonary host-defense is often characterized by either acute or chronic inflammatory response to a given challenge. Acute inflammation is the host's stereotyped response to a variety of insults such as infection, trauma, allograft rejection and ischemia-reperfusion. This defense mechanism may be summarized by the following, often overlapping, "steps": (a) injury to the host and local generation of mediators; (b) recruitment of inflammatory cells (e.g., neutrophils, eosinophils) at the site of injury, an event always associated with increased microvascular permeability; (c) destruction and/or phagocytosis of invading organisms and tissue debris; (d) removal of inflammatory cells and debris; (e) tissue remodeling and repair, leading to resolution and restoration of normal lung function. Modern perceptions have emphasized the detrimental effects of inflammation and its associa-tion with disease. Indeed, inflammation is the basis of a variety of pulmo-nary diseases such as asthma and acute respiratory distress syndrome

(ARDS). In these pathologies, persisting stimuli or injury results in tissue destruction and ongoing (chronic) inflammation that fails to resolve and that may culminate in end-stage fibrosis. Although the specific molecular mechanisms that result in acute or chronic inflammation have not been fully elucidated, it is now clear that humoral mediators and interactions between immune and non-immune cells are integral to our understanding of the pathogenesis of pulmonary inflammation.

Recent observations indicate that enhanced intrapulmonary expression and production of endothelins (ETs), in particular endothelin-1 (ET-1), occur in a variety of pulmonary diseases, including pulmonary hypertension, asthma, certain pulmonary tumors and ARDS (for reviews see [1–5]). Since ETs are potent constrictors of vascular and airway smooth muscle cells, both *in vitro* and *in vivo*, much of the research on ETs has concentrated on these actions. The sites of intrapulmonary synthesis of ETs and distribution of their receptors have, to a large extent, been characterized and are discussed in other chapters of this book. An increasing body of evidence indicates that ETs may also function as proinflammatory mediators. This chapter will summarize the possible proinflammatory actions of ETs in the lung with a particular emphasis on ET-1. Where possible, we have focused on studies performed in humans or human cells and tissues.

2. Proinflammatory Actions of Endothelins

2.1. Effects on Inflammatory Cells

2.1.1. Neutrophil granulocytes: Extravasation and activation of neutrophil granulocytes are essential in the inflammatory response. The pulmonary microvascular sequestration and activation of neutrophil and mononuclear phagocytes are associated with enhanced microvascular injury, leukocyte migration, hemorrhage and parenchymal injury. These are characteristic events in a number of lung diseases, in particular ARDS and chronic obstructive pulmonary disease (COPD). Microvascular injury often leads to substantial and permanent pulmonary dysfunction.

Non-activated human neutrophils rapidly converts big ET-1 to ET-1, *via* a pathway blocked by phosphoramidon (presumably due to inhibition of ECE) [6], whereas activated neutrophils can destroy high amounts (up to 100 nmol/l) of ET-1 within a few minutes, probably *via* the release of a neutral peptidase(s) [7]. These observations suggest the possibility that neutrophils modulate local levels of ET-1. Indeed, inhibition of granulocyte derived proteases by Eglin C reduced the increase in plasma ET-1 associated with myocordial ischemia [8].

Although controversy exists about whether ETs by themselves activate human neutrophils, ET-1 appears to be a potent modulator of neutrophil responses. While ET-1 by itself does not stimulate the neutrophil respiratory burst [9] and superoxide production [10], it primes neutrophils to increase superoxide production in response to the bacterial chemotaxin f-Met-Leu-Phe [10, 11]. ET-1 stimulates elastase release from human neutrophils [12] and promotes neutrophil aggregation that is mediated by platelet-activating factor (PAF) [13,14]. A recent clinical study raised the possibility that ET-1-mediated activation of neutrophils may occur in patients undergoing cardiopulmonary bypass surgery [15].

ET-1, ET-2 and ET-3 are chemoattractants for human neutrophils *in vitro* (maximum responses at 100 nM) [9]. ET-1-induced neutrophil migration is blocked with BQ-123, indicating involvement of ET_A receptors, and involves activation of protein kinase C, protein tyrosine kinase and phosphatase [16]. Unlike ET-1, ET-2 at high concentrations inhibits f-Met-Leu-Phe-activated migration, *via* activation of both ET_A and ET_B receptors [17]. ET-1-activated neutrophils migrate from the venous lumen into tissue matrix of the human umbilical cord, and induce a massive tissue destruction [12]. When injected into experimental animals, ET-1 causes a dose-dependent selective neutrophil leukocytopenia due to intravascular sequestration in the lung and spleen [18], and induces intrapulmonary accumulation of neutrophils [19, 20], suggesting that ET-1 may promote neutrophil adhesion *in vivo*. This action of ET-1 is predominantly mediated via ET_A receptors [18]. ET-1 stimulates surface expression of the adhesion molecule CD11b/CD18 on human neutrophils and augments adhesion of human neutrophils to cultured bovine endothelial cells [21]. ET-1 was also shown to upregulate expression of adhesion molecules on endothelial cells (see below).

ET-1 may function as a modulator and/or mediator of interactions between inflammatory cells (Fig. 1). For instance, ET-1-activated monocytes stimulate neutrophil superoxide production via soluble mediators, possibly including interleukin-8 (IL-8) [22]. The presence of ET-1 counterbalances the inhibitory action of neutrophils on platelet activation, therefore acting as an indirect pro-aggregatory agent [14]. This action of ET-1 is probably mediated through production of PAF by neutrophils [14]. These observations indicate an important role for neutrophils in mediating vascular or tissue injury elicited by ET-1, though neutrophil-independent mechanisms might be also involved in certain circumstances [23].

2.1.2. Monocytes/macrophages: Mononuclear phagocytes are well equipped to play a central role in the amplification of the inflammatory response by virtue of their distribution in the airway, and by their capacity to generate a variety of pro-inflammatory mediators and cytokines that modulate the function of other cells [24]. Recruitment of monocytes from the circulation and their maturation to macrophages are important processes for the removal of inflammatory cells and debris at the inflammatory locus. Macrophages are thought to be the key effector cells in COPD.

ET-1 and ET-3 are produced and secreted by monocytes and macrophages in a manner similar to that of cytokines [25]. Two distinct preproET-1

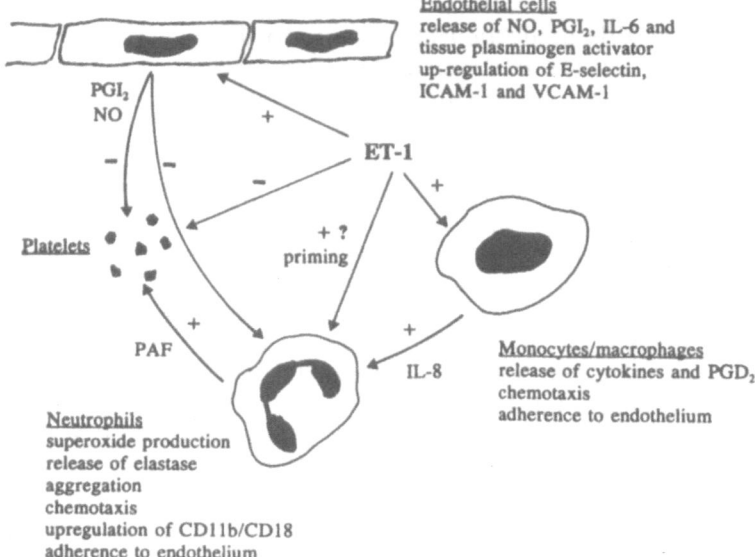

Figure 1. Potential roles for ET-1 in mediating cross-talk between inflammatory cells and the endothelium. ET-1 is produced by endothelial cells, neutrophils and monocytes (for further details see text). Adhesion of platelets and neutrophils to endothelial cells stimulates ET-1 production (not shown). Abbreviations: NO, nitric oxide; PGI_2, prostacyclin; IL-6, interleukin-6; IL-8, interleukin-8; PAF, platelet-activating factor; PGD_2, prostaglandin D_2.

mRNAs produced by a single gene are expressed in human pulmonary epithelial cells and monocytes [26]. The mechanisms controlling the production of these alternative transcripts might be cell-specific. ET-1 increases intracellular calcium concentration in human monocytes [27] and stimulates production of IL-6, IL-8, tumor necrosis factor-α (TNF-α), transforming growth factor β, granulocyte-macrophage colony stimulating factor and prostaglandin (PG) E_2 [22, 27, 28]. The amounts of IL-6 produced by monocytes incubated with 1 nM ET-1 were comparable to those produced in response to endotoxin [27]. ET-1 is a strong chemoattractant for monocytes [29] and promotes monocytes adherence to human umbilical vein endothelial cells [30]. The ET-1 receptor-response coupling in endothelial cells involves the activation of p60SRC and JAK1-like kinases [30]. In contrast, exposure of human peripheral blood monocytes to ET-1 markedly reduces the adhesion of these cells to human saphenous vein [31]. This can be attributed to ET_B receptor-dependent production of NO by the monocytes [31]. NO has been demonstrated to inactivate cytokine-stimulated cells and impair chemotaxis [32, 33], suggesting that stimulatory and inhibitory signal molecules released in response to ET-1 may provide a mechanism for autocrine regulation of monocyte adhesion. Of interest, ET_B receptor blockade with BQ-788 upregulated ET-1-induced monocyte adherence to umbilical vein endothelial cells [30] and restored monocyte adherence to control levels in the saphenous vein [31].

ET-1 may serve as an autocrine modulator of activation of macrophages. ET-1 raises intracellular calcium (EC_{50}, 50 nM), superoxide production (EC_{50}, 100 nM) and induces rapid phosphorylation of a 48 kDa and a 35 kDa protein in human alveolar macrophages [34]. However, single-cell experiments revealed that only a subpopulation of human alveolar macrophages responds to ET-1 [34]. While ET-1 (up to a concentration of 100 nM) neither increases free calcium concentration nor augments superoxide formation in guinea-pig alveolar macrophages, it markedly potentiates superoxide production in response to f-Met-Leu-Phe or PAF, but not to phorbol myristate acetate (PMA), without affecting f-Met-Leu-Phe or PAF-induced changes in intracellular free calcium concentration [18]. This potentiating action of ET-1 was completely reversed by FR 139317, indicating the involvement of ET_A receptors [18]. ET-1 transiently increases thromboxane A_2 (TXA_2) production by guinea-pig alveolar macrophages [35].

Accumulation of monocytes/macrophages could alter vascular responsiveness to ET-1. For instance, ET-1-induced contractions of guinea-pig coronary artery strips are markedly augmented in the presence of human peripheral blood monocytes or guinea-pig alveolar macrophages [36]. The enhanced response requires ET-1 binding to monocytes/macrophages that is not inhibited by BQ-123 and appears to be independent from mediator release. Magazine et al. [36] speculated that monocytes/macrophages may more effectively present ET-1 to ET receptors on vascular smooth muscle, leading to an enhanced contraction. Mononuclear cells activated by acute rejection of allotransplanted lung in dogs cause contraction of pulmonary arteries that are mediated, in part, by oxygen radicals and ETs [37].

2.1.3. Cells obtained from bronchoalveolar lavage: Intravenous bolus injection of ET-1 to rats results in increases in the bronchoalveolar lavage (BAL) fluid concentration of 15-HETE and augments generation of free oxygen radicals by lavage cells in response to PMA [38]. The effect of ET-1 appears to be specific for 15-lipoxygenase, since neither 5-lipoxygenase nor cyclooxygenase pathway was activated by ET-1. On the other hand, ET-1 (10 nM to 1 μM) augments release of TXA_2 and PGD_2 from cells obtained from canine airways by lavage [39]. Although these studies have not identified the cells responsible for mediator synthesis, both 15-HETE (precursor of lipoxins) and free radicals could contribute to damage of pulmonary tissues.

2.1.4. Platelets: Platelets are generally considered primarily for their active participation in thrombosis and hemostasis. Activated platelets contribute to microvascular injury, and consequently to pulmonary dysfunction (see above). In addition, platelets may play a role in neutrophil accumulation [40] as well as in allergen-induced eosinophil infiltration [41], and thus may have a role in the pathogenesis of asthma.

Although ET-1 may act as a pro-aggregatory agent under certain circumstances, as described above, numerous studies demonstrated that ETs either have no direct effect on human or animal platelets or inhibit aggregation *in vitro*. Both ET-1 and ET-3 inhibit serotonin and thrombin-induced aggregation and calcium mobilization in human platelets [42–44]. ET-3 reduces calcium uptake and Ca^{2+} content of platelet internal stores, suggesting that ET-3 is functionally coupled to Ca^{2+} pumps of the dense tubular system, thereby lowering Ca^{2+} release in response to agonists [45]. Incubation of human platelets with ET-1 activates protein kinase C [44], raises cytosolic pH [44] and causes accumulation of cGMP, but not cAMP [46]. These actions are mediated via ET_A and ET_B receptors, respectively [44, 46].

Injection of ET-1 to experimental animals results in a marked anti-aggregatory action, as measured *in vivo* [47–49] or *ex vivo* [50]. This action of ET-1 is mediated through stimulation of prostacyclin and nitric oxide (NO) release from endothelial cells [47, 49, 51, 52]. Of interest, ET-1 and sarafotoxin S6b inhibits repetitive platelet thrombus formation in the stenosed canine coronary artery, demonstrating the potential of an *in vivo* antithrombotic activity of these peptides [53]. By contrast, ET-1 (6 nmol/kg) injected into the jugular vein promoted trapping of platelets (loosely aggregated and adhered partly to the endothelium) in the rat pulmonary microcirculation [20].

2.1.5. Mast cells: The mast cell has been proposed as the initiator of inflammatory responses to allergic stimuli [54]. Mast cells are abundant in lung tissue, and they release a wide array of mediators, including histamine, lipid mediators and proteolytic enzymes that may contribute to smooth muscle contraction, mucus secretion and recruitment of neutrophils and eosinophils.

ET-1 releases histamine from guinea-pig lung mast cells, but not peritoneal mast cells through ET_B receptors [55]. However, only a subset of pulmonary mast cells appear to possess ET_B receptors, since ET-1 maximally releases approximately 30% of total histamine content, and about one-fourth of pulmonary mast cells bind radiolabeled ET-1 [55]. ET-1 can directly act as a histamine and serotonin secretagogue and as a stimulator of leukotriene C_4 (LTC_4) production in mouse bone-marrow-derived mast cells, *via* ET_A receptor activation [56]. In the course of Th_2-dominated immune response, IL-4 renders mast cells functionally responsive to ET-1 [47]. ET-1-induced mast cell activation and subsequent release of potent mediators (amines, PGD_2, LTC_4) [55, 56] may have clinical importance in allergic diseases. However, it is not known at present whether the ETs activate human lung mast cells.

2.1.6. Eosinophils: Eosinophil granulocytes are predominant among the cells involved in the inflammation in asthma, and eosinophilic inflammation of the airways correlates with the severity of the disease [57]. Infiltrating eosinophils release various cytotoxic substances such as major basic protein, that contribute to epithelial damage. However, little is known at present about the mechanisms of eosinophil recruitment into the lung. A recent study suggested that ET-1 may function as an eosinophil chemoattractant. In a rat model of eosinophilic inflammation of the lower airways (produced by intratracheal instillation of Sephadex beads), the increases in the BAL fluid ET level were correlated with considerable increases in total cell count, eosinophils and neutrophils. Treatment with the ET_A/ET_B antagonist bosentan inhibited the increase in eosinophils in BAL fluid and reduced the extent of the inflammatory regions around the Sephadex beads [58]. ET-1, up to 1 µM, alone has no stimulatory effect on guinea pig or human eosinophils (unpublished observations). BQ-788 (ET_B receptor antagonist) or SB 209670 (mixed ET_A/ET_B receptor antagonist) inhibited antigen-induced infiltration of eosinophils, and also neutrophils, in BAL in mice [59].

2.1.7. Lymphocytes: The number of lymphocytes and their activation state in the airways is increased in patients with asthma and COPD [60]. Both $CD4^+$ and $CD8^+$ cells are present and probably increased number of Th_2 cells. To date there have been no reports on the effects of ETs on T-cell function. Preliminary results in our laboratory inducate that ET-1 can have either stimulatory or inhibitory effects on the proliferation of T-cells, depending upon the individual donor (unpublished observations).

2.2. Effects on Vascular Endothelial Cells

Endothelial cells respond to injury with acute alterations in mediator generation and surface molecule expression [61]. These responses generally mediate or ameliorate the local tissue response. Endothelial cells may respond by altering the release of mediators and expression of adhesion molecules, or by altering the expression of various genes whose products are central to endothelial cell matrix remodeling and interactions with inflammatory cells and to angiogenesis.

The vascular endothelium is not only a major site of ET-1 production, but is a target for ETs. Vasoactive agents and cytokines are important regulators of the ET-1 gene expression in rat pulmonary endothelial cells with different responses at the mRNA and protein levels depending on the mediator involved [62]. Both ET-1 and ET-3 were reported to stimulate the release of IL-6 [63], arachidonic acid [64], prostacyclin [51, 52], PAF [65] and tissue plasminogen activator from cultured endothelial cells [66]. ET-1, ET-2 and ET-3 upregulate the expression of the adhesion molecules, E-selectin, ICAM-1 and VCAM-1 on cerebro-microvascular endothelial cells [67], thereby regulating interactions between circulating leukocytes and endothelium. Since cytokines, such as IL-1 and TNF-α, can increase expression of these adhesion molecules and ET-1 production at the same time, it is possible that ET-1 may mediate some of the actions of these cyto-kines.

2.3. Effects on Vascular Permeability

The general features of an inflammatory response include vascular dilatation and increased microvascular permeability with the formation of an exudate consisting of both plasma proteins and migrating inflammatory cells. Capil-lary engorgement and/or leakage in the pulmonary vascular bed could directly alter airway geometry, because of its anatomical arrangement rela-tive to the airway wall [68]. Airway lumen narrows as a result of a combina-tion of external airway compression from dilatation of the pulmonary vessels within the confines of the bronchovascular sheath (with or without local edema) and internal bronchial narrowing secondary to mucosal vascular congestion and edema. In addition, airway reactivity increases under these circumstances; this might be related to a mechanical increase in mucosal size, for example mucosal edema [69]. Thus, mucosal swelling produced by vascular effects alone can have profound effects upon airway function.

ET-1 is capable of inducing pulmonary edema in experimental animals. ET-1 elevates pulmonary capillary hydrostatic pressure in isolated perfused lungs, thereby leading to enhanced fluid filtration rate [70, 71]. Intravenous administration of ET-1 to rats or guinea-pigs increases microvascular permeability in the large airways [18, 72–74]. These actions of ET-1 are mediated predominantly through activation of ET_A receptors [18, 74], and subsequent release of PAF [72] and TXA_2 [75].

Controversy exist about whether ET-1 by itself facilitates albumin escape into the pulmonary parenchyma. While most studies have found no evi-dence that ET-1 promotes albumin extravasation in the parenchyma [18, 70, 72–76] or increases pulmonary vascular fluid filtration coefficient [77, 78], several other studies in guinea-pig and rat perfused lungs reported increased edema formation [19, 70, 77, 79] and albumin escape in the rat lung [80]. However, ET-1 does not produce edema in blood-perfused lungs [70, 77]. While ET-1 does not enhance albumin movements across cultured pulmonary endothelial monolayers [77, 78], together with other mediators it can promote pulmonary albumin escape in a synergistic fashion. Indeed, simultaneous administration of ET-1 and PAF [72] or leukotoxin [81] cau-ses parenchymal edema. ET_A-receptor dependent elevation of pulmonary artery and capillary pressure [81] and increases in vascular permeability [73, 74] contribute to lung edema. Furthermore, inhibition of endogenous NO synthesis also leads to a marked albumin escape into the parenchyma in response to exogenous ET-1 in conscious rats [82]. Acute NO blockade results in increases in albumin escape in the large airways and parenchyma

[83]. These are, in part, the consequence of unmasking the actions of endogenous ETs, which are mediated predominantly *via* ET_A receptors [83].

2.4. Mucus Secretion

Inflammatory reactions that affect mucous membranes such as those of the airways are characterised by mucus hypersecretion and shedding of the epithelial lining cells into the lumen. Mucosal inflammation can be detected even in patients with newly diagnosed asthma [84]. Immunoreactive ET and ET-1 mRNA was detected primarily in the vascular endothelium and the serous cells in human nasal submucosal gland [85, 86]. ET-1 and ET-2, but not ET-3, stimulate lactoferrin and mucuos glycoprotein secretion from serous and mucuos cells [86] and from feline isolated tracheal submucosal glands [87], although high peptide concentrations $(0.1-10 \ \mu M)$ were required to observe a significant effect. By contrast, ET-1 decreases secretion from tracheal explants [87] and inhibits phenylephrine- or methacholine-induced secretion from serous cells in the ferret trachea [88]. Intranasal ET-1 stimulates nasal secretions and triggers rhinorrhea, itching and sneezing in both allergic and nonallergic individuals, being more potent in allergic individuals [89]. This would suggest that allergic inflammation increases responsiveness of the nasal airways to ET-1. ET-1 induces prostanoid release from human nasal mucosa [90].

2.5. Airway Epithelial Cells

Alterations in the structural and functional properties of airway epithelium have been proposed to contribute to airway inflammation and to underlie the pathophysiology of several lung disorders such as asthma and cystic fibrosis [91]. The airway epithelium is a rich source of ET-1 and ET-3 [92–95]. The ETs likely exert both paracrine and autocrine effects. LPS and various cytokines [4] as well as immune complexes [96] have the potential to enhance ppET-1 mRNA expression and ET-1 release. ET-1, but not ET-2 or ET-3, increases ciliary beat frequency and chloride secretion from canine tracheal epithelial cells [97, 98], and consequently ET-1 may also augment mucociliary clearance. ET-1 stimulates secretion of phosphatidylcholine and diacylglycerol from rat alveolar type II cells [99], suggesting a role for ET-1 in the regulation of surfactant secretion.

2.6. Tissue Remodeling

Tissue repair and resolution are essential components of acute inflammation. Alterations in airway architecture, including airway smooth muscle hypertrophy/hyperplasia and fibrosis, are characteristic features of chronic airway diseases. Among the most interesting effects of ETs in the lung are their potential effects on tissue remodeling. ET-1 stimulates proliferation of rabbit and sheep cultured airway smooth muscle cells [100–102]. Salbutamol pretreatment does not affect the mitogenic response to ET-1, while it inhibits the proliferative effects of thrombin and epidermal growth factor (EGF) [103]. ET-1 and ET-3 stimulate collagen secretion from pulmonary fibroblasts [104] and may therefore be involved in the increased collagen formation observed in asthmatic airways [105]. ET-1 stimulates proliferation of vascular smooth muscle cells and fibroblasts through activation of protein kinase C [106–108], and subsequent activation of MEK (mitogen-activated protein or ERK kinase), which, in turn, activates ERK2 (extracellular regulated kinase 2) [109]. Activation of ERK kinase appears to be the critical endpoint on ET signal transduction pathway, since inhibition of this kinase blocks ET-1-induced proliferation of airway smooth muscle cells [109]. In addition, ET-1 induces type V collagen production by cultured vascular smooth muscle cells [110]. The ETs also promote transformation of fibroblasts into myofibroblast that can evolve into smooth muscle cells [111]. This process may contribute to the extension of smooth muscle cells into normally nonmuscular small blood vessels that characteristically occurs with chronic hypoxia. ET-1 is a potent stimulus of DNA synthesis and proliferation of human pulmonary vascular smooth muscle cells in culture [112]. ET-1 is considerably more potent than ET-3 and sarafotoxin S6c, and its effect is blocked by BQ-123, indicating that an ET_A receptor is involved [112]. ET-1 increases expression of the proto-oncogenes *c-fos* and *c-myc* [106, 113, 114], and early growth response gene-1 [115]. Demoly et al. [116] have found *c-fos* expression in bronchial biopsies from 8 out of 12 patients with asthma, but not in specimens from 10 nonasthmatic individuals, although the *c-fos* expression was localized to the airway epithelium rather than smooth muscle cells. The regulation of ET-1 gene transcription by *fos* and *jun* oncogenes [117] may create a positive feedback loop, with ET-1 stimulating expression of oncogenes resulting higher ET-1 production. However, ET-1 or ET-3 generally acts as comitogen with serum or other growth factors, including EGF, platelet-derived growth factor (PDGF) and insulin for airway and vascular smooth muscle cells and fibroblasts [104, 112, 118]. In addition, both ET-1 and ET-3 induce chemotaxis of fibroblasts prepared from rat pulmonary arteries [104]. The ET concentrations which induced chemotaxis (1 pM–0.1 nM) were much lower than those required to produce fibroblast replication [104]. Of interest, a recent study has shown that the ET_A receptor antagonist BQ-610 effectively blocks cigarette smoke-induced mitogenesis in rat airways and blood vessels [119], suggesting the involvement of ET-1 in smoking-related remodeling of airways.

2.7. Cytotoxicity

One mechanism by which ETs might contribute to tissue injury is modulating activation of inflammatory cells, as described. Another possibility is that ET-1 may enhance the cytotoxic activity of other agents. Indeed, ET-1, although non-toxic by itself, potentiates ischemia-induced damage to myocardial cells [120]. Furthermore, ET-1 secreted from endothelial cells enhanced NO-induced cytotoxicity in cultured rat vascular smooth muscle cells through activation of ET_A receptors [121].

3. Potential Proinflammatory Roles of Endothelins in Pulmonary Diseases

Based on their wide range of proinflammatory actions which are highlighted above, endothelins, in particular ET-1, have been implicated in the pathophysiology of various human pulmonary diseases (see Chapters 12 and 13). Figure 2 summarizes potential mechanisms by which ETs may contribute to the inflammatory process in the lung. However, these assumptions are predominantly based on circumstantial evidence. For instance, numerous studies have reported an association between development of

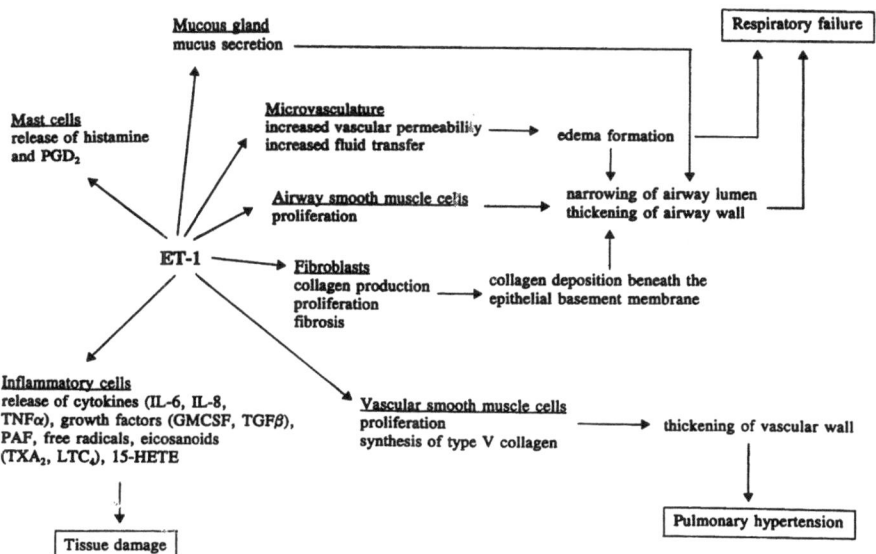

Figure 2. Potential proinflammatory actions of endothelin-1 in the lung. Abbreviations: PGD_2, prostaglandin D_2; IL-6, interleukin-6; IL-8, interleukin-8; $TNF\alpha$, tumor necrosis factor α; GMCSF, granulocyte/macrophage colony stimulating factor; $TGF\beta$, transforming growth factor β; PAF, platelet-activating factor; TXA_2, thromboxane A_2; LTC_4, leukotriene C_4.

Table 1. Potential proinflammatory actions of endothelin-1 in pulmonary diseases

Disease	Species	Proinflammatory actions	Receptor	References
Asthma	human	subepithelial fibrosis	?	[104, 105]
	human	airway wall thickening	?	[104, 105]
	sheep	airway hyperreactivity	ET_A	[125]
Air embolism	rat	increased protein in BAL fluid	?	[126]
	rat	edema	?	[126]
Neurogenic edema	rat	edema	ET_A	[127]
Lung ischemia-reperfusion	rat	endothelial injury	ET_A/ET_B	[128]
	rat	accumulation of neutrophils	ET_A/ET_B	[128]
	rat	edema	ET_A	[128, 129]
Lung transplant rejection	dog	activation of mononuclear cells ?	?	[37, 130]
Monocrotaline-induced pulmonary hypertension	rat	increased collagen production	?	[110]
	rat	right ventricular hypertrophy	ET_A	[131]
Pulmonary hypoxia-induced pulmonary hypertension	rat	right ventricular hypertrophy	ET_A	[132–134]
High-altitude-induced pulmonary hypertension	human	edema?, neutrophil influx ?	?	[135]
Cryptogenic fibrosing alveolitis	human	young granulation tissue, fibrosis	?	[136]
	human	type II cell proliferation	?	[136]
Systemic sclerosis	human	pulmonary fibrosis, mitogen	ET_A	[137]
Disseminated intravascular coagulation	human	multiorgan system failure	?	[138, 139]
	human	pro-coagulant actions	?	[66, 140, 141]
Epithelial carcinoma	cell culture	autocrine/paracrine growth factor	?	[142]

acute lung injury both in patients and experimental animals and increases in ET-1 gene expression and production, elevated ET levels in the BAL fluid, plasma and tissues and augmented binding and/or action of ET-1. However, one should keep in mind that elevated ET-1 concentrations would not necessarily indicate a mediator role for this peptide, as they may also be a marker of tissue damage [122, 123]. On the other hand, because release of ET-1 by endothelial and probably other cells is polarized toward the basolateral side [124], ET-1 levels in biological fluids may not correctly represent local production rate, and local concetrations of ET-1 might be much higher than those detected in biological fluids. To understand the pathological significance of endogenous ETs, it is necessary to demonstrate the beneficial effects of selective ET receptor antagonists or inhibitors of its production on the respective pathology in animal models of the disease, and, ultimately, in humans. Table 1 lists pulmonary pathologies where the proinflammatory actions of ET-1 have been documented or implicated. The pathologies where ET receptor antagonists have been successfully employed against are also presented in this table.

4. Conclusions

The ability of ETs, in particular ET-1, to stimulate inflammatory cells, promote their adhesion to endothelial cells and migration, modulate interactions between inflammatory cells, enhance vascular permeability, trigger cytokine and growth factor release, alter gene expression, augment proliferation of airway and vascular smooth muscle cells and fibroblasts illustrates their potential to contribute to ongoing inflammation and tissue remodeling. Despite these provocative findings, further studies are necessary to critically assess the requirement of ET peptides in the development and progression of pulmonary diseases, including asthma, pulmonary hypertension, pulmonary fibrosis and acute lung injury. An important problem to be addressed is defining the ET receptor subtypes that should be targeted in humans based on experiments using animal models. The marked species differences in ET receptor subtypes mediating the same effects of exogenous ETs [1–5] complicates predicting the efficacy of ET receptor antagonists in human diseases. The required profile will likely depend on the specific disorder. The recent development of nonpeptide, orally active ET receptor antagonists makes these types of preclinical and clinical investigations feasible for the first time.

References

1 Hay DWP, Henry PJ, Goldie RG (1993) Endothelin and the respiratory system. *Trends Pharmacol Sci* 14: 29–32

2 Filep JG (1994) Endothelin peptides: Biological actions and pathophysiological significance in the lung. *Life Sci* 52: 119–133

3 Rubanyi GM, Polokoff MA (1994) Endothelins: Molecular biology, biochemistry, pharmacology, physiology, and pathophysiology. *Pharmacol Rev* 46: 325–415

4 Michael JR, Markewitz BA (1996) Endothelins and the lung. *Am J Respir Crit Care Med* 154: 555–581

5 Hay DWP, Henry PJ, Goldie RG (1996) Is endothelin-1 a mediator in asthma? *Am J Respir Crit Care Med* 154: 1594–1597

6 Sessa WC, Kaw S, Hecker M, Vane JR (1991) The biosynthesis of endothelin-1 by human polymorphonuclear leukocytes. *Biochem Biophys Res Comm* 174: 613–618

7 Patrignani P, Del Maschio A, Bazzoni G, Daffonchio L, Hernandez A, Modica R, Montesanti L, Volpi D, Patrono C, Dejana E (1991) Inactivation of endothelin by polymorphonuclear leukocyte-derived lytic enzymes. *Blood* 78: 2715–2720

8 Tonnessen T, Ilebekk A, Naess PA, Christensen G (1996) Inhibition of granulocyte-derived proteases reduces the increase in plasma endothelin associated with myocardial ischemia in the pig. *Basic Res Cardiol* 91: 289–295

9 Wright CD, Cody WL, Dunbar JB, jr, Doherty AM, Hingorani GP, Rapandalo ST (1994) Characterization of endothelins as chemoattractants for human neutrophils. *Life Sci* 55: 1633–1641

10 Ishida K, Takeshige K, Minakami S (1990) Endothelin-1 enhances superoxide generation of human neutrophils stimulated by the chemotactic peptide N-formyl-methionyl-leucyl-phenylalanine. *Biochem Biophys Res Comm* 173: 496–500

11 Hafström I, Ringertz B, Lundberg T, Palmblad J (1993) The effect of endothelin, neuropeptide Y, calcitonin gene-related peptide and substance P on neutrophil functions. *Acta Physiol Scand* 148: 341–346

12 Halim A, Kanayama N, El Maradny E, Maehara K, Terao T (1995) Activated neutrophil by endothelin-1 caused tissue damage in human umbilical cord. *Thromb Res* 77: 321–327

13 Gómez-Garré D, Guerra M, Gonzalez E, López-Farré A, Riesco A, Escanero J, Egido J (1992) Aggregation of human polymorphonuclear leukocytes by endothelin: role of platelet-activating factor. *Eur J Pharmacol* 224: 167–172

14 López-Farré A, Caramelo C, Esteban A, Alberola ML, Millás I, Montón M, Casado S (1995) Effects of aspirin on platelet-neutrophil interactions: role of nitric oxide and endothelin-1. *Circulation* 91: 2080–2088

15 Lundblad R, Moen O, Fosse E (1997) Endothelin-1 and neutrophil activation during heparin-coated cardiopulmonary bypass. *Ann Thorac Surg* 63: 1361–1367

16 Elferink JGR, de Koster BM (1994) Endothelin-induced activation of neutrophil migration. *Biochem Pharmacol* 48: 865–871

17 Elferink JGR, de Koster BM (1996) The effect of endothelin-2 (ET-2) on migration and changes in cytosolic free calcium of neutrophils. *Naunyn Schmiedeberg's Arch Pharmacol* 353: 130–135

18 Filep JG, Fournier A, Földes-Filep E (1995) Acute pro-inflammatory actions of endothelin-1 in the guinea-pig lung: involvement of ET_A and ET_B receptors. *Br J Pharmacol* 115: 227–236

19 Helset E, Ytrehus K, Tveita T, Kjaeve J, Jorgensen L (1994) Endothelin-1 causes accumulation of leukocytes in the pulmonary circulation. *Circ Shock* 44: 201–209

20 Helset E, Lindal S, Olsen R, Hyklebust R, Jorgensen L (1996) Endothelin-1 causes sequential trapping of platelets and neutrophils in pulmonary microcirculation in rats. *Am J Physiol* 271: L538–L546

21 López-Farré A, Riesco A, Espinosa G, Digiuni E, Cernadas MR, Alvarez V, Montón M, Rivas F, Gallego MJ, Egido J, Casado S, Caramelo C (1993) Effect of endothelin-1 on neutrophil adhesion to endothelial cells and perfused heart. *Circulation* 88: 1166–1171

22 Huribal M, Kumar R, Cunningham ME, Sumpio BE, McMillen MA (1994) Endothelin-stimulated monocyte supernatants enhance neutrophil superoxide production. *Shock* 1: 184–187

23 Lopez-Belmonte J, Whittle BJR (1995) Endothelin-1 induces neutrophil-independent vascular injury in the rat gastric microcirculation. *Eur J Pharmacol* 278: R7–R9

24 Lee TH, Lane SJ (1992) The role of macrophages in the mechanisms of airway inflammation in asthma. *Am Rev Respir Dis* 145: S27–S30

25 Ehrenreich H, Anderson RW, Fox CH, Rieckmann P, Hoffman GS, Travis WD, Coligan JE, Kehrl JH, Fauci AS (1990) Endothelins, peptides with potent vasoactive properties, are produced by human macrophages. *J Exp Med* 172: 1741–1748

26 Sun G, De Angelis G, Nucci F, Ackerman V, Bellini A, Mattoli S (1996) Functional analysis of the preproendothelin-1 gene promoter in pulmonary epithelial cells and monocytes. *Biochem Biophys Res Comm* 221: 647–652

27 McMillen MA, Huribal M, Cunningham ME, Kumar R, Sumpio BE (1995) Endothelin-1 increases intracellular calcium in human monocytes and causes production of interleukin-6. *Crit Care Med* 23: 34–40

28 McMillen MA, Huribal M, Kumar R, Sumpio BE (1993) Endothelin-stimulated human monocytes produce prostaglandin E_2 but not leukotriene B_4. *J Surg Res* 54: 331–335

29 Achmad TH, Rao GS (1992) Chemotaxis of human blood monocytes toward endothelin-1 and the influence of calcium channel blockers. *Biochem Biophys Res Comm* 189: 994–1000

30 Chisholm LJ, Dovgan PS, Agrawal DK (1996) Modulation of monocyte adherence to endothelial cells by endothelin-1: involvement of Src (p60src) and JAK1-like kinases. *J Vasc Surg* 23: 288–300

31 King JK, Srivastava KD, Stefano GB, Bilfinger TV, Bahou WF, Magazine HI (1997) Human monocyte adhesion is modulated by endothelin B receptor-coupled nitric oxide release. *J Immunol* 158: 880–886

32 Stefano GB, Rodriguez M, Glass R, Cesares F, Hughes TK, Bilfinger TV (1995) Hyperstimulation of leukocytes by plasma from cardiopulmonary by-pass patients is diminished by morphine and IL-10 pretreatment. *J Cardiovasc Surg* 36: 25–30

33 Makman MH, Bilfinger TV, Stefano GB (1995) Human granulocytes contain an opiate alkaloid-selective receptor mediating inhibition of cytokine-induced activation and chemotaxis. *J Immunol* 154: 1323–1330

34 Haller H, Schaberg T, Lindschau C, Lode H, Distler A (1991) Endothelin increases $[Ca^{2+}]_i$, protein phosphorylation, and O_2^- production in human alveolar macrophages. *Am J Physiol* 261: L478–L484

35 Millul V, Lagente V, Gillardeux O, Boichot E, Dugas B, Mencia-Huerta JM, Bereziat G, Braquet P, Masliah J (1991) Activation of guinea pig alveolar macrophages by endothelin-1. *J Cardiovasc Pharmacol* 17 (Suppl 7): S233–S235

36 Magazine HI, Andersen TT, Bruner CA, Malik AB (1994) Vascular contractile potency of endothelin-1 is increased in the presence of monocytes and macrophages. *Am J Physiol* 266: H1620–H1625

37 Cale AR, Ricagna F, Wiklund L, McGregor CG, Miller VM (1994) Mononuclear cells from dogs with acute lung allograft rejection cause contraction of pulmonary arteries. *Circulation* 90: 952–958

38 Nagase T, Fukuchi Y, Jo C, Teramoto S, Uejima Y, Ishida K, Shimizu T, Orimo H (1990) Endothelin-1 stimulates arachidonate 15-lipoxygenase activity and oxygen radical formation in the rat distal lung. *Biochem Biophys Res Comm* 168: 485–489

39 Ninomiya H, Yu XY, Hasegawa S, Spannhake EW (1992) Endothelin-1 induces stimulation of prostaglandin synthesis in cells obtained from canine airways by bronchoalveolar lavage. *Prostaglandins* 43: 401–411

40 Merhi Y, Guidoin R, Provost P, Leung TK, Lam JYT (1995) Increase of neutrophil adhesion and vasoconstriction with platelet deposition after deep arterial injury by angioplasty. *Am Heart J* 129: 445–451

41 Page CP (1989) Platelets and asthma. *Agents Actions* (Suppl 28): 75–84

42 Pietraszek MH, Takada Y, Takada A (1992) Endothelins inhibit serotonin-induced platelet aggregation via a mechanism involving protein kinase C. *Eur J Pharmacol* 219: 289–293

43 Asterie-Dequeker C, Ktuzalen L, David-Dufilho M, Devynck MA (1992) *In vitro* inhibition by endothelins of thrombin-induced platelet aggregation and Ca^{2+} mobilization in human platelets. *Br J Pharmacol* 106: 966–971

44 Touyz RM, Lariviere R, Schiffrin EL (1995) Endothelin influences pH_i of human platelets through protein kinase C mediated pathways. *Thromb Res* 78: 55–65

45 Astarie-Dequeker C, Pernollet MG, Le Breton G, Devynck MA (1995) Endothelin-3 reduces Ca^{2+}-uptake and Ca^{2+} content of platelet internal stores. *Biochem Biophys Res Comm* 210: 889–897

46 Dockrell MEC, Webb DJ, Williams BC (1996) Activation of the endothelin B receptor causes a dose-dependent accumulation of cyclic GMP in human platelets. *Blood Coag Fibrinolysis* 7: 178–180

47 Hermán F, Magyar K, Chabrier PE, Braquet P, Filep J (1989) Prostacyclin mediates antiaggregatory and hypotensive actions of endothelin in anaesthetized beagle dogs. *Br J Pharmacol* 98: 38–40

48 Thiemermann C, May GR, Page CP, Vane JR (1990) Endothelin-1 inhibits platelet aggregation *in vivo*: a study with [111]indium-labelled platelets. *Br J Pharmacol* 99: 303–308

49 Filep J, Hermán F, Battistini B, Chabrier PE, Braquet P, Sirois P (1991) Antiaggregatory and hypotensive effects of endothelin-1 in beagle dogs: Role for prostacyclin. *J Cardiovasc Pharmacol* 17 (Suppl 7): S216–S218

50 Thiemermann C, Lidbury P, Thomas R, Vane J (1988) Endothelin inhibits *ex vivo* platelet aggregation in the rabbit. *Eur J Pharmacol* 158: 181–182

51 Filep JG, Battistini B, Côté YP, Beaudoin AR, Sirois P (1991) Endothelin-1 induces prostacyclin release from bovine aortic endothelial cells. *Biochem Biophys Res Comm* 1771: 171–176

52 Kato K, Sawada S, Toyoda T, Kobayashi K, Shirai K, Yamamoto K, Tamagaki T, Yamagami M, Yoneda M, Takeda O (1992) Influence of endothelin on human platelet aggregation and prostacyclin generation from human vascular endothelial cells in culture. *Jpn Circ J* 56: 422–431

53 Leadley RJ, jr, Lee P, Erickson LA, Shebuski RJ (1993) The snake venom peptide sarafotoxin S6b inhibits repetitive platelet thrombus formation in the stenosed canine coronary artery. *J Cardiovasc Pharmacol* 22 (Suppl 8): S199–S203

54 Chung KF (1986) Role of inflammation in the hyperreactivity of the airways in asthma. *Thorax* 41: 657–662

55 Uchida Y, Ninomiya H, Sakamoto T, Lee JY, Endo T, Nomura A, Hasegawa S, Hirata F (1992) ET-1 released histamine from guinea pig pulmonary but not peritoneal mast cells. *Biochem Biophys Res Comm* 189: 1196–1201

56 Egger D, Geuenich S, Denzlinger C, Schmitt E, Mailhammer R, Ehrenreich H, Dörmer P. Hültner L (1995) IL-4 renders mast cells functionally responsive to endothelin-1. *J Immunol* 154: 1830–1837

57 Bousquet J, Chanez P, Lacoste JY, Barnéon G, Ghavanian N, Enander I, Venge P, Ahlstedt S, Simony-Lafontaine J, Godard P, Michel FB (1990) Eosinophilic inflammation in asthma. *N Engl J Med* 323: 1033–103⁹

58 Finsnes F, Skjønsberg OH, Tønnessen T, Naess O, Lyberg T, Christensen G (1997) Endothelin production and effects of endothelin antagonism during experimental airway inflammation. *Am J Respir Crit Care Med* 155: 1404–1412

59 Fujitani Y, Trifileff A, Tsuyuki S, Coyle AJ, Bertrand C (1997) Endothelin receptor antagonists inhibit antigen-induced lung inflammation in mice. *Am J Respir Crit Care Med* 155: 1890–1894

60 Ollerenshaw S, Woolcock AJ (1992) Characteristics of the inflammation in biopsies from large airways of subjects with asthma and subjects with chronic airflow limitation. *Am Rev Respir Dis* 145: 922–927

61 Gerritsen ME, Bloor CM (1993) Endothelial gene expression in response to injury. *FASEB J* 7: 523–532

62 Golden CL, Kohler JP, Nick HS, Visner GA (1995) Effects of vasoactive and inflammatory mediators on endothelin-1 expression in pulmonary endothelial cells. *Am J Respir Cell Mol Biol* 12: 503–512

63 Xin X, Cai Y, Matsumoto K, Agui T (1995) Endothelin-1-induced interleukin-6 production by rat aortic endothelial cells. *Endocrinology* 136: 132–137

64 Stanimirovic DB, McCarron R, Bertrand N, Spatz M (1993) Endothelins release ^{51}Cr from cultured human cerebro microvascular endothelium. *Biochem Biophys Res Comm* 191: 1–8

65 Catalán RE, Martinez AM. Aragonés D, Martinez A, Diaz G (1996) Endothelin stimulates phosphoinositide hydrolysis and PAF synthesis in brain microvessels. *J Cereb Blood Flow Metab* 16: 1325–1334

66 Yamamoto C, Kaji T, Sakamoto M, Koizumi F (1992) Effect of endothelin on the release of tissue plasminogen activator and plasminogen activator inhibitor-1 from cultured human endothelial cells and interaction with thrombin. *Thromb Res* 67: 619–624

67 McCarron RM, Wang L, Stanimirovic DB, Spatz M (1993) Endothelin induction of adhesion molecule expression on human brain microvascular endothelial cells. *Neurosci Lett* 156: 31–34

68 McFadden ER, jr (1992) Microvasculature and airway response. *Am Rev Respir Dis* 145: S42–S43

69 Hogg JC, Paré PD, Moreno R (1987) The effect of submucosal edema on airway resistance. *Am Rev Respir Dis* 135: 554–556

70 Rodman DM, Stelzner TJ, Zamora MR, Bonvallet ST, Oka M, Sato K, O'Brien RF, McMurtry IF (1992) Endothelin-1 increases the pulmonary microvascular pressure and causes pulmonary edema in salt solution but not blood-perfused rat lungs. *J Cardiovasc Pharmacol* 20: 658–663

71 Helset E, Kjaeve J, Hauge A (1993) Endothelin-1-induced increases in microvascular permeability in isolated, perfused rat lungs requires leukocytes and plasma. *Circ Shock* 39: 15–20

72 Filep JG, Sirois MG, Rousseau A, Fournier A, Sirois P (1991) Effects of endothelin-1 on vascular permeability in the conscious rat: interactions with platelet-activating factor. *Br J Pharmacol* 104: 797–804

73 Filep JG, Sirois MG, Földes-Filep E, Rousseau A, Plante GE, Fournier A, Yano M, Sirois P (1993) Enhancement by endothelin-1 of microvascular permeability via the activation of ET_A receptors. *Br J Pharmacol* 109: 880–886

74 Filep JG, Clozel M, Fournier A, Földes-Filep E (1994) Characterization of receptors mediating vascular responses to endothelin-1 in the conscious rats. *Br J Pharmacol* 113: 845–852

75 Sirois MG, Filep JG, Rousseau A, Fournier A, Plante GE, Sirois P (1992) Endothelin-1 enhances vascular permeability in conscious rats: role of thromboxane A_2. *Eur J Pharmacol* 214: 119–125

76 Macquin-Mavier I, Levame M, Istin N, Harf A (1989) Mechanisms of endothelin-mediated bronchoconstriction in the guinea pig. *J Pharmacol Exp Ther* 250: 740–745

77 Barnard JW, Barman SA, Adkins WK, Longenecker GL, Taylor AE (1991) Sustained effects of endothelin-1 on rabbit, dog, and rat pulmonary circulations. *Am J Physiol* 261: H479–H486

78 Horgan MJ, Pinheiro JMB, Malik AB (1991) Mechanism of endothelin-1-induced pulmonary vasoconstriction. *Circ Res* 69: 157–164

79 Ercan ZS, Kiling M, Yazar Ö, Korkosuz P, Türker RK (1993) Endothelin-1-induced oedema in rat and guinea-pig isolated perfused lungs. *Arch Int Pharmacodyn* 323: 74–84

80 Zimmerman RS, Martinez AJ, Maymind M, Barbee RW (1992) Effect of endothelin on plasma volume and albumin escape. *Circ Res* 70: 1027–1034

81 Ishizakai T, Shigemori K, Nakai T, Miyabo S, Hayakawa M, Ozawa T, Voelkel NF, Chang SW (1995) Endothelin-1 potentiates leukotoxin-induced edematous lung injury. *J Appl Physiol* 79: 1106–1111

82 Filep JG, Földes-Filep E, Rousseau A, Sirois P, Fournier A (1993) Vascular responses to endothelin-1 following inhibition of nitric oxide synthesis in the conscious rat. *Br J Pharmacol* 110: 1213–1221

83 Filep JG (1997) Endogenous endothelin modulates blood pressure, plasma volume and albumin escape after systemic nitric oxide blockade. *Hypertension* 30: 22–28

84 Laitinen LA, Laitinen A, Haahtela T (1993) Airway mucosal inflammation even in patients with newly diagnosed asthma. *Am Rev Respir Dis* 147: 697–704

85 Casasco A, Benazzo M, Casasco M, Cornaglia AI, Springall DR, Calligaro A, Mira E, Polak JM (1993) Occurrence, distribution and possible role of the regulatory peptide endothelin in the nasal mucosa. *Cell Tissue Res* 274: 241–247

86 Mullol J, Chowdhury BA, White MV, Ohkubo K, Rieves RD, Baraniuk J, Hausfeld JN, Shelhamer JH, Kalinef MA (1993) Endothelin in human nasal mucosa. *Am J Respir Cell Mol Biol* 8: 393–402

87 Shimura S, Ishihara H, Satoh M, Masuda T, Nagaki N, Sasaki H, Takishima T (1992) Endothelin regulation of mucus glycoprotein secretion from feline tracheal submucosal glands. *Am J Physiol* 262: L208–L213

88 Webber SE, Yurdakos E, Woods AJ, Widdicombe JG (1992) Effects of endothelin-1 on tracheal submucosal gland secretion and epithelial function in ferret. *Chest* 101: 63S–67S

89 Riccio MM, Reynolds CJ, Hay DWP, Proud D (1995) Effects of intranasal administration of endothelin-1 to allergic and nonallergic individuals. *Am J Respir Crit Care Med* 152: 1757–1764

90 Wu T, Mullol J, Rieves RD, Logun C, Hausfield J, Kalimer MA, Shelhamer JH (1992) Endothelin-1 stimulates eicosanoid production in cultured human nasal mucosa. *Am J Respir Cell Mol Biol* 6: 168–174

91 Hay DWP (1995) Endothelins. In: D Raebum, Ma Giembycz (eds): *Airways smooth muscle: peptide receptors, ion channels and signal transduction*. Birkhäuser Verlag, Basel, Switzerland, 1–50

92 Black PN, Ghatei MA, Takahashi K, Bretherton-Watt D, Krausz T, Dollery CT, Bloom SR (1989) Formation of endothelin by cultured airway epithelial cells. *FEBS Lett* 255: 129–132

93 Mattoli S, Soloperto M, Marini M, Fasoli A (1991) Levels of endothelin in the broncho-alveolar lavage fluid of patients with symptomatic asthma and reversible airflow obstruction. *J Allergy Clin Immunol* 88: 376–384

94 Ninomiya H, Uchida Y, Ishii Y, Nomura A, Kameyama M, Saotome M, Endo T, Hasegawa S (1991) Endotoxin stimulates endothelin release from cultured epithelial cells of guinea-pig trachea. *Eur J Pharmacol* 203: 299–302

95 Rozengurt N, Springall DR, Polak JM (1990) Localization of endothelin-like immuno-reactivity in airway epithelium of rats and mice. *J Pathol* 160: 5–8

96 Ehrenreich H, Burd PR, Rottem M, Hultner L, Hylton JB, Garfield M, Coligan JE, Metcalfe DD, Fauci AS (1992) Endothelins belong to the assortment of mast cell-derived and mast cell-bound cytokines. *New Biol* 4: 147–156

97 Satoh M, Shimura S, Ishihara H, Nagaki M, Sasaki H, Takishima T (1992) Endothelin-1 stimulates chloride secretion across canine tracheal epithelium. *Respiration* 59: 145–150

98 Tamaoki J, Kanemura T, Sakai N, Isono K, Kobayashi K, Takizawa T (1991) Endothelin stimulates ciliary beat frequency and chloride secretion in canine cultured tracheal epithelium. *Am J Respir Cell Mol Biol* 4: 426–431

99 Sen M, Grunstein NM, Chander A (1994) Stimulation of lung surfactant secretion by endothelin-1 from rat alveolar type II cells. *Am J Physiol* 266: L225–L262

100 Glassberg MK, Ergul A, Winner A, Puett D (1994) Endothelin-1 promotes mitogenesis in airway smooth muscle cells. *Am J Respir Cell Mol Biol* 10: 316–324

101 Noveral JP, Rosenberg RA, Anbar NA, Grunstein MM (1992) Role of endothelin-1 in regulating proliferation of cultured rabbit airway smooth muscle cells. *Am J Physiol* 263: L317–L324

102 Stewart AG, Grigoriadis G, Harris T (1994) Mitogenic actions of endothelin-1 and epidermal growth factor in cultured airway smooth muscle. *Clin Exp Pharmacol Physiol* 21: 277–285

103 Tomlinson PR, Wilson JW, Stewart AG (1994) Inhibition by salbutamol of the proliferation of human airway smooth muscle cells grown in culture. *Br J Pharmacol* 111: 641–647

104 Peacock AJ, Dawes KE, Shock A, Gray AJ, Reeves JT, Laurent GJ (1992) Endothelin-1 and endothelin-3 induces chemotaxis and replication of pulmonary artery fibroblasts. *Am J Respir Cell Mol Biol* 7: 492–499

105 Brewster CEP, Howarth PH, Djukanovic R, Wilson J, Holgate ST, Roche WR (1990) Myofibroblasts and subepithelial fibrosis in bronchial asthma. *Am J Respir Cell Mol Biol* 3: 507–511

106 Komuro I, Kurimara H, Sugiyama T, Takaku F, Yazaki Y (1989) Endothelin stimulates c-fos and c-myc expression and proliferation of vascular smooth muscle cells. *FEBS Lett* 238: 249–252

107 Muldoon L, Rodland RD, Forsythe ML, Magun BE (1989) Stimulation of phosphatidyl-inositol hydrolysis, diacylglycerol release, and gene expression in response to endothelin, a potent new agonist for fibroblasts and smooth muscle cells. *J Biol Chem* 264: 8529–8536

108 Takuwa N, Takuwa Y, Yanagisawa M, Yamashita K, Masaki T (1989) A novel vasoactive peptide endothelin stimulates mitogenesis through inositol lipid turnover in Swiss 3T3 fibroblasts. *J Biol Chem* 264: 7856–7861

109 Welchel A, Evans J, Posada J (1997) Inhibition of ERK activation attenuates endothelin-stimulated airway smooth muscle cell proliferation. *Am J Respir Cell Mol Biol* 16: 589–596

110 Mansoor AM, Honda M, Saida K, Ishinaga Y, Kuramochi T, Maeda A, Takabatake T, Mitsui Y (1995) Endothelin induced collagen remodeling in experimental pulmonary hypertension. *Biochem Biophys Res Comm* 215: 981–986

111 Villaschi S, Nicosia RF (1994) Paracrine interactions between fibroblasts and endothelial cells in a serum-free co-culture model: modulation of angiogenesis and collagen gel contraction. *Lab Invest* 71: 291–299

112 Zamora MR, Dempsey EC, Walchak SJ, Stelzner TJ (1993) BQ 123, an ET_A receptor antagonist, inhibits endothelin-1 mediated proliferation of human pulmonary artery smooth muscle cells. *Am J Respir Mol Cell Biol* 9: 429–433

113 Simonson MS, Jones JM, Dunn MJ (1992) Differential regulation of *fos* and *jun* gene expression and AP-1 *cis*-element activity by endothelin isopeptides: possible implications for mitogenic signaling by endothelin. *J Biol Chem* 267: 8643–8649

114 Bobik A, Grooms A, Millar JA, Mitchell A, Grinpukel S (1990) Growth factor activity of endothelin on vascular smooth muscle. *Am J Physiol* 258: C408–C415

115 Neyses L, Nouskas J, Vetter H (1991) Inhibition of endothelin-1-induced myocardial protein synthesis by an antisense oligonucleotide against the early growth response gene-1. *Biochem Biophys Res Comm* 181: 22–27

116 Demoly P, Basset-Seguin N, Chanez P, Campbell AM, Gauthier-Rouviére C, Godard P, Michel FB, Bousquet J (1992) *c-fos* proto-oncogene expression in bronchial biopsies of asthmatics. *Am J Respir Cell Mol Biol* 7: 128–133

117 Lee ME, Dhadly MS, Temizer DH, Clifford JA, Yoshimuzi M, Quertermous T (1991) Regulation of endothelin-1 gene expression by *fos* and *jun*. *J Biol Chem* 266: 19034–19039

118 Janakidevi K, Fisher MA, Del Vecchio PJ, Tiruppathi C, Figge J, Malik AB (1992) Endothelin-1 stimulates DNA synthesis and proliferation of pulmonary artery smooth muscle cells. *Am J Physiol* 263: C1295–C1301

119 Dadmanesh F, Wright JL (1997) Endothelin-A receptor antagonist BQ-610 blocks cigarette smoke-induced mitogenesis in rat airways and vessels. *Am J Physiol Lung Cell Mol Physiol* 16: L614–L618

120 Stawski G, Olsen UB, Grande P (1991) Cytotoxic effect of endothelin-1 during "simulated" ischaemia in cultured rat myocytes. *Eur J Pharmacol* 201: 123–124

121 Nakahashi T, Fuluo K, Inoue T, Morimoto S, Hata S, Yano M, Ogihara T (1995) Endothelin-1 enhances nitric oxide-induced cytotoxicity in vascular smooth muscle. *Hypertension* 25: 744–747

122 Stewart DJ, Levy RD, Cernacek P, Langleber. D (1991) Increased plasma endothelin-1 in pulmonary hypertension: marker or mediator of disease? *Ann Intern Med* 114: 464–469

123 Filep JG, Bodolay E, Sipka S, Gyimesi E, Csipö I, Szegedi G (1995) Plasma endothelin correlates with antiendothelial antibodies in patients with mixed connective tissue disease. *Circulation* 92: 2969–2974

124 Wagner OF, Christ G, Wojta J, Vierhapper H, Parzer S, Nowotny PJ, Schneider B, Waldhausl W, Binder BR (1992) Polar secretion of endothelin-1 by cultured endothelial cells. *J Biol Chem* 267: 16066–16068

125 Noguchi K, Ishikawa K, Yano M, Ahmed A, Cortes A, Abraham WM (1995) Endothelin-1 contributes to antigen-induced airway hyperresponsiveness. *J Appl Physiol* 79: 700–705

126 Wang D, Li MH, Hsu K, Shen CY, Chen HI, Lin YC (1992) Air embolism-induced lung injury in isolated rat lungs. *J Appl Physiol* 72: 1235–1242

127 Herbst C, Tippler B, Shams H, Simmet Th (1995) A role for endothelin in bicuculline-induced neurogenic pulmonary oedema in rats. *Br J Pharmacol* 115: 753–760

128 Khimenko PL, Moore TM, Taylor AE (1996) Blocked ET_A receptors prevent ischemia and reperfusion injury in rat lungs. *J Appl Physiol* 80: 203–207

129 Okada M, Yamashita C Okada K (1995) Contribution of endothelin-1 to warm ischemia/reperfusion injury of the rat lung. *Am J Respir Crit Care Med* 152: 2105–2110

130 Ricagna F, Miller VM, Tazelaar HD, McGregor CGA (1996) Endothelin-1 and cell proliferation in lung organ cultures. Implications for lung allografts. *Transplantation* 62: 1492–1498

131 Miyauchi T, Yorikane R, Sakai S, Sakurai T, Okada M, Nishikibe M, Yano M, Yamaguchi I, Sugishita Y, Goto K (1993) Contribution of endogenous endothelin-1 to the progression of cardiopulmonary alterations in rats with monocrotaline-induced pulmonary hypertension. *Circ Res* 73: 887–897

132 Bonvallet ST, Morris KG, Yano M, Zamora MR, McMurtry IF, Stelzner TJ (1994) BQ123, an ET$_A$-receptor antagonist, attenuates hypoxic pulmonary hypertension in rats. *Am J Physiol* 266: H1327–H1331

133 Eddahibi S, Raffestin B, Clozel M, Levame M, Adnot S (1995) Protection from pulmonary hypertension with an orally active endothelin receptor antagonist in hypoxic rats. *Am J Physiol* 268: H828–H835

134 DiCarlo VS, Chen SJ, Meng QC, Durand J, Yano M, Chen YF, Oparil S (1995) ET$_A$-receptor antagonist prevents and reverses chronic hypoxia-induced pulmonary hypertension in rat. *Am J Physiol* 269: L690–L697

135 Droma Y, Hayano T, Takabayeshi Y, Koizumi T, Kubo K, Kobayashi T. Sekiguchi M (1996) Endothelin-1 and interleukin-8 in high altitude pulmonary oedema. *Eur Respir J* 9: 1947–1949

136 Giaid A, Michel RP, Stewart DJ, Sheppard M, Corrin B, Hamid Q (1993) Expression of endothelin-1 in lungs of patients with cryptogenic fibrosing alveolitis. *Lancet* 341: 1550–1554

137 Cambrey AD, Harrison NK, Dawes KE, Southcott AM, Black CM, du Bois RM, Laurent GJ, McAnulty RJ (1994) Increased levels of endothelin-1 in bronchoalveolar lavage fluid from patients with systemic sclerosis contribute to fibroblast mitogenic activity *in vitro*. *Am J Respir Cell Mol Biol* 11: 439–445

138 Asakura H, Jokaji H, Saito M, Uotani C, Kumabashiri I, Morishita E, Yamazaki M, Matsuda T (1992) Role of endothelin in disseminated intravascular coagulation. *Am J Hematol* 41: 71–75

139 Ishibashi M, Ito N, Fujita M, Furue H, Yamaji T (1994) Endothelin-1 as an aggravating factor of disseminated intravascular coagulation associated with malignant neoplasms. *Cancer* 73: 191–195

140 Halim A, Kanayama N, El Maradny E, Maehara K, Terao T. Coagulation *in vivo* microcirculation and *in vitro* caused by endothelin-1. *Thromb Res* 72: 203–209

141 Ruschitzka F, Schrader J, Luders S, Schulz E, Groneau C, Talartschik J, Eisenhauer T, Verwiebe R, Warneke G, Scheler F (1993) Effects of endothelin on coagulation, prostaglandins and hemodynamics. *Contrib Nephrol* 101: 30–36

142 Shichiri M, Hirata Y, Nakajima T, Ando K, Imai T, Yanagisawa M, Masaki T, Marumo F (1991) Endothelin-1 is an autocrine/paracrine growth factor for human cancer cell lines. *J Clin Invest* 87: 1867–1871

Pulmonary Actions of the Endothelins
ed. by R. G. Goldie and D. W. P. Hay
© 1999 Birkhäuser Verlag Basel/Switzerland

CHAPTER 11
Endothelin-induced Neuromodulation

Karen O. McKay

Children's Chest Research Centre, Department of Respiratory Medicine, Royal Alexandra Hospital for Children, Westmead NSW 2145, Australia

1   Introduction
2   Endothelin as a Neurotransmitter
3   Endothelin as a Neuromodulator
3.1   Central Neurotransmission
3.2   Peripheral Autonomic Neurotransmission
3.2.1   Sensory nerves
3.2.2   Sympathetic nerves
3.2.3   Parasympathetic nerves
3.2.4   Non-adrenergic non-cholinergic nerves
4   Neuromodulatory Substances Released by Endothelin
5   Summary
        References

1. Introduction

Despite the relative paucity of information with regard to the effects of endothelin on neurotransmission, there are indications that this peptide has functions and effects that extend beyond that of a potent contractor of smooth muscle, and may include that of a neuroactive mediator. Binding sites for endothelin have been found on neural structures, immunoreactive endothelin has been demonstrated within nerves, and endothelin has been shown to have effects on neural functioning, leading to the proposal that endothelin may be a neuropeptide or, alternatively, may be a modulator of neural function [1, 2].

Endothelin is released locally within airways [3, 4], and it is therefore possible that endothelin may have significant effects on neural structures within the lung. Endothelin-like immunoreactivity has been detected in the lung as well as in the associated autonomic nervous supply, in particular, the vagus nerve [5]. Binding sites for endothelin have been localized over nerves in airway connective tissue [6] and a number of studies have recently demonstrated the effect of the endothelin isopeptides on neural activity within the lung. In this chapter, these studies and others well be examined and the suggestion that endothelin is a significant neuromodulator within the respiratory system will be addressed.

2. Endothelin as a Neurotransmitter

The notion that endothelin may be associated with neuronal function was first suggested at the William Harvey Workshop on Endothelin in London in 1989 [1]. Early studies indicated that endothelin-3, may in fact, be a neural form of the peptide [7]. An endothelin-3-like immunoreactive material was shown to be present in porcine brain homogenates and studies of cDNA libraries demonstrated endothelin-3 expression in the human hypothalamus [8]. The detection of endothelin-1-like immunoreactivity in motor neurons and sensory neurons in the spinal cord, and mRNA for endothelin-1 in the human spinal cord and brain [2, 9] suggested that not only endothelin-3, but all members of the endothelin family may be neurotransmitters.

There is, at present, no study in which endothelin has been shown to be released within the lung in response to nerve stimulation induced by application of an electrical field. Significant release of endothelin in response to high frequency electrical stimulation has however been demonstrated in rat iris sphincter. As this release was abolished by tetrodotoxin, this was interpreted as being due to the neural release of the peptide [10]. In addition, endothelin is secreted by hypothalamic neurons in culture [11]. It is therefore not possible to eliminate a neurotransmitter role for endothelin within the peripheral nervous system, but it is more likely that endothelin released from other sites within the lung acts as a local modulator of peripheral nervous function.

3. Endothelin as a Neuromodulator

While there is no evidence to date that the endothelins act as neurotransmitters in the lung, there is evidence that endothelin may amplify or dampen neuronal activity, or the effect of neural activation, therefore satisfying the characteristics required for classification as a neuromodulator. The proximity of the sites of synthesis and release of the endothelin isopeptides to neural structures both in the central and peripheral nervous systems, suggest that the endothelins may influence functioning within these structures. The brain is a richly vascularised structure and endothelin released from vascular endothelial cells may reach sufficiently high local concentrations to alter central neurotransmission. In the lung, transmission through nerve fibres and ganglia may be influenced by endothelin released from epithelial and endothelial cells. Neuromodulators, which do not function as a neurotransmitters, may alter the synthesis, release, interactions with receptors, uptake and metabolism of neurotransmitters, and the results of studies with the endothelins suggest that they fulfil this function.

3.1. Central Neurotransmission

Quantitative receptor autoradiography has revealed the presence of specific endothelin-1 binding sites in the brain. High density sites for radio-labelled endothelin-1 and endothelin-3 were visualized in the rat brain [12–14] and have been localized to the hippocampus as well as the cerebellum [15]. *In situ* hybridization and immunocytochemical studies of human brain have displayed mRNA for endothelin-1 and endothelin-1 labelled neuronal cell bodies and fibres in the cortex, the hippocampus, the medullary portion of the brainstem and the cerebellum [2, 9]. This finding is of significance as the cell bodies of the phrenic and intercostal nerves, which control respiratory function, are located in the medulla. These binding sites for endothelin indicate the potential for modulation of central respiratory control by endothelin, although it is not clear if there are binding sites on the inspiratory neurons within the medulla, or on adjacent neural processes with input to these neurons. These binding sites are likely to however, be primarily located on the cardiovascular control centres which are also situated in the medulla, and may mediate the increases in arterial blood pressure resulting from infusion of endothelin-1 into the central nervous system of rats [16]. Administration of endothelin-1 to the venterolateral medulla of the rat has, however, been shown to result in a 30% increase in respiratory rate. This increase was seen 15 min after administration and the effect was no longer obvious after 30 min. Most importantly, the response was not accompanied or preceded by changes in pH, P_{CO_2} or P_{O_2} indicating it was unlikely to be due to reflex changes in respiratory rate [17].

3.2. Peripheral Autonomic Neurotransmission

3.2.1. Sensory nerves: In recent years it has become obvious that local irritation of the airway mucosa and resultant sensory nerve activation may have a significant impact on airway functioning. Presently, there is no information as to the effect of endothelin on sensory nerve transmission in the lung. It has however been suggested that endothelin may be a sensory neuropeptide in the guinea pig [18] and evidence for this suggestion is provided by a study demonstrating high concentrations of endothelin-like immunoreactivity in porcine peripheral sensory ganglia [19].

3.2.2. Sympathetic nerves: Despite rich innervation of the airway vasculature and glandular structures, the sympathetic nervous system is considered to have little direct influence on the control of airway calibre. Indeed, in many species including the human, studies have been unable to demonstrate the presence of sympathetic nerve fibre supply to the airway smooth muscle [20]. It is however, important to acknowledge the presence of adrenoceptors on airway parasympathetic ganglia [21], a prime site for neuromodulation. In addition to potential modulation via these ganglionic receptors, alteration of sympathetic neurotransmitter release by endothelin in the airway vasculature, changes in bronchovascular size, and possible diffusion of adrenoceptor agonists to the nearby airway smooth muscle, may alter airway calibre.

At this time, there has been only one report examining the effect of endothelin on sympathetic neurotransmission in tissue isolated from the lung, despite a significant body of evidence demonstrating a marked modulatory role for endothelin on sympathetic neurotransmission in general. In guinea pig isolated pulmonary arterial tissue, endothelin concentration-dependently enhanced sympathetic nerve-induced contractile responses. This adrenergic modulation was found to occur at two sites; by inhibition at pre-junctional sites (the stimulation-evoked release of tritiated noradrenaline was decreased in the presence of endothelin), and, by stimulation at post-junctional sites (the contractile response to exogenously administered noradrenaline was increased after endothelin pretreatment) [22]. In addition, endothelin and sarafotoxin S6b (one of the endothelin-like peptides of the sarafotoxin family) potentiated nerve-stimulation induced vasoconstriction in the isolated perfused mesenteric bed of the rat. This alteration was found to occur via action of the peptides at a postsynaptic site [23, 24]. Conversely, intravenous endothelin-1 has been shown to have a significant pre-junctional stimulatory effect on the sympathetic nervous system, enhancing the release of adrenal catecholamines induced by splanchnic nerve stimulation in the anaesthetised dog [25] and augmenting the response to nerve stimulation in rabbit isolated ear arteries [26].

Endothelin has therefore been shown to have both stimulatory and inhibitory effects on sympathetic nervous transmission, and these effects occur via action at both pre-junctional and post-junctional sites. These differences in the modulatory effect of endothelin may be due to the well established species differences in the actions of this family of peptides [27], the concentration of endothelin used, and the parameters employed for nerve stimulation in the various studies. At this time, in the single study reported in pulmonary vascular tissue, endothelin has been shown to have a stimulatory effect on the contractile response to sympathetic neural stimulation in the guinea pig pulmonary vascular bed, an influence involving an enhancement of the vascular smooth muscle responsiveness to the neurotransmitter noradrenaline.

3.2.3. Parasympathetic nerves: Studies in a number of species have displayed modulation of airway parasympathetic neurotransmission by the endothelins. Parasympathetic neurotransmission may be modified pre-synaptically, postsynaptically, or postjunctionally, as well as via direct action at the parasympathetic ganglion [28, 29]. Receptors for the neuropeptide substance P have been localised on rabbit parasympathetic ganglia

[28] and application of substance P has been shown to augment the response to both pre- and postganglionic nerve stimulation, indicating that this modulation may be due to an action of substance P on the ganglion [30]. Thus, identification of receptors on the ganglia, as well as nerve fibres, may indicate a neuromodulatory role for various mediators.

The demonstration of binding sites for the endothelin isopeptides on human airway nerve fibres [6], rat tracheal nerve trunks [31], guinea pig vagus nerve [32] and airway parasympathetic ganglia [33, 34], coupled with the demonstration of parasympathetic neuromodulation by all three endothelins in the vas deferens and ileum [35, 36], was considered suggestive of a similar neuromodulatory role for the endothelins in the lung. The unexpected finding of immunoreactive sarafotoxin S6b co-localised with endothelin-3 in feline and rodent airway nerves and ganglia, also suggested that isoforms of this endothelin-related peptide may influence pulmonary neural functioning [37]. Further, as a result of studies of the electromechanical effects of endothelin on ferret airways, Lee and co-workers [38] concluded that endothelin may stimulate neurotransmitter release from nerve terminals. The findings reported in studies since that time have provided strong evidence to support a neuromodulatory role for endothelin in the lung, and include the recent demonstration of altered transmission in human airways [39].

The first endothelin isopeptide shown to produce effects on parasympathetic neurotransmission in the airways was the purported neural form, endothelin-3. Endothelin-3 induced a concentration-dependent increase in the response to electrical field stimulation in rabbit isolated bronchus [40]. At a concentration of 10 nM (which is subcontractile in the absence of neutral endopeptidase inhibition in this tissue [41]), endothelin-3 increased the contractile response to nerve stimulation to 205% of the initial response, while a ten-fold increase in the concentration of endothelin-3 further increased the response to 315% of the initial response. This potentiation was independent of the direct contractile effect of endothelin-3 on the smooth muscle, as the greatest potentiation of the response to stimulation occurred when the rise in tone had abated. While also augmenting the response to electrical field stimulation in rabbit bronchus, a greater concentration of endothelin-1 was required for potentiation and the magnitude of the effect was not as great as that produced by endothelin-3 [42]. This suggested that the mechanism of the potentiation may involve activation of the non isopeptide-selective ET_B receptor.

Endothelin-1 also induced potentiation of cholinergically mediated contractions of mouse trachea [43]. As in rabbit airways, the potentiation was concentration-dependent, but the effect was complicated by the direct contractile action of endothelin-1. At subcontractile concentrations (0.1 nM), endothelin-1 caused a 1.35 fold increase in response to electrical field stimulation, while a 1.95 fold increase was apparent when the concentration was increased to 1 nM. Further increases of the concentration of endothelin-1 produced large direct contractile responses of the smooth muscle and an inhibition of the response to nerve stimulation. When the endothelin$_A$ receptor antagonist BQ123 was present, preventing the direct contractile response to endothelin-1, higher concentrations of the peptide also produced marked enhancement of the response to electrical field stimulation. This apparent endothelin-induced inhibition of the response to nerve stimulation in the presence of large contractions of airway smooth muscle (which approach the maximal contractile capacity of the tissue) seems to be a consistent finding, as it has also been reported in human bronchus [39]. A similar facilitation of the response to nerve stimulation by sarafotoxin S6c and endothelin-1 has been reported in rat trachea [44].

The demonstration of binding sites for endothelin on airway parasympathetic ganglia (Fig. 1) and the report of endothelin-induced modulation of synaptic transmission in feline colonic [45] and rabbit pelvic [46] parasympathetic ganglia, suggested that the potentiation of the response to electrical field stimulation may have been due to an action of endothelin at the cholinergic ganglia. Modulation of neurotransmission through parasympathetic ganglia by beta$_2$ and alpha adrenergic agonists has been demonstrated in ferret trachea *in vitro* [47]. In 1991, Kushiku et al. reported inhibition of sympathetic ganglionic transmission by endothelin-3 [48]. If the endothelins inhibited pulmonary sympathetic neurotransmission, the modulating influence of adrenergic mediators may have been attenuated, and an increase in cholinergic responsiveness observed. This was not the case in the rabbit airways, as neither phentolamine, nor propranolol had a significant effect on the potentiation to neural stimulation induced by endothelin-3 (unpublished observations). Further, as hexamethonium had no effect upon the potentiation in either rabbit or mouse airways, the augmentation of the cholinergic responsiveness in these tissues was considered a post-ganglionic phenomenon.

Locations distal to the ganglia at which parasympathetic neurotransmission can be enhanced include the postganglionic nerve ending (an increase in the release of acetylcholine) and the airway smooth muscle (postjunctional enhancement of the response to acetylcholine). Contrary to the potentiation of postjunctional cholinergic responses in the gastrointestinal system [35], the response of airway tissue to exogenous acetylcholine is not altered by concentrations of endothelins having a potentiating effect on the response to nerve stimulation [40, 43, 44]. The concentration-dependent facilitation of parasympathetic cholinergic neurotransmission by the endothelins in the airways is, therefore, likely to be due to increased neurotransmitter release.

The proposition that the neuromodulatory effect of endothelin in the airways occurs as a result of enhanced release of acetylcholine from postganglionic prejunctional nerve fibres is supported by the results of a study carried out on airway nerve and muscle cells in co-culture. Endothelin$_B$

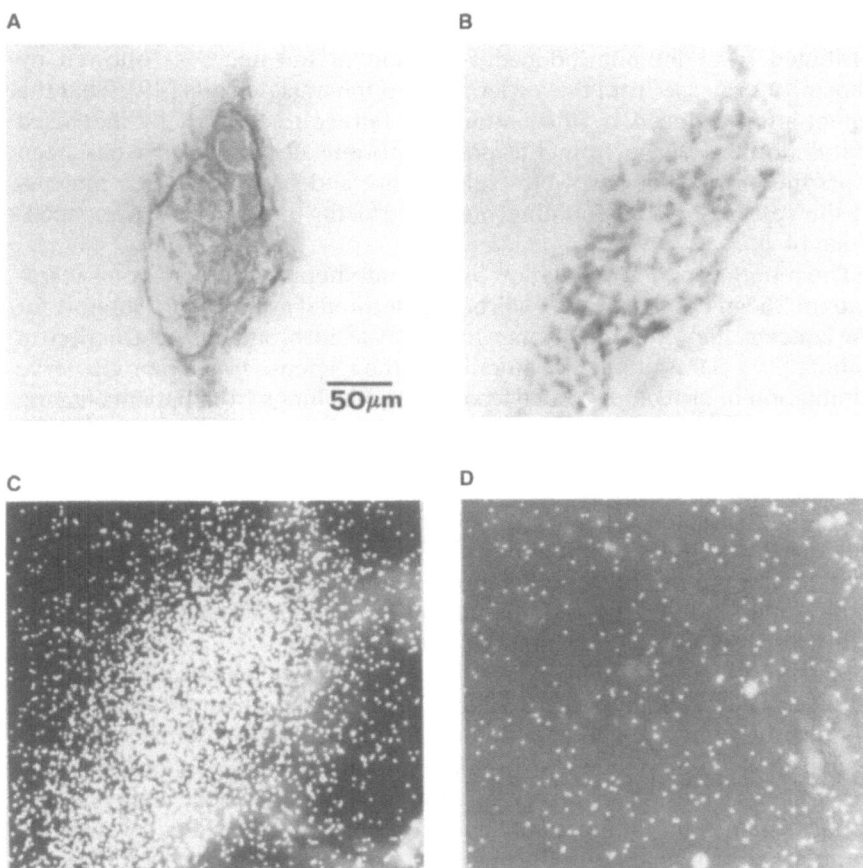

Figure 1. Autoradiographic localization of binding sites for [¹²⁵I] endothelin-2 on human airway parasympathetic ganglia. **A** shows the light-field view of ganglia stained for acetylcholinesterase; **B** is the adjacent section stained with methyl-green pyronin. "Total binding" is shown in plate **C** and "nonspecific binding" is seen in **D**. Originally published in *J Cardiovasc Pharmacol* (1991) 17: S206–209 (ref. [33]) and reproduced with permission.

receptors were identified (by radiolabelled endothelin-1 and endothelin-3 in the presence of the endothelin$_A$ receptor antagonist, BQ-123) on cell bodies, processes and varicosities of cholinergic neurons in primary cultures of guinea pig tracheal smooth muscle. Application of endothelin-1 and endothelin-3 to the culture dish resulted in large increase in intracellular calcium concentration in the cholinergic neurons. After stimulation of the neurons, a delayed contraction of the smooth muscle cells, which did not occur in the presence of the neurotoxin tetrodotoxin, was seen. As the application of the endothelins was not accompanied by an immediate rise in intracellular calcium in the muscle cells (as is associated with a direct contractile action on the muscle), the contraction of the muscle cells was attributed to endothelin-induced activation of the neurons followed by transmitter release from the varicosities to the muscle cells [49]. That the potentiation induced by the endothelin family is a result of increased acetylcholine release from the postganglionic nerve endings has been confirmed. In rat trachea, both endothelin-1 and sarafotoxin S6c increased the release of acetylcholine measured using the tritium efflux technique [44].

The report of neuromodulation by the endothelin$_B$ receptor agonist sarafotoxin S6c in human airways has been interpreted as providing support for the concept that endothelin may have a role in bronchial obstruction in asthma [39]. Sarafotoxin S6c augmented the response to cholinergic nerve stimulation in bronchial tissues from nine of 12 lungs from patients having resections for pulmonary carcinoma. In tissue from the remaining three lungs, no potentiation by sarafotoxin was seen, and this was attributed to the fact that the baseline nerve-induced contraction approximated the maximal contractile capacity of the tissue. This finding, coupled with the lack of effect of sarafotoxin S6c on the response to acetylcholine, provides additional support for a prejunctional enhancement of the response to nerve stimulation by the endothelins rather than a postjunctional increase in the responsiveness of the smooth muscle.

In contrast to the findings in other species, the neuromodulatory influence of the endothelins in ovine isolated trachea is an inhibitory one. Both sarafotoxin S6c and endothelin-1 markedly inhibited the contractile response to cholinergic nerve stimulation in this tissue. As in airway tissue from other species, the neuromodulation was found to occur via a prejunctional response as the postjunctional response to exogenous cholinergic stimulation was unaffected by the presence of sarafotoxin S6c [50]. As discussed by these investigators, this finding highlights the significant species differences in the activity of the endothelins in the airways.

The evidence to date is consistent with the observation that endothelin-induced neuromodulation in the airways occurs via activity at endothelin$_B$ receptors. This was first suggested by the finding of a greater potentiation of the parasympathetic response by endothelin-3 than endothelin-1 in rabbit isolated airways [40]. In addition, the endothelin$_A$ receptor antagonist, BQ-123, has been shown to have no effect upon the potentiation of the contractile response to electrical field stimulation induced by endothelin-1. This proposition is further supported by the observed abolition of potentiation after desensitization of the endothelin$_B$ receptor in mouse trachea [43] and the absence of effect of the endothelin$_B$ receptor-selective antagonist, BQ-788 on the endothelin-1 and sarafotoxin-induced attenuation in sheep trachealis [50]. The receptor responsible for the facilitation of the cholinergic response in human airways is also of the endothelin$_B$ type as this facilitation was absent in the presence of the endothelin$_B$ receptor-selective antagonist, BQ-788 39]. This confirmation of endothelin$_B$ receptor-mediated enhancement of neurotransmission by the endothelins was somewhat unexpected as autoradiographic studies have shown a predominance of endothelin$_A$ receptors on human bronchial neuronal tissues [51]. At present, there is little indication of a significant role for endothelin$_A$ receptor-mediated alteration of neurotransmission in bronchial tissue although, in the reports detailing the neuromodulatory effect of the endothelins in rabbit and rat trachea, it was shown that this receptor subtype was involved to some extent in facilitation of parasympathetic neurotransmission [44, 52].

The augmented response to nerve stimulation in airway tissue is reported to be prolonged, and is present at 45 [43] and 90 min [41] after the application of endothelin. This may be due to the binding characteristics of endothelin in neural tissue. In rat cerebellar neurons, the dissociation of endothelin from its binding site was slow, with approximately $90-95\%$ binding remaining after 14 h [53].

Other structures innervated by parasympathetic nerve endings within the airways include the submucosal glands and the airway vasculature. While endothelin has been shown to have a marked effect on parasympathetic nerve stimulation-induced airway contraction, the same does not appear to be true for neurogenic vascular leakage or neurally mediated mucus secretion, although this statement is based on ancillary findings in studies of endothelin in the airways. Aerosol administration of endothelin to guinea pigs did not result in alteration of lung permeability, although, as mentioned, this was a comment made in the report of a study examining the potential of endothelin to induce airways hyperresponsiveness in sensitized guinea pigs [54].

Endothelin-1, but not endothelin-2 or endothelin-3, has been shown to dose-dependently enhance the release of mucus glycoprotein from submucosal glands isolated from feline trachea. This increased release was not neurally mediated, as prior treatment with the muscarinic receptor antagonist atropine in addition to adrenoceptor antagonists did not affect release. It is therefore likely that this increase is due to postjunctional binding of endothelin to receptors on secretory cells within the mucus glands. Interestingly, when epithelial cells were present in these preparations, the application of endothelin resulted in decreased glandular secretion, implying that airway epithelial disruption and the absence of epithelial derived modulatory factors may result in induction of mucous secretion by endothelin [55].

In summary therefore, in airway tissue, the evidence is strongly suggestive of prolonged facilitation of parasympathetic neurotransmission by the endothelins. The mechanism responsible for this neuromodulation is likely to involve occupation of endothelin$_B$ receptors on postganglionic nerve endings causing a partial depolarization of the vessicle membrane, increased release of the neurotransmitter acetylcholine, and a significantly greater contractile response of the smooth muscle. While still effecting a neuromodulatory response in sheep trachealis, the modulation by endo-thelin-1 in this tissue results in inhibition of the contractile response to cholinergic nerve stimulation [50].

3.2.4. Non-adrenergic non-cholinergic nerves: During development, neuroblasts from the vagal nuclei establish myenteric ganglia in the gut and ganglia within the airway wall [56] and so it is likely that some of the mechanisms involved in neural control may be shared by the lung and the gut. The effect of the endothelins on inhibitory non-adrenergic non-cholinergic (iNANC) neural function in the lung has not been reported, therefore, an examination of endothelin-induced neuromodulation in the gastrointestinal tract may provide information as to the likely effects of the endothelins on pulmonary iNANC neurotransmission. When the iNANC nervous system is absent in the gut, as is the case in Hirschsprung's disease [57], the gastrointestinal smooth muscle develops uncontrolled myogenic activity resulting in spasm of the gastrointestinal system [58]. The finding that knockout of the gene for endothelin-3 in mice resulted in animals with an aganglionic colon and other features of Hirschsprung's disease therefore suggests that endothelin-3 may be important in the development of neural crest-derived structures, in particular non-adrenergic non-cholinergic nerves. This action is likely mediated by endothelin$_B$ receptors, as deletion of these receptors resulted in similar changes in the gut in homozygous animals [59].

Again, by examination of the effects of the endothelin isopeptides on neurotransmission in other organs, indications of the potential effects of these peptides on pulmonary excitatory non-adrenergic non-cholinergic (eNANC) neural functioning in the lung can be gleaned. Endothelin-1 potentiated, in a concentration-dependent fashion, the contractile response to eNANC nerve stimulation in rat urinary bladder [60], indicating that the endothelins may also have effects on this nervous system in the lung.

4. Neuromodulatory Substances Released by Endothelin

Endothelin may also have indirect effects on pulmonary neurotransmission by the release of neuroactive mediators. Evidence relating to this proposition includes the demonstration of relaxation of precontracted guinea pig isolated trachea in response to application of endothelin-1. This relaxation was due to the release of nitric oxide from epithelial cells, as the response was abolished by epithelial denudation and the presence of the nitric oxide synthetase inhibitor L-NMMA [61]. Not only has NO been shown to be the predominant neurotransmitter of the iNANC nervous system [62], it has also been touted as a neuromodulator. Cholinergic responsiveness in human [63] and guinea pig [62] isolated airways is augmented in the presence of a nitric oxide synthetase inhibitor, suggesting that NO released by

endothelin from epithelial cells has the potential to modulate airway cholinergic neurotransmission.

As a second example, endothelin has also been shown to release substance P from the hypothalamus [64], and from the spinal cord via activation of dihydropyridine-sensitive calcium channels [65]. As mentioned
above, substance P [30] has a significant neuromodulatory role in the pulmonary parasympathetic nervous system. It is possible therefore, that
endothelin may have indirect effects on neurotransmission in the airways
via release of neuromodulatory substances.

5. Summary

As a result of studies in the adrenergic, cholinergic and non-adrenergic
non-cholinergic nervous systems in various organs isolated from a number
of species, Wiklund and coworkers [66] concluded that the endothelins
(both endothelin-1 and endothelin-3) induced a general pattern of neuromodulation – a prejunctional inhibitory effect and a stimulatory postjunctional effect. On the basis of the studies to date in the respiratory system,
the actions of endothelins on the pulmonary parasympathetic nervous
system differ from this general pattern. While inducing a contractile
response postjunctionally, the endothelins are without effect on the post-
junctional response to acetylcholine. Moreover, the endothelins have a
significant prejunctional stimulatory effect in airway tissue. This effect is
likely to be mediated by the occupation of receptors of the endothelin$_B$ subtype and an enhanced release of the neurotransmitter acetylcholine.

In addition to contractile activity on the blood vessels and airways of the
lung, the endothelins therefore act on the neural pathways within the lung.
There are specific binding sites for the endothelins on pulmonary neural
structures. In addition, endothelin-1, endothelin-3, as well as the related
peptide sarafotoxin S6c, significantly modulate the response to parasympathetic nerve stimulation in isolated airways. The neuromodulation effect-
ed by the endothelin isopeptides in this tissue occurs at a site that is both
prejunctional and postganglionic and is not a result of a decrease in sympathetic input to the airway parasympathetic nervous system. As pulmonary cholinergic nerves supply submucosal glandular tissue and airway
vasculature and associated structures, in addition to airway smooth muscle,
this modulatory effect of endothelin on cholinergic neurotransmission may
have effects on additional structures relevant to normal and abnormal airway functioning.

Although there are a considerable number of studies reporting the activity of endothelin on neural functioning in the lung and other organs, a significant role for endothelin in the function or modulation of function of *in
vivo* pulmonary neurotransmission has yet to be demonstrated. The body of
literature pertaining to endothelin-induced alterations in neurotransmission

is based upon the application of exogenous synthetic endothelins or endothelin receptor agonists, to various preparations. To substantiate a neuromodulatory role for endothelin, demonstration of altered neural function in the presence of endothelin synthesis inhibition or receptor blockade is necessary.

References

1 Hiley CR (1989) Functional studies on endothelin catch up with molecular biology. *TiPS* 10: 47–49
2 Giaid A, Gibson SJ, Herrero MT, Gentleman S, Legon S, Yanagisawa M, Masaki T, Ibrahim NBN, Roberts GW, Rossi ML, Polak JM (1991) Topographical localisation of endothelin mRNA and peptide immunoreactivity in neurones of the human brain. *Histochemistry* 95: 303–314
3 Black PN, Ghatei MA, Takahashi K, Bretherton-Watt D, Krausz T, Dollery CT, Bloom SR (1989) Formation of endothelin by cultured airway epithelial cells. *FEBS Lett* 255: 129–132
4 Ninomiya H, Uchida Y, Ishii Y, Nomura A, Kameyama M, Saotome M, Endo T, Hasegawa S (1991) Endotoxin stimulates endothelin release from cultured epithelial cells of guinea-pig trachea. *Eur J Pharmacol* 203: 299–302
5 Franco-Cereceda A, Matran R, Lou Y-P, Lundberg JM (1990) Occurrence and effects of endothelin in guinea-pig cardiopulmonary tissue. *Acta Physiol Scand* 138: 539–547
6 Power RF, Wharton J, Zhao Y, Bloom SR, Polak JM (1989) Autoradiographic localization of endothelin-1 binding sites in the cardiovascular and respiratory systems. *J Cardiovasc Pharmacol* 13: S50–56
7 Yanagisawa M, Masaki T (1989) Molecular biology and biochemistry of the endothelins. *TiPS* 10: 374–378
8 Bloch KD, Eddy RL, Shows TB, Quertermous T (1989) cDNA cloning and chromosomal assignment of the gene encoding endothelin 3. *J Biol Chem* 264: 18156–18161
9 Lee ME, de la Monte SM, Ng S-C, Block KD, Quertermous ST (1990) Expression of the potent vasoconstrictor endothelin in the human central nervous system. *J Clin Invest* 86: 141–147
10 Shinkai M, Tsuruoka H, Wakabayashi S, Yamamoto Y, Takayanagi I (1994) Pre- and post-junctional actions of endothelin in the rat iris sphincter preparation. *Naunyn-Schmiedeberg's Arch Pharmacol* 350: 63–67
11 Krsmanovic LZ, Stojilkovic SS, Balla T, Al-Damluji S, Weiner RI, Catt KJ (1991) Receptors and neurosecretory actions of endothelin in hypothalamic neurons. *Proc Natl Acad Sci USA* 88: 11124–11128
12 Ambar I, Kloog Y, Schvartz I, Hazum E, Sokolovsky M (1989) Competitive interaction between endothelin and sarafotoxin: Binding and phosphoinositides hydrolysis in rat atria and brain. *Biochem Biophys Res Com* 158: 195–201
13 Jones CR, Hiley C, Pelton JT, Mohr M (1989) Autoradiographic visualization of the binding sites for [^{125}I] endothelin in rat and human brain. *Neurosci Lett* 97: 276–279
14 Fuxe K, Änggård E, Lundgren K, Cintra A, Agnati LF, Galton S, Vane J (1989) Localization of [^{125}I]endothelin-1 and [^{125}I]endothelin-3 binding sites in the rat brain. *Acta Physiol Scand* 137: 563–564
15 Niwa M, Kawaguchi T, Fujimoto M, Kataoka Y, Taniyama K (1991) Receptors for endothelin in the central nervous system. *J Cardiovasc Pharmacol* 17: S137–139
16 Rubanyi GM, Parker Botelho LH (1991) Endothelins. *FASEB J* 5: 2713–2720
17 Mosqueda-Garcia R, Inagami T, Appalsamy M, Sugiura M, Robertson RM (1992) Endothelin as a neuropeptide. Cardiovascular effects in the brainstem of normotensive rats. *Circ Res* 72: 20–35
18 Franco-Cereceda A, Rydh M, Lou Y-P, Dalsgaard C-J, Lundberg JM (1991) Endothelin as a putative sensory neuropeptide in the guinea pig: different roperties in comparison with calcitonin gene-related peptide. *Regul Pept* 32: 253–265

19 Hemsén A, Lundberg JM (1991) Presence of endothejlin-1 and endothelin-3 in peripheral tissues and the central nervous system of the pig. *Regul Pept* 36: 71–83

20 Richardson JB, Beland J (1976) Nonadrenergic inhibitory nervous system in human airways. *J Appl Physiol* 41: 764–771

21 Richardson JB, Ferguson CC (1979) Neuromuscular structure and function in the airways. *Federation Proc* 38: 202–208

22 Wiklund NP, Öhlén A, Cederqvist B (1989) Adrenergic neuromodulation by endothelin in the guinea pig pulmonary artery. *Neurosci Letters* 101: 269–273

23 Han S-P, Kneupfer MM, Trapani AJ, Fok KF, Westfall TC (1990) Endothelin and sarafotoxin S6b have similar vasoconstrictor effects and postsynaptically mediated mechanisms. *Eur J Pharmacol* 177: 29–34

24 Tabuchi Y, Nakamaru M, Rakugi H, Nagano M, Mikami H, Ogihara T (1989) Endothelin inhibits presynaptic adrenergic neurotransmission in rat mesenteric artery. *Biochem Biophys Res Comm* 161: 803–808

25 Takeuchi A, Kimura T, Satoh S (1992) Enhancement by endothelin-1 of the release of catecholamines from the canine adrenal gland in response to splanchnic nerve stimulation. *CEPP* 19: 633–666

26 La M, Wong-Dusting HK, Rand MJ (1991) Endothelin-1 enhances responses to sympathetic nerve stimulation in endothelium-denuded and endothelium-intact rabbit ear arteries. *Neurochem Int* 18: 465–469

27 McKay KO, Armour CL, Black JL (1996) Endothelin receptors and activity differ in human, dog and rabbit lung. *Am J Physiol* 270: L37–43

28 Lundberg JM, Fahrenkrug J, Hökfelt T, Martling CR, Larsson O, Takemoto K, Ånggärd A (1984) Coexistence of peptide histidine isoleucine (PHI) and VIP in nerves regulating blood flow and bronchial smooth muscle tone in various mammals including man. Peptides 5: 593–606

29 Martin JG, Wang A, Zacour M, Biggs DF (1990) The effects of vasoactive intestinal polypeptide on cholinergic neurotransmission in isolated innervated guinea pig tracheal preparations. *Respir Physiol* 79: 111–122

30 Hall AK, Barnes PJ, Meldrum LA, Maclagan J (1989) Facilitation by tachykinins of neurotransmission in guinea-pig pulmonary parasympathetic nerves. *Br J Pharmacol* 97: 274–280

31 Turner NC, Power RF, Polak JM, Bloom SR, Dollery CT (1989) Endothelin-induced contractions of tracheal smooth muscle and identification of specific endothelin binding sites in the trachea of the rat. *Br J Pharmacol* 98: 361–366

32 Franco-Cereceda A, Matran R, Lou Y-P, Lundberg JM (1990) Occurrence and effects of endothelin in guinea-pig cardiopulmonary tissue. *Acta Physiol Scand* 138: 539–547

33 McKay KO, Black JL, Diment LM, Armour CL (1991) Functional and autoradiographic studies of endothelin-1 and endothelin-2 in human bronchi, pulmonary arteries, and airway parasympathetic ganglia. *J Cardiovasc Pharmacol* 17: S206–209

34 Kobayashi M, Ihara M, Sato N, Saeki T, Ozaki S, Ikemoto F, Yano M (1993) A novel ligand, [^{125}I]BQ-3030, reveals the localization of endothelin ETB receptors. *Eur J Pharmacol* 235: 95–100

35 Wiklund NP, Wiklund CU, Öhlén A, Gustafsson LE (1989) Cholinergic neuromodulation by endothelin in guinea pig ileum. *Neurosci Lett* 101: 342–346

36 Mattera GG, Eglezos A, Renzetti AR, Mizrahi J (1993) Comparison of the cardiovascular and neural activity of endothelin-1, -2, -3 and respective proendothelins: effects of phosphoramidon and thiorphan. *Br J Pharmacol* 110: 331–337

37 Seldeslagh KA, Lauweryns JM (1993) Sarafotoxin expression in the bronchopulmonary tract: immunohistochemical occurrence and colocalization with endothelins. *Histochemistry* 100: 257–263

38 Lee H-K, Leikauf GD, Sperelakis N (1990) Electromechanical effects of endothelin on ferret bronchial and tracheal smooth muscle. *J Appl Physiol* 68: 417–420

39 Fernandes LB, Henry PJ, Rigby PJ, Goldie RG (1996) Endothelin$_B$ (ETB) receptor-activated potentiation of cholinergic nerve-mediated contraction in human bronchus. *Br J Pharmacol* 118: 1873–1874

40 McKay KO, Armour CL, Black JL (1993) Endothelin-3 increases transmission in the rabbit pulmonary parasympathetic nervous system. *J Cardiovasc Pharmacol* 22: S181–184

41 McKay KO, Black JL, Armour CL (1992) Phosphoramidon potentiates the contractile response to endothelin-3, but not to endothelin-1 in isolated airway tissue. *Br J Pharmacol* 105: 929–932

42 Black JL, Carey D, Johnson PRA, Armour CL (1995) Potentiation of neural responses via the endothelin$_B$ receptor in rabbit but not human bronchus. *Am Rev Respir Dis* 151: A824

43 Henry PJ, Goldie RG (1995) Potentiation by endothelin-1 of cholinergic nerve-mediated contractions in mouse trachea via activation of ET$_B$ receptors. *Br J Pharmacol* 114: 563–569

44 Knott PG, Fernandes LB, Henry PJ, Goldie RG (1996) Influence of endothelin-1 on cholinergic nerve-mediated contractions and acetylcholine release in rat isolated tracheal smooth muscle. *J Pharmacol Exp Ther* 279: 1142–1147

45 Nishimura T, Krier J, Akasu T (1991) Endothelin causes prolonged inhibition of nicotinic transmission in feline colonic parasympathetic ganglia. *Am J Physiol* 261: G628–633

46 Nishimura T, Akasu T, Krier J (1991) Endothelin modulates calcium channel current in neurones of rabbit pelvic parasympathetic ganglia. *Br J Pharmacol* 103: 1242–1250

47 Skoogh BE, Ullman A (1991) Modulation of cholinergic neurotransmission to the airways. *Am Rev Respir Dis* 143: 1427–1428

48 Kushiku K, Ohjimi H, Yamada H, Tokunaga R, Furukawa T (1991) Endothelin-3 inhibits ganglionic transmission at preganglionic sites through activation of endogenous thromboxane A$_2$ production in dog cardiac sympathetic ganglia. *J Cardiovasc Pharmacol* 17: S197–199

49 Takimoto M, Inui T, Okada T, Urade Y (1993) Contraction of smooth muscle by activation of endothelin receptors on autonomic neurons. *FEBS Letters* 324: 277–282

50 Henry PJ, Goldie RG (1995) Endothelin-1, via activation of an ET$_B$ receptor, inhibits cholinergic nerve-mediated contractions in sheep trachea. *J Cardiovasc Pharmacol* 26: S117–119

51 Knott PG, D'Aprile AC, Henry PJ, Hay DWP, Goldie RG (1995) Receptors for endothelin-1 in asthmatic human peripheral lung. *Br J Pharmacol* 114: 1–3

52 Yoneyama T, Hori M, Tanaka T, Matsuda Y, Karaki H (1995) Endothelin ET$_A$ and ET$_B$ receptors facilitating parasympathetic neurotransmission in the rabbit trachea. *J Pharmacol Exp Ther* 275: 1084–1089

53 Hiley CR, Jones CR, Pelton JT, Miller RC (1990) Binding of [^{125}I]-endothelin-1 to rat cerebellar homogenates and its interactions with some analogues. *Br J Pharmacol* 101: 319–324

54 Lagente V, Touvey C, Mencia-Huerta J-M, Chabrier P-E, Braquet P (1990) Bronchopulmonary effects of endothelin. *Clin Exptl Allergy* 20: 343–348

55 Shimura S, Ishihara H, Satoh M, Masuda T, Nagaki N, Sasaki H, Takishima T (1992) Endothelin regulation of mucus glycoprotein secretion from feline tracheal submucosal glands. *Am J Physiol* 262: L208–213

56 Richardson JB, Ferguson CC (1979) Neuromuscular structure and function in the airways. *Federation Proc* 38: 202–208

57 Frigo GM, del Tacca M, Lecchini S, Crema A (1973) Some observations on the intrinsic nervous mechanism in Hirschsprung's disease. *Gut* 14: 35–40

58 Hukuhara T, Kotami S, Sato G (1961) Effects of obstruction of intramural ganglion cells on colon motility: possible genesis of congenital megacolon. *Jap J Physiol* 11: 635–640

59 Battistini B, Botting R, Warner TD (1995) Endothelin: a knockout in London. *TiPS* 16: 217–222

60 Donoso MV, Salas C, Sepulveda G, Lewin J, Fournier A, Huidobro-Toro JP (1994) Involvement of ETA receptors in the facilitation by endothelin-1 of non-adrenergic non-cholinergic transmission in the rat urinary bladder. *Br J Pharmacol* 111: 473–482

61 Filep JG, Battistini B, Sirois P (1993) Induction by endothelin-1 of epithelium-dependent relaxation *in vitro*: role for nitric oxide. *Br J Pharmacol* 109: 637–644

62 Belvisi MG, Stretton CD, Barnes PJ (1992) Nitric oxide is the endogenous neurotransmitter of bronchodilator nerves in human airways. *Eur J Pharmacol* 210: 221–222

63 Ward JK, Fox AJ, Miura M, Tadjkarimi S, Yacoub MH, Barnes PJ, Belvisi MG (1993) Modulation of cholinergic neurotransmission by nitric oxide and VIP in human airways *in vitro*. *J Clin Invest* 92: 736–742

64 Calvo JJ, Gonzalez R, Carvalho LF, Takahashi K, Kanse SM, Hart GR, Ghatei MA, Bloom SR (1990) Release of substance P from rat hypothalamus and pituitary by endothelin. *Endocrinology* 126: 2288–2295

65 Yoshizawa T, Kimura S, Kanazawa I, Uchiyama Y, Yanagisawa M, Masaki T (1989) Endothelin localizes in the dorsal horn and acts on the spinal neurones: possible involvement of
   dyhydropyridine-sensitive calcium channels and substance P release. *Neuroscience Letters*
   102: 179–184
66 Wiklund NP, Öhlén A, Wiklund CU, Cederqvist B, Hedqvist P, Gustafsson LE (1989) Neuromuscular actions of endothelin on smooth, cardiac and skeletal muscle from guinea-pig,
   rat and rabbit. *Acta Physiol Scand* 137: 399–407

Pulmonary Actions of the Endothelins
ed. by R. G. Goldie and D. W. P. Hay
© 1999 Birkhäuser Verlag Basel/Switzerland

CHAPTER 12
Role of Endothelin in Pulmonary Hypertension

Bernadette Raffestin[2], Saadia Eddahibi[1] and Serge Adnot[1]

[1] Département de physiologie et INSERM U492, Hôpital Henri Mondor, 94010 Créteil, France
[2] Département de Physiologie, Hôpital Ambroise Paré, Université René Descartes, 92104 Boulogne, France

1   Introduction
1.1 Pulmonary Hypertension – Clinical Aspects
1.2 Vascular Remodeling Associated with Pulmonary Hypertension
1.3 Treatment of Pulmonary Hypertension
1.4 Animal Models of Pulmonary Hypertension
2   Vasoactive Effects of Endothelin in the Pulmonary Circulation,
     Changes Associated with Pulmonary Hypertension
2.1 Vasoreactivity of the Normal Pulmonary Circulation to Endothelin
2.2 Changes in Vasoreactivity to ET During Pulmonary Hypertension
3   Mitogenic Effects of ET-1
4   Evidence for Increased Production of ET-1 During Pulmonary Hypertension
4.1 Experimental Models of Pulmonary Hypertension
4.2 Pulmonary Hypertension in Humans
5   Treatment with ET-Receptor Antagonists in Pulmonary Hypertension
5.1 Hypoxic Pulmonary Hypertension
5.2 Monocrotaline-induced Pulmonary Hypertension
5.3 Neonatal Pulmonary Hypertension
5.4 Congestive Heart Failure
6   Conclusion
     References

1. Introduction

1.1. Pulmonary Hypertension – Clinical Aspects

Pulmonary hypertension, defined as a mean pulmonary artery pressure above 20 mmHg or a pulmonary artery systolic pressure above 30 mmHg, is characterized by an increase in pulmonary vascular resistance that impedes ejection of blood by the right ventricle. Right ventricular hypertrophy occurs in response to the prolonged pressure overload. As a consequence, cardiac output may not increase appropriately with exercise or may even be markedly decreased at rest.

Hypoxic pulmonary hypertension is the most common form of pulmonary hypertension. Of variable severity, it affects patients of all ages and occurs in subjects who live at high altitude and in patients with lung diseases such as chronic obstructive pulmonary disease. Although pulmonary hypertension is usually moderate in most patients with stable hypoxic chronic obstructive pulmonary disease and progresses slowly, its presence infers a poor prognosis [1]. Cases of pulmonary hypertension related to hypoxic episodes during upper airway obstruction from hypertrophied tonsils and adenoids have also been reported in children. In patients with lung diseases, factors other than hypoxia may contribute to aggravate pulmonary hypertension: increased blood viscosity due to polycythemia, disappearance of small arteries and capillaries, occlusion of arteries by thrombi or inflammatory infiltrates.

Pulmonary hypertension may also be the consequence of recurrent thromboembolic disease or increased left atrial pressure secondary to acquired heart diseases. In congestive heart failure, the severity of pulmonary hypertension is an important determinant of the prognosis.

Congenital heart disease with increased pulmonary blood flow due to left to right cardiac shunt may also lead, if surgical correction is not performed early enough, to progressive and irreversible pulmonary vascular remodeling causing severe pulmonary hypertension and secondary right to left shunting of blood through the heart defect, the so-called Eisenmenger syndrome. In such cases, corrective surgery is usually not advised.

The idiopathic variant of pulmonary hypertension, called primary pulmonary hypertension, is a rare and often fatal disease for which there is no identifiable cause, although associations have been found with portal hypertension, infection with the human immunodeficiency virus, and use of appetite suppressants. Cases of familial primary pulmonary hypertension have also been reported, suggesting that a genetic susceptibility may also be involved.

In newborns, failure to achieve the normal decline of pulmonary vascular resistance after birth can lead to persistent pulmonary hypertension. Neonatal pulmonary hypertension is often associated with severe parenchymal lung disease which further complicates the clinical course and response to treatment.

1.2. Vascular Remodeling Associated with Pulmonary Hypertension

Normal pulmonary vessels are characterized by their low tone and sparcity of smooth muscle cells. The transition of the larger elastic pulmonary arteries to muscular arteries takes place in adults at a diameter between 1000 and 500 μm. In infants, this transition is usually at a smaller diameter. These muscular arteries have normally a thin media (in the range of 5% of the external diameter) and a relatively wide lumen. More distally, arteries with a diameter ranging from 100 to 70 μm tend to lose their muscular coat gradually, so that small arteries become partially muscular and non muscular. Functional and structural abnormalities contribute to the increased pulmonary vascular resistance seen in secondary and primary pulmonary hypertension. Abnormal vasoconstriction, increased interaction between circulating blood cells and vessels walls promoting blood clotting, proliferation of smooth muscle cells, and increased extracellular matrix production, all contribute to progressive obliteration of the distal pulmonary arteries.

The most common and earliest lesion in pulmonary hypertension is medial hypertrophy with thickening of muscular arteries and progressive muscularization of normally partially muscular and non muscular distal arteries [2]. At a later stage, cellular intimal proliferation develops with appearance of the onion-skin lesions due to concentric-laminar intimal fibrosis which may progressively lead to obliteration of vessels. Plexiform lesions, consisting of dilated arterial segments with destruction of their wall and a plexus of capillary-like channels are characteristic of advanced pulmonary hypertension, either primary or secondary to congenital heart disease with a shunt.

1.3. Treatment of Pulmonary Hypertension

In hypoxic patients suffering from chronic obstructive pulmonary disease, long-term administration of oxygen improves survival and quality of life although this benefit is not clearly linked to changes in pulmonary hemodynamics. In patients with severe pulmonary hypertension, survival has been improved using medical therapies such as oral anticoagulation [3], vasodilator drugs, including calcium channel blockers [4, 5]. The major limitation of vasodilators is their effect on the systemic circulation. Patients with severe pulmonary hypertension have low systemic blood pressure due to reduced cardiac output. A further lowering with a vasodilator drug may compromise perfusion of vital organs and cause right to left shunting of blood through a heart defect or a patent foramen ovale. Furthermore, the ability of vasodilators or calcium channel blockers to decrease pulmonary vascular resistance varies with the severity of the disease and the initial benefit obtained in some patients may not be sustained. The progressive downhill course of patients suffering from severe pulmonary hypertension has prompted the pursuit of more aggressive therapy such as continuous infusion of prostacyclin or heartlung transplantation. Although prostacyclin improves survival and quality of life, this treatment is costly and not always well tolerated [6]. Inhaled prostacyclin and nitric oxide have been tried in a few patients, but larger studies are needed to document their effectiveness [7, 8]. In transplant recipients, one year survival rate does not exceed 65% [9]. Due to limitations and failure of treatments, novel therapies are there fore badly needed in patients with severe pulmonary hypertension.

1.4 Animal Models of Pulmonary Hypertension

Real advances in therapy come usually from the understanding of the disease process. Pulmonary hypertension is no exception and various animal models of pulmonary hypertension have been developed to obtain a better understanding of its causal mechanisms.

Among models of pulmonary hypertension, the hypoxic model has been widely used. Continued exposure of rats to hypoxia, either hypobaric or normobaric (by reducing atmospheric pressure to 0.5 atm or oxygen fraction to 10%) results in a steady increase in pulmonary artery pressure which is related in part to a vasoconstrictive component but also to polycythemia and structural remodeling of pulmonary arteries [10]. Indeed, thickness of the media from muscularized arteries increases (Fig. 1) and new muscle cells arise in the distal precapillary arteries, so that after 2 weeks of hypoxia, 70 to 80% of arteries at the level of alveolar duct and about 50% of those at the alveolar wall level contain some muscle, whereas only 30% at alveolar duct and 10% at alveolar wall levels do in normoxic rats [11]. In this model, migration and proliferation of immature smooth muscle cells are not observed in the intima. After return to normoxic conditions, recovery from pulmonary hypertension and regression of muscularization occurs progressively over a period of several weeks [12].

Pulmonary hypertension can be produced in several species by administration of monocrotaline (Fig. 1), a pyrrolizidine alkaloid derived from the plant crotalaria spectabilis. The alkaloid is converted in the liver to an active pyrrole that causes early and diffuse injury to the endothelium in various organs. After administration of the drug either daily in the food, or acutely as a single subcutaneous injection, increased pulmonary vascular permeability precedes pulmonary hypertension and medial thickening which are first apparent 2 weeks after onset of treatment [13, 14]. By 4 weeks, pulmonary artery pressure has doubled and occlusion of numerous precapillary arteries is observed concomitantly with abnormalities of lung parenchyma, including interstitial edema and increased numbers of mast

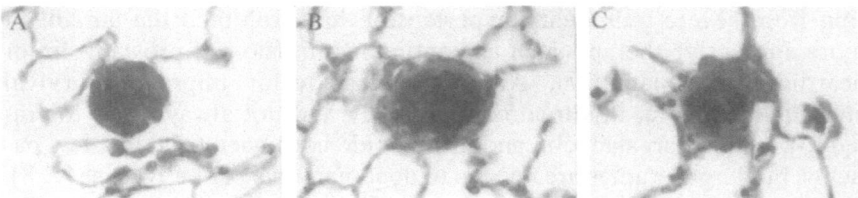

Figure 1. Histological aspect of pulmonary arterioles (diameter: 40–50 µm) in normal rats (A), chronically hypoxic rats (B), and monocrotaline-treated rats (C). Vascular remodeling results from medial hypertrophy in hypoxic rats and from both intimal and medial hypertrophy in monocrotaline-treated rats.

cells and leukocytes. After a single subcutaneous injection of monocrotaline (60 mg), no recovery of pulmonary hypertension is observed and mortality was around 33% at 4 weeks in one of our studies.

There have been numerous attempts to reproduce in animals the structural alterations of the pulmonary vasculature observed in the clinical conditions of increased blood flow due to congenital heart lesions. However, when a surgical anastomosis of the aorta to the pulmonary artery is performed in animals, only the very young, such as the piglet, develop mild pulmonary hypertension with muscularization of distal vessels [15].

Animal models have also been used to study structural and functional changes associated with perinatal pulmonary hypertension. Much of our understanding of the adaptation of the pulmonary circulation at birth comes from work done in fetal, chronically instrumented lamb. At birth, there is a sudden transition for gas exchange from the placenta to the lung. This requires a dramatic increase in blood flow and decrease in resistance in the pulmonary circulation. Failure of this adaptation leads to the clinical syndrome of persistent pulmonary hypertension of the newborn characterized by an elevated pulmonary vascular resistance, resulting in right to left shunting across the foramen ovale and ductus arteriosus with severe hypoxemia [16]. These features can be experimentally reproduced in the lamb. Ligation of the ductus arteriosus in late-gestation fetus causes marked elevation of intrauterine pulmonary artery pressure, right ventricular hypertrophy, hypertensive lung structural changes and failure to achieve the normal decline in pulmonary artery pressure at birth [17].

There is accumulating evidence from these experimental models that the pulmonary vascular endothelial cell plays a central part in the pathogenesis of secondary and primary pulmonary hypertension. Under normal circumstances, the endothelial cell elaborates a variety of substances such as prostacyclin and nitric oxide that promote vasodilation, inhibit platelets adhesion and activation. Endothelial cells have also an important role in preventing the propensity of blood to clot. They express at their surface thrombomodulin, a protein necessary for activation of proteine C by thrombin, synthesize heparan-sulfate and tissue plasminogen activator inhibitor. Hypoxia, oxygen free radicals, increased shear forces and pressure, all environmental stimuli that arise in the setting of pulmonary hypertension, may promote endothelial dysfunction.

Among the various substances produced by endothelial cells, endothelin (ET) induces vasoconstriction and proliferation of smooth muscle cells. Increased production of endothelin may play a part in the pathogenesis of pulmonary hypertension since ET-1 release by endothelial cells is stimulated by thrombin, hypoxia and various inflammatory mediators, all stimuli which can be encouffered in the various conditions associated with pulmonary hypertension [18]. In the past few years, numerous studies have been devoted to examining the role of ET in the regulation of pulmonary vascular tone and development of pulmonary vascular remodeling associated with pulmonary hypertension. Their results suggest new therapeutic approaches involving control of the ET system. For example, development of ET receptor antagonists may lead to a significant advance in the treatment of pulmonary hypertension.

2. Vasoactive Effects of Endothelin in the Pulmonary Circulation, Changes Associated with Pulmonary Hypertension

Both ET-1 and ET-3 are abundantly expressed in the lung but only ET-1 is synthetized by endothelial cells [19, 20]. Expression of preproET-1 mRNA and ET-1 immunostaining has also been demonstrated under basal conditions in smooth muscle cells isolated from normoxic rats pulmonary arteries [21].

2.1. Vasoreactivity of the Normal Pulmonary Circulation to Endothelin

The vasoactive effects of ET-1 are mediated by receptor subtypes ET_A and ET_B; there is controversial evidence for the existence of ET_B receptor subtypes, ET_{B1} and ET_{B2} (see chapter 5) [22, 23]. ET_A is found on vascular smooth muscle cells and mediates the major part of vasoconstriction elicited by ET. The putative ET_{B1} receptor is found on endothelial cells and mediates ET-1-induced vasodilation, whereas the purported ET_{B2} receptor subtype present on smooth muscle cells, is also involved in vasoconstriction [24]. Results of studies in various species are consistent with this dual vascular action of ET-1 on the pulmonary circulation. In intact animals, short-term ET-1 infusion has been shown to reverse the increased pulmonary pressor response to acute hypoxia [25, 26]. Similarly, low doses of ET-1 caused vasodilation in rat isolated lungs preconstricted with acute hypoxia [27] or infusion of the endoperoxide analog U46619 (Fig. 2) [11]. This effect is mimicked by the selective ET_B receptor agonist IRL-1620, unaffected by the selective ET_A receptor antagonist BQ-123 and blocked by the combined ET_A- and ET_B- receptor antagonist, bosentan. In contrast, higher doses of ET-1 or IRL-1620 cause vasoconstriction in rat isolated lung, an effect which is more marked with ET-1 than with similar doses of IRL-1620. This effect is potentiated by NO synthesis inhibitors, suggesting that NO released in response to ET-1 may attenuate the pulmonary vasoconstrictor effect of ET-1. Pretreatment with bosentan inhibits vasoconstriction to both ET-1 and IRL-1620, whereas pretreatment with BQ-123 inhibitis only vasoconstriction induced by ET-1.

NORMOXIA

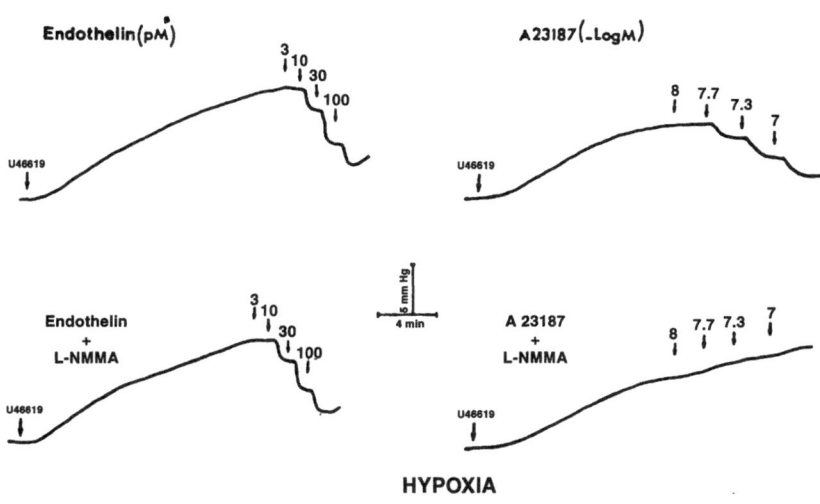

HYPOXIA

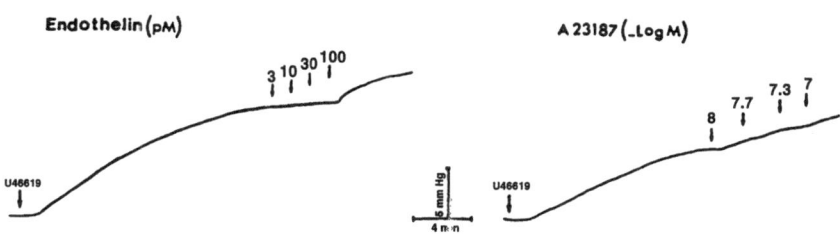

Figure 2. Vasodilator responses to ET-1 and ionophore A23187 in isolated lungs from control normoxic (normoxia) and hypoxic rats (hypoxia). Baseline pulmonary artery pressure was elevated by a continuous infusion of U46619 (50 pmol/min). ET-1 or ionophore A23187 were administered into the perfusate as a 50 μl bolus of increasing doses in the absence or presence of NG-monomethyl-L-arginine (L-NMMA) (5×10^{-4} M) as indicated. Each tracing represents one of five separate experiments. (Reproduced with permission from J Cardiovasc Pharmacol, 17 (Suppl. 7): S358–S361, 1991.)

2.2. Changes in Vasoreactivity to ET During Pulmonary Hypertension

Important changes in vasoreactivity to ET-1 have been found in various experimental models of pulmonary hypertension. Pulmonary hypertension caused by exposure of rats to chronic hypoxia results in loss of the vasodilator effect of ET-1 (Fig. 2) and in enhancement of its vasoconstrictor effects [28]. Vasodilation in response to a selective ET_B receptor agonist is also abolished, whereas the vasoconstrictor effect of the agonist is potentiated [11]. Interestingly, the vasoconstrictor actions of ET-1 and IRL-1620 in lungs from chronically hypoxic rats are almost completely inhibited in the presence of bosentan, a combined ET_A and ET_B receptor antagonist, whereas they are only partly inhibited by BQ-123. This finding suggests that if endogenous ET-1 contributes to the increased pulmonary vascular tone observed in chronic hypoxic pulmonary hypertension, this effect may be mediated by both ET_A and ET_B receptors. These alterations in the vaso-reactivity to ET-1 do not appear due to a change in the ratio of ET_B to ET_A receptor number. No quantitative changes in binding of ET-1 to ET_A or ET_B receptors have been detected in the lungs of chronically hypoxic rats [28]. Evidence of increased ET_A and ET_B-receptor mRNA levels in the lungs from chronically hypoxic rats also does not support a decrease in the num-ber of either of the ET receptor subtypes [29]. However, the alterations in functional responses to ET seen in chronic hypoxic pulmonary hyperten-sion may be partly ascribable to a change in the distribution of ET_B re-ceptors between smooth muscle and endothelial cells. In a recent study, Barber et al. demonstrated upregulation of ET_B receptors in arterial smooth muscle in response to a chronic increase in blood flow produced by a surgical fistula [30].

Loss of the vasodilator effect of ET-1 during chronic hypoxic pulmonary hypertension may also be related to functional alterations distal to the endothelial ET_B receptor. In the normal pulmonary circulation, transduc-tion mechanisms in response to activation of endothelial ET_B receptors have been shown to vary according to species, age and type of vessels. In isolated large-conduit pulmonary arteries from normal adult rats, vasodila-tion in response to ET-3 is abrogated by endothelial denudation or inhibi-tion of NO synthesis [31], whereas ET-1 fails to produce vasodilation [27]. In the pulmonary vasculature from fetal and newborn animals, the vasodi-lator action of ET-1 has also been shown to be primarily mediated by NO formation [32–34]. In contrast, in isolated lung preparations from adult cats and rats, vasodilation to ET-1 or ET-3 remains unaffected by NO synthesis inhibitors (Fig. 2) but is reduced by glybenclamide, an antagonist of ATP-sensitive K^+ channels and by tetraethylammonium, a non selective K^+-channel blocker, suggesting that in adults of these species, vasodilation of pulmonary resistance vessels in response to ET-1 is not mediated by NO [27, 28, 35]. Vasodilation in response to the K^+ channel opener pinacidil is preserved in lungs from chronically hypoxic rats, indicating that loss of the vasodilatory effect of ET may not be related to unresponsiveness of smooth-muscle K channels.

Apart from nitric oxide, endothelium may produce an unidentified factor that activates K^+ channels of the underlying smooth muscle, causing it to hyperpolarize, and consequently to relax. It is conceivable that endothelin-induced vasodilation is mediated by the release of an endothelium-depend-ent hyperpolarization factor   that its loss is among the functional ab-normalities of the pulmonary endothelium associated with chronic hypoxic pulmonary hypertension. Supporting this possibility is our finding that vasodilation induced by the endothelium-dependent agents acetylcholine and ionophore A23187 is also abolished in lungs from chronically hypoxic rats (Fig. 2) [36, 37]. Moreover, compared with lungs from normoxic rats, the vasoconstrictor response to ET-1 is enhanced, although it is not sensitive to potentiation by NO synthesis inhibitors. This last finding suggests that decreased ability of the endothelium to release NO may contribute to the increased pulmonary vasoconstrictor response to ET-1 seen in pulmonary hypertension related to chronic hypoxia.

Pulmonary hypertension induced by ductus ligation in the fetal lamb during late gestation has also been shown to be associated with altered reactivity to ET-1. In this experimental model, ET_B receptor-mediated vasodilation was diminished and ET_A receptor-mediated vasoconstriction enhanced, concomitantly with impairment of endothelium-dependent vasodilation and endogenous NO activity [38]. Similarly, in lambs with pulmonary hypertension secondary to a surgical creation of a shunt between ascending aorta and main pulmonary artery during late gestation, loss of vasodilation and increased vasoconstriction in response to ET-1 has also been observed [39].

3. Mitogenic Effects of ET-1

In addition to its vasoconstricting properties, ET-1 has been shown to be mitogenic or co-mitogenic with serum or other growth factors for pulmonary vascular smooth muscle cells and fibroblasts from various species [40]. Growth promoting effects of ET have been described in smooth muscle cells from the bovine and porcine pulmonary circulation [41, 42]. Moreover, ET-1 demonstrated significant co-mitogenic effects in presence of trace serum on smooth muscle cells isolated from human proximal pulmonary arteries [43]. The concentrations required for threshold and maximal growth responses were similar to those required to elicit contractile responses. This growth-promoting effect may be mediated by ET_A receptor activation, since it is inhibited by the selective ET_A antagonist, BQ-123, and is not seen with the selective ET_B agonist, S6c. The exact mechanism leading from receptor activation to growth enhancement is unknown. Binding of ET-1 to the ET_A receptor stimulates phospholipase C through activation of a G protein. However ET-1 has also been shown to increase tyrosyl-phosphorylation and to stimulate the activity of mitogen-activated protein kinases in smooth muscle cells (44), an effect which is abolished in presence of the ET_A antagonist BQ-123 [45]. In addition, ET-1 has been shown to induce the expression of other growth factors, such as platelet-derived growth factor-A and transforming growth factor-β in cultured vascular smooth muscle cells [42, 46]. Because of its ability to mimic its two features, pulmonary vasospasm as well as proliferation of pulmonary vascular smooth muscle, ET-1 may contribute significantly to the development of pulmonary hypertension.

4. Evidence for Increased Production of ET-1 During Pulmonary Hypertension

4.1. Experimental Models of Pulmonary Hypertension

Numerous reports suggest increased production of ET-1 in humans and animals with pulmonary hypertension. Hypoxic pulmonary hypertension has been shown to be associated with increased ET-1 synthesis. There are several reports of increased plasma levels of ET-1 in rats exposed to acute [47] and chronic hypoxia [29, 48]. Furthermore, concomitant increases in gene transcript levels for ET-1 and the ET_A and ET_B receptors, as well as increases in ET-1, have been reported in lung tissue from chronically hypoxic rats [29]. Stelzner et al. also found an increase in preproET-1 mRNA and ET-1 peptide in whole lung homogenates and pulmonary artery smooth muscle cells from fawn-hooded rats exposed at Denver altitude [21, 49]. This strain develops pulmonary hypertension in an environment with mild reduction of oxygen tension whereas control Sprague-Dawley exhibit normal pulmonary artery pressure at this altitude. However, these authors failed to find evidence of an increased ET-1 production in the lung of Sprague-Dawley rats with pulmonary hypertension in response to a 3-week exposure to severe hypobaric hypoxia. Hypoxia has been shown to stimulate ET-1 gene expression and ET-1 synthesis in cultured endothelial cells [50]. Moreover, diminished nitric oxide production in endothelial cells, another event associated with hypoxia [51], may also lead to an increase in ET-1 production [52]. Increased production of ET-1 in the lung of rats with hypoxic pulmonary hypertension may therefore be a direct consequence of hypoxia. However, ET-1 production by endothelial cells has also been shown to increase when cells are cultured on a membrane exposed to cyclical elongation [53]. We therefore cannot rule out that increased production of ET-1 represents an adaptative process to the increased wall stress and shear stress associated with pulmonary hypertension. The fact that preproendothelin-1 mRNA is increased in the lung and staining for ET-1 is more intense in the pulmonary artery endothelial cells from rats with pulmonary hypertension secondary to congestive heart failure lends further support to this idea [54]. Experimental data also suggest that endogenous ET-1 may be invoved in perinatal pulmonary hypertension. Increased ET-1 content has been demonstrated in the lung tissue from neonate lambs with pulmonary hypertension secondary to partial ligation of the ductus arteriosus [38].

4.2. Pulmonary Hypertension in Humans

Numerous reports also suggest an increased production of ET-1 in patients with pulmonary hypertension. Elevated immunoreactive plasma levels have been found in patients with pulmonary hypertension either primary [55] or secondary, newborns with persistent pulmonary hypertension [56], children with congenital heart disease [57], adults with respiratory distress syndrome [58, 59] and patients with chronic congestive heart failure [60]. Since Dupuis et al. have recently demonstrated that the normal human lung is a site for both clearance and production of ET-1 [61], the possibility remains that reduced lung clearance may contribute to the increase in ET-1 plasma levels seen in some cases of pulmonary hypertension. However, in contrast to normal subjects, patients with pulmonary hypertension due to various causes exhibit higher levels of ET-1 in arterial than in venous plasma, suggesting increased production of ET-1 by the lung [62]. Moreover, increased ET-1-like immunoreactivity has been observed in the vascular endothelium of lung tissue obtained both from patients with primary and from patients with secondary pulmonary hypertension [63]. Expression of ET-1 was found in the vessels that are most severely affected by the morphologic abnormalities of pulmonary hypertension, and pre-proendothelin mRNA as demonstrated by *in situ* hybridization closely matched the distribution of ET-1 like immunoreactivity, providing strong evidence for an increase in the local production of ET-1 by the pulmonary vascular endothelium in pulmonary hypertension. The greatest degree of immunostaining occurred in the endothelium of elastic and muscular pulmonary arteries, which also demonstrated severe medial thickening and intimal proliferation.

5. Treatment with ET-Receptor Antagonists in Pulmonary Hypertension

Reports of the beneficial effect of chronic treatment with specific ET-receptor antagonists in various models of pulmonary hypertension gives further support to the view that ET-1 plays an important role in the initiation or progression of pulmonary vessel lesions.

5.1. Hypoxic Pulmonary Hypertension

In intact animals from various species, pretreatment with the ET_A receptor antagonist BQ-123 or compounds with mixed antagonist properties for ET_A and ET_B receptors produced a profound decrease in the pulmonary pressor response to acute hypoxia [48, 64, 65]. During exposure of rats to hypoxia for 2 weeks, continuous iv infusion of the ET_A receptor antagonist

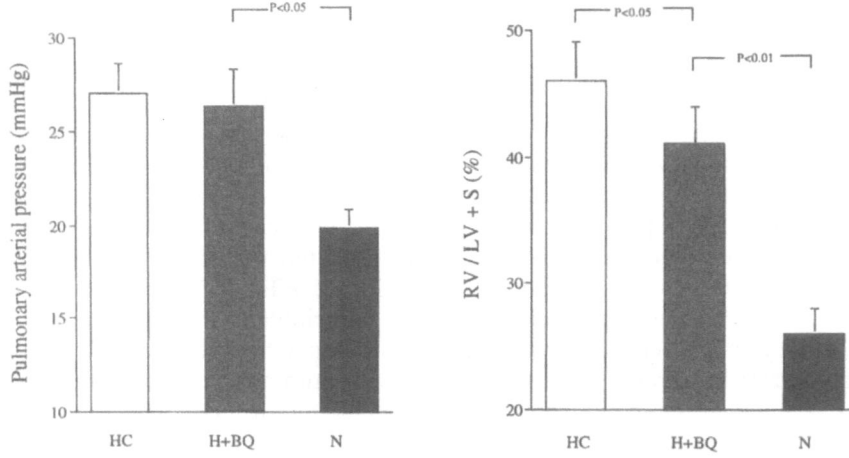

Figure 3. Pulmonary arterial pressure, ratio of right ventricle-to-left ventricle + septum weight (RV/LV + S) and muscularization of pulmonary arteries in rats exposed to hypoxia (10% O_2) for 15 days and treated with BQ-123 (H + BQ), or vehicle (HC). Values are mean ± SE. Values in untreated rats exposed to normoxia (N) are also indicated.

BQ-123 (0.4 mg/h) (Fig. 3), or oral administration of a non-peptide compound with mixed antagonist properties for ET_A and ET_B receptors, bosentan (100 mg/kg/day), attenuated the development of pulmonary hypertension and reduced the degree of right ventricular hypertrophy without altering systemic artery pressure [11, 66, 67]. Treatment with BQ-123, A-127722 (10 mg/kg/day), a non peptide ET_A receptor antagonist [68] or bosentan also attenuated the increase in wall thickness of distal pulmonary arteries in response to chronic hypoxia in these animals. Attenuation of the muscularization of arteries at both the alveolar duct and alveolar wall levels was also clearly demonstrated with BQ-123 (Fig. 4) or bosentan. Furthermore, when BQ-123, A-127722 or bosentan treatment was started 2 weeks after the beginning of exposure to hypoxia and continued for 4 weeks, progression of pulmonary hypertension and right heart hypertrophy was prevented and significant reversal of pulmonary vascular remodeling occurred despite continued hypoxic exposure.

5.2. Monocrotaline-induced Pulmonary Hypertension

In dogs, acute infusion of FR-13917, an ET_A receptor antagonist, 4 and 8 weeks after a single injection of dehydromonocrotaline via the right atrium, causes a significant reduction of pulmonary hypertension [69]. In rats treated with monocrotaline, chronic treatment with bosentan has been shown to attenuate the progressive increase in pulmonary artery pressure

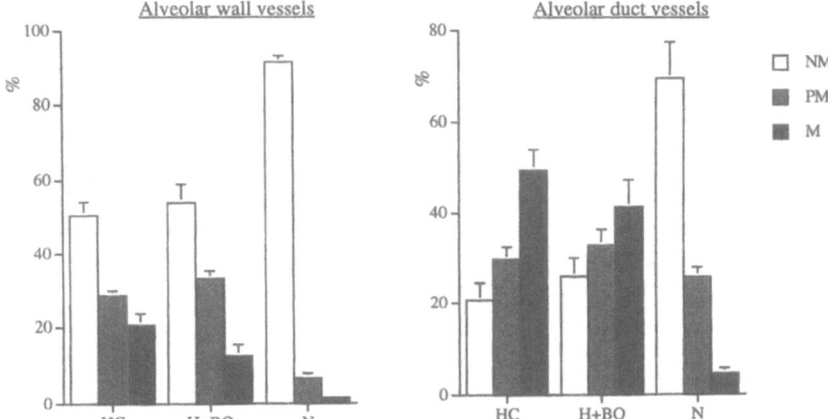

Figure 4. Bar graphs showing the percentage of muscularized (M), partially muscularized (PM), and non muscularized (NM) arteries at the alveolar duct and alveolar wall levels in rats exposed to hypoxia (10% O_2) for 15 days and treated with BQ123 (H + BQ, n = 10), or vehicle (HC, n = 9). Values in untreated rats exposed to normoxia are also indicated (N, n = 10). Results are expressed as means ± S.E. Degree of muscularization of distal pulmonary arteries was less severe at both the alveolar duct and alveolar wall levels in the BQ123 than in the vehicle-treated group.

[70], and chronic treatment with BQ-123 decreased pulmonary artery medial thickening and the development of right ventricular hypertrophy [71]. In the latter study, the beneficial effect of treatment with the ET_A receptor antagonist was observed despite the fact that ET-1 synthesis was decreased in the lungs from the monocrotaline-treated animals. However, plasma levels of ET-1 were increased. This finding is perhaps related to the increased synthesis of the peptide observed in the heart and kidneys, although reduced pulmonary clearance of ET-1 may also have played a role. In the monocrotaline model, ET-1 may contribute to the pulmonary vascular leak and lung inflammation which precede pulmonary hypertension. Indeed, reports have shown that ET-1 increased microvascular permeability, an effect mediated by PAF [72] and stimulated neutrophil adhesion to endothelial cells by promoting the expression of adhesion molecules on the neutrophil surface [73]. It is noteworthy that PAF receptor antagonists have also been shown to attenuate pulmonary vascular leak and development of pulmonary hypertension in monocrotaline-treated rats [74].

5.3. Neonatal Pulmonary Hypertension

Beneficial effects of BQ-123 have also been shown in neonatal pulmonary hypertension induced by partial ligation of the ductus arteriosus in fetal lambs [75]. Acute infusion of BQ-123 after delivery potentiated the decrease in pulmonary vascular resistance caused by ventilation with 100% O_2 and NO, suggesting that ET_A-receptor mediated vasoconstriction may contribute to the residual high tone of the pulmonary vessels in neonatal pulmonary hypertension. Moreover, when the BQ-123 infusion was started at the time of ductus ligation and continued until delivery, fetal pulmonary artery pressure was lower, vasodilation at delivery was more noticeable, and right ventricular hypertrophy and distal muscularization of small pulmonary arteries were less marked. These findings support the hypothesis that endogenous ET-1 may contribute to pulmonary vascular remodeling and failure of the pulmonary circulation to adapt at birth.

5.4. Congestive Heart Failure

Some experimental data also suggest that treatment with ET receptor antagonists reduces pulmonary hypertension secondary to congestive heart failure. In rats, infusion with BQ-123, started the day after coronary ligation and continued for 2 weeks, significantly reduced right ventricular systolic pressure and right ventricular hypertrophy without affecting left ventricular hemodynamics [54]. Since treatment with ET receptor antagonists has been reported to prevent cardiac hypertrophy *in vivo*, it is possible that a longer treatment would also ameliorate left ventricular remodeling or left heart failure [76].

6. Conclusion

Studies demonstrating that blocking the action of endogenous ET protects against the development of this disease not only suggest that ET may play an important role in the development of pulmonary hypertension but also provide a rationale for a new therapeutic approach for pulmonary hypertension due to various causes. Prior to the development of ET receptor antagonists, the pharmacological tools used to reduce progression of pulmonary artery hypertension were usually designed to potentiate vasodilator mechanisms rather than to inhibit vasoconstrictor and growth-stimulating processes. Results obtained with ET receptor antagonists in experimental pulmonary hypertension, suggest hope that ET receptor antagonists or ET production inhibitors may specifically reduce the progression of vascular narrowing in patients with pulmonary hypertension, particularly those with primary pulmonary hypertension. However, clinical studies need to be conducted to confirm the significance of these preclinical findings. It has been shown clearly that ET_B receptor activa ion can mediate pulmonary vasoconstriction, but the effects of selective ET_B receptor blockade, in particular in humans, is not currently known. We actually do not know whether selective ET_A receptor or non-selective ET_A and ET_B receptor antagonism is the best therapeutic strategy in pulmonary hypertension. Characterization of actions of ET receptor antagonists outside the pulmonary circulation, in particular in the kidney, may also be important.

References

1 MacNee W (1994) Pathophysiology of cor pulmonale in chronic obstructive pulmonary disease. *Am Respir Crit Care Med* 150: 833–852
2 Heath D, Edwards JE (1958) The pathology of hypertensive pulmonary vascular changes: a description of six grades of structural changes in the pulmonary arteries with special reference to congenital cardiac defects. *Circulation* 18: 533–547
3 Fuster V, Steele PM, Edwards WD, Gersh BJ, McGoon MD, Frye RL (1984) Primary pulmonary hypertension: natural history and the importance of thrombosis. *Circulation* 70: 580–587
4 Rich S, Kaufman E, Levy PS (1992) The effect of high doses of calcium-channel blockers on survival in primary pulmonary hypertension. *N Eng J Med* 327: 76–81
5 Rich S (1995) Calcium channel blockers for the treatment of primary pulmonary hypertension. *Chest* 29: 252–254
6 Barst RJ, Rubin LJ, Long WA, McGoon MD, Rich S, Badesh DB, Groves BM, Tapson VF, Bourge RC, Brundage BH, Koerner SK, Langleben D, Keller CA, Murali S, Uretsky BF, Clayton LM, Jöbsis MM, Blackburn SD, Shortino JW, Crow JW (1996) A comparison of continuous intravenous epoprostenol (prostacyclin) with conventional therapy for primary pulmonary hypertension. *N Engl Med* 334: 296–301
7 Olschewski H, Walmwrath D, Schermuly R, Ghofrani A, Grimminger F, Seeger W (1996) Aerosolized prostacyclin and iloprost in severe pulmonary hypertension. *Ann Intern Med* 124: 820–824
8 Channick RN, Newhart JW, Johnson FW, Williams PJ, Auger WR, Fedullo PF, Moser KM (1996) Pulsed delivery of inhaled nitric oxide to patients with primary pulmonary hypertension. *Chest* 109: 1545–1549
9 Trulock EP (1997) Lung transplantation. *Am J Respir Crit Care Med* 155: 789–818
10 Rabinovitch M, Gamble M, Nadas AS, Miettinen OS, Reid L (1979) Rat pulmonary circulation after chronic hypoxia: hemodynamic and structural features. *Am J Physiol* 236: H818–827
11 Eddahibi S, Raffestin B, Clozel M, Levame M, Adnot S (1995) Protection from pulmonary hypertension with an orally active endothelin receptor antagonist in hypoxic rats. *Am J Physiol* 268: H828–H835
12 Fried R, Reid L (1984) Early recovery from hypoxic pulmonary hypertension: a structural and functional study. *J Appl Physiol: Respir Environ Exercise Physiol* 57: 1247–1253
13 Meyrick B, Gamble W, Reid L (1980) Development of crotalaria pulmonary hypertension: hemodynamic and structural study. *Am J Physiol* 239: H692–H702
14 Sugita T, Hyers TM, Dauber IM, Wagner WW, McMurtry IF, Reeves JT (1983) Lung vessel leak precedes right ventricular hypertrophy in monocrotaline-treated rats. *J Appl Physiol: Respir Environ Exercise Physiol* 54: 371–374
15 Reid L, Geggel R, Fried R, Langleben D (1986) Anatomy of pulmonary hypertensive states. In: EH Bergofsky (ed.) *Abnormal Pulmonary Circulation*. Churchill Livingstone, New York, 221–263
16 Levin DL, Heyman MA, Kitterman JA, Gregory GA, Phibbs RH, Rudolph AM (1976) Persistent pulmonary hypertension of the newborn infant. *J Pediatr* 89: 626–630
17 Morin FC (1989) Ligating the ductus arteriosus before birth causes persistent pulmonary hypertension in the newborn lamb. *Pediatr Res* 25: 245–250
18 Golden CL, Kohler JP, Hick HS, Visner GA (1995) Effects of vasoactive and inflammatory mediators on endothelin-1 expression in pulmonary endothelial cells. *Am J Respir Cell Mol Biol* 12: 503–512
19 Firth JD, Ratcliffe PJ (1992) Organ distribution of the three rat endothelin messenger RNAs and the effects of ischemia on renal gene expression. *J Clin Invest* 90: 1023–1031

20 Marciniak SJ, Plumpton C, Barker PJ, Huskisson NS, Davenport AP (1992) Localization of immunoreactive endothelin and proendothelin in the human lung. *Pulm Pharmacol* 5: 175–182

21 Zamora MR, Stelzner TJ, Webb S, Panos RJ, Ruff L, Dempsey EC (1996) Overexpression of endothelin-1 and enhanced growth of pulmonary artery smooth muscle cells from fawn-hooded rats. *Am J Physiol* 270: L101–L109

22 Arai H, Hori S, Aramori I, Ohkubo H, Nakanishi S (1990) Cloning and expression of a cDNA encoding and endothelin receptor. *Nature Lond* 348: 730–732

23 Sakurai T, Yanagisawa M, Takuwa Y, Miyazaki H, Kimura S, Goto K, Masaki T (1990) Cloning of cDNA encoding a non-isopeptide-selective subtype of the endothelin receptor. *Nature Lond* 348: 732–735

24 Warner TD (1993) Characterization of endothelin synthetic pathways and receptor subtypes: physiological and pathophysiological implications. *Eur Heart J* 14: 42–47

25 Deleuze PH, Adnot S, Shiiya N, Thoraval FR, Eddahibi S, Braquet P, Chabrier PE, Loisance DY (1992) Endothelin dilates bovine pulmonary circulation and reverses hypoxic pulmonary vasoconstriction. *J Cardiovasc Pharmacol* 19: 354–360

26 Liska J, Holm P, Owall A, Franco-Cereceda A (1995) Endothelin infusion reduces hypoxic pulmonary hypertension in pigs *in vivo*. *Acta Physiol Scand* 154: 489–498

27 Hasunuma K, Rodman DM, O'Brien RF, Mc Murtry IF (1990) Endothelin 1 causes pulmonary vasodilation in rats. *Am J Physiol* 259: H48–H54

28 Eddahibi S, Springall D, Mannan M, Carville C, Chabrier PE, Levame M, Raffestin B, Polak J, Adnot S (1993) Dilator effect of endothelins in pulmonary circulation: changes associated with chronic hypoxia. *Am J Physiol* 265: L571–L580

29 Li H, Chen S-J, Chen Y-F, Cheng Meng Q, Durand J, Oparil S, Elton TS (1994) Enhanced endothelin-1 and endothelin receptor gene expression in chronic hypoxia. *J Appl Physiol* 77: 1451–1459

30 Barber DA, Michener SR, Ziesmer SC, Miller VM (1996) Chronic increases in blood flow upregulate endothelin-B receptors in arterial smooth muscle. *Am J Physiol* 270: H65–H71

31 Carville C, Raffestin B, Eddahibi S, Blouquit Y, Adnot S (1993) Loss of endothelium-dependent relaxation of large pulmonary arteries from rats exposed to chronic hypoxia: effects of *in vitro* and *in vivo* supplementation with L-arginine. *J Cardiovasc Pharmacol* 22: 889–896

32 Perreault T, De Marte J (1993) Maturational changes in endothelium-derived relaxations in newborn piglet pulmonary circulation. *Am J Physiol* 264: H302–309

33 Pinheiro JM, Malik AB (1993) Mechanisms of endothelin-1 induced pulmonary vasodilation in neonatal pigs. *J Physiol* (Lond) 469: 739–752

34 Ivy DD, Kinsella JP, Abman S (1994) Physiologic characterization of endothelin A and B receptor activity in the ovine fetal pulmonary circulation. *J Clin Invest* 93: 2141–2148

35 Lippton HL, Cohen GA, McMurtry IF, Hyman AL (1991) Pulmonary vasodilation to endothelin isopeptides *in vivo* is mediated by potassium channel activation. *J Appl Physiol* 70: 947–952

36 Adnot S, Raffestin B, Eddahibi S, Braquet P, Chabrier P-E (1991) Loss of endothelium-dependent relaxant activity in the pulmonary circulation of rats exposed to chronic hypoxia. *J Clin Invest* 87: 155–162

37 Eddahibi S, Adnot S, Carville, C Blouquit Y, Raffestin B (1992) L-arginine restores endothelium-dependent relaxation in pulmonary circulation of chronically hypoxic rats. *Am J Physiol* 263: L194–L200

38. Ivy DD, Ziegler JW, Fox JJ, Kinsella JP, Abman SH (1996) Chronic intrauterine pulmonary hypertension alters endothelin receptor activity in the ovine fetal lung. *Pediatr Res* 39: 435–442

39 Wong J, Reddy VM, Hendriks-Munoz K, Liddicoat JR, Gerrets R, Fineman JR (1995) Endothelin vasoactive responses in lambs with pulmonary hypertension and increased pulmonary blood flow. *Am J Physiol* 269: H1965–1972

40 Peacock AJ, Dawes KE, Shock A, Gray AJ, Reeves JT, Laurent GJ (1992) Endothelin-1 and endothelin-3 induce chemotaxis and replication of pulmonary artery fibroblasts. *Am J Respir Cell Mol Biol* 7: 492–499

41 Hassoun PM, Thappa V, Landman MJ, Fanburg BL (1992) Endothelin-1: mitogenic activity on pulmonary artery smooth muscle cells and release from hypoxic endothelial cells. *Proc Soc Exp Biol Med* 199: 165–170

42 Janakidevi K, Fisher MA, Del Vecchio PJ, Tiruppathi C, Figge J, Malik AB (1992) Endo-thelin-1 stimulates DNA synthesis and proliferation of pulmonary artery smooth muscle cells. *Am J Physiol* 263: C1295–1301

43 Zamora MA, Dempsey EC, Walchak SJ, Stelzner TJ (1993) BQ123, an ETA receptor an-tagonist, inhibits endothelin-1-mediated proliferation of human pulmonary artery smooth muscle cells. *Am J Respir Cell Mol Biol* 9: 429–433

44 Koide W, Kawahara Y, Tsuda T, Ishida Y, Shii K, Yokohama M (1992) Endothelin-1 stimu-lates tyrosine phoshorylation and the activities of two mitogen-activated protein kinases in cultured vascular smooth muscle cells. *J Hypertension* 10: 1173–1182

45 Guo X, Okada K, Fujita N, Ishikawa S, Komatsu N, Saito T (1996) Inhibitory effect of BQ-123 on endothelin – stimulated mitogen-activated protein kinase and cell growth of rat vas-cular smooth muscle cells. *Hypertens Res* 19: 23–30

46 Hahn AWA, Resink TJ, Scott-Burden T, Powell J, Dohi Y, Bühler FR (1990) Stimulation of endothelin mRNA and secretion in rat vascular smooth muscle cells: a novel autocrine function. *Cell Regul* 1: 649–659

47 Shirakami G, Nakao K, Saito Y, Magaribuchi T, Jougasaki M, Mukoyama M, Arai H, Hosoda K, Suga S-I, Ogawa Y, Yamada T, Mori K, Imura H (1991) Acute pulmonary alveolar hy-poxia increases lung and plasma endothelin-1 levels in conscious rats. *Life Sci* 48: 969–976

48 Oparil S, Chen S-J, Cheng Meng C, Elton TS, Mitsuo Y, CY-F (1995) Endothelin-A recep-tor antagonist prevents acute hypoxia-induced pulmonary hypertension in the rat. *Am J Phy-siol* 268: L95–L100

49 Stelzner TJ, O'Brien RF, Yanagisawa M, Sakurai T, Sato K, Webb S, Zamora M, McMurtry IF, Fisher JH (1992) Increased lung endothelin-1 production in rats with idiopathic pulmo-nary hypertension. *Am J Physiol* 262: L614–620

50 Kourembanas S, Marsden PA, McQuillan LP, Faller DV (1991) Hypoxia induces endothelin gene expression and secretion in cultured human endothelium. *J Clin Invest* 88: 1054–1057

51 Liao JK, Zulueta JJ, Yu F-S, Peng H-B, Cote CG, Hassoun PM (1995) Regulation of bovine endothelial constitutive nitric oxide synthase by oxygen. *J Clin Invest* 96: 2661–2666

52. Kourembanas S, McQuillan LP, Leung GK Faller DV (1993) Nitric oxide regulates the expression of vasoconstrictors and growth factors by vascular endothelium under both normoxia and hypoxia. *J Clin Invest* 92: 99–104

53 Carosi JA, Eskin SG, Mcintire LV (1992) Cyclical strain effects on production of vasoactive materials in cultured endothelial cells. *J Cell Physiol* 151: 29–36

54 Sakai S, Miyauchi T, Sakurai T, Yamaguchi I, Kobayashi M, Goto K, Sugishita Y (1996) Pulmonary hypertension caused by congestive heart failure is ameliorated by long-term application of an endothelin receptor antagonist. *J Am Coll Cardiol* 28: 1580–1588

55 Cacoub P, Dorent R, Maistre G, Nataf P, Carayon A, Piette JC, Godeau P, Cabrol C, Grand-jbakhch I (1993) Endothelin-1 in primary pulmonary hypertension and the Eisenmenger syndrome. *Am J Cardiol* 71: 448–450

56 Rosenberg AA, Kennaugh J, Koppenhafer SL, Loomis M, Abman SH (1993) Elevated immunoreactive endothelin-1 levels in infants with persistent pulmonary hypertension of the newborn. *J Pediatr* 123: 109–114

57 Allen S, Chatfield BA, Koppenhafer SL, Schaffer MS, Wolfe RR, Abman SH (1993) Circulating immunoreactive endothelin-1 levels in children with pulmonary hyperten-sion: association with acute hypoxic pulmonary vasoreactivity. *Am Rev Respir Dis* 148: 519–522

58 Druml W, Steltzer H, Waldhäusl W, Lenz K, Hammerle A, Vierhapper H, Gasic S, Wagner OF (1993) Endothelin-1 in adult respiratory distress syndrome. *Am Rev Respir Dis* 148: 1169–1173

59 Langleben D, Demarchie M, Laporta D, Spanier A, Schlesinger RD, Stewart DJ (1993) Endothelin-1 in acute lung injury and the adult respiratory distress syndrome. *Am Rev Respir Dis* 148: 1646–1650

60 Cody RJ, Haas JL, Capers Q, Kelley R (1992) Plasma endothelin correlates with the extent of pulmonary hypertension in patients with chronic congestive heart failure. *Circulation* 85: 504–509

61 Dupuis J, Stewart DJ, Cernacek P, Gosselin G (1996) Human pulmonary circulation is an important site for both clearance and production of endothelin-1. *Circulation* 94: 1578–1584

62 Stewart DJ, Levy RD, Cernack P, Langleben D (1991) Increased plasma endothelin-1 in pul-monary hypertension: marker or mediator of disease? *Ann Intern Med* 114: 464–469

63 Giaid A, Yanagisawa M, Langleben D, Levy R, Shennib H, Kimura S, Masaki T, Duguid WP, Stewart DJ (1993) Expression of endothelin-1 in the lungs of patients with pulmonary hypertension. *N Engl J Med* 328: 1732–1739

64 Holm P, Liska J, Clozel M, Franco-Cereceda A (1996) The endothelin antagonist bosentan: hemodynamic effects during normoxia and hypoxic pulmonary hypertension. *J Thorac Cardiovasc Surg* 112: 890–897

65 Willette RN, Ohlstein EH, Mitchell MP, Sauermelch CF, Beck GR, Luttmann MA, Hay DW (1997) Nonpeptide endothelin receptor antagonists. VIII: attenuation of acute hypoxia-induced pulmonary hypertension in the dog. *J Pharmacol Exp Ther* 280: 695–701

66 DiCarlo VS, Chen S-J, Cheng Meng Q, Durand J, Yano M, Chen Y-F, Oparil S (1995) ETA-receptor antagonist prevents and reverses chronic hypoxia induced pulmonary hypertension in rat. *Am J Physiol* 269: L690–697

67 Chen SJ, Chen YF, Meng QC, Durand J, Dicarlo VS, Oparil S (1995) The endothelin-receptor antagonist bosentan prevents and reverses hypoxia-induced pulmonary hypertension in the rat. *J Appl Physiol* 79: 2122–2131

68 Chen SJ, Chen YF, Opgenorth TJ, Wessale JL, Meng QC, Durand J, Dicarlo VS, Oparil S (1997) The orally active nonpeptide endothelin A-receptor antagonist A-127722 prevents and reverses hypoxias -induced pulmonary hypertension and pulmonary vascular remodeling in Sprague-Dawley rats. *J Cardiovasc Pharmacol* 29: 713–725

69 Okada M, Yamashita C, Okada M, Okada K (1995) Role of endothelin-1 in beagles with dehydromonocrotaline-induced pulmonary hypertension. *Circulation* 92: 114–119

70 Hess P, Clozel M, Clozel JP (1996) Telemetry monitoring of pulmonary arterial pressure in freely moving rats. *J Appl Physiol* 81: 1027–1032

71 Miyauchi T, Yorikane R, Sakai S, Sakurai T, Okada M, Nishikibe M, Yano M, Yamaguchi M, Sugishita Y, Goto K (1993) Contribution of endogenous endothelin-1 to the progression of cardiopulmonary alterations in rats with monocrotaline-induced pulmonary hypertension. *Circ Res* 73: 887–897

72 Filep JG, Sirois MG, Rousseau A, Fournier A, Sirois P (1991) Effects of endothelin-1 on vascular permeability in the conscious rat: interactions with platelet-activating factor. *Br J Pharmacol* 104: 797–804

73 Farre AL, Riesco A, Espinoza G, Digiuni E, Cernadas MR, Alvarez V, Monton M, Rivas M, Callego MJ, Edigo J, Casado S, Caramelo C (1993) Effect of endothelin-1 on neutrophil adhesion to endothelial cells and perfused heart. *Circulation* 88: 1166–1171

74 Ono S, Voelkel NF (1991) PAF antagonists inhibit monocrotaline-induced lung injury and pulmonary hypertension. *J Appl Physiol* 71: 2483–2492

75 Ivy DD, Parker TA, Ziegler JW, Galan HL, Kinsella JP, Tuder RM, Abman SH (1997) Prolonged endothelin A receptor blockade attenuates chronic pulmonary hypertension in the ovine fetus. *J Clin Invest* 99: 1179–1186

76 Ito H, Hiroe M, Hirata Y (1994) Endothelin ETA receptor antagonist blocks cardiac hypertrophy provoked by hemodynamic overload. *Circulation* 2198–2203

Pulmonary Actions of the Endothelins
ed. by R. G. Goldie and D. W. P. Hay
© 1999 Birkhäuser Verlag Basel/Switzerland

CHAPTER 13
Endothelin as a Putative Mediator in Asthma and Other Respiratory Diseases

Anthony E. Redington

Dept. of Respiration Medicine, 2nd floor, Thomas Guy House, Guy's Hospital, London SE1 9RT, UK

1   Introduction
2   Endothelin Receptors
3   Biological Properties of Endothelin in the Pulmonary System
3.1   Airway Smooth Muscle Contractile Responses
3.2   Neuromodulation
3.3   Bronchial Hyperresponsiveness
3.4   Microvascular Permeability
3.5   Mucociliary Function
3.6   Proinflammatory Properties
3.7   Mitogenic and Fibrogenic Properties
4   Endothelin as a Putative Mediator in Asthma
4.1   Pathophysiology of Asthma
4.2   Expression of Endothelin in Asthma
4.2.1   Peripheral blood
4.2.2   Bronchial tissue
4.2.3   Bronchoalveolar lavage fluid
4.2.4   Induced sputum
4.3   Endothelin Receptor Expression in Asthma
4.4   ECE Expression in Asthma
4.5   Conclusions
5   Endothelin as a Putative Mediator in Other Respiratory Diseases
5.1   Chronic Obstructive Pulmonary Disease (COPD)
5.2   Idiopathic Pulmonary Fibrosis
5.3   Systemic Sclerosis
5.4   Lung Cancer
5.5   Acute Lung Injury
5.6   Sickle Cell Disease
5.7   Behçet's Syndrome
5.8   Pulmonary Embolism
        References

1. Introduction

In the decade since the discovery of endothelins (ETs), there has been a rapid expansion of knowledge concerning their molecular biology, biological effects, and their expression in health and disease. Significant attention has focused on the lung, and a role for ETs in the pathophysiology of a number of respiratory diseases has been proposed. Research into ETs is now moving into a new phase as a result of the availability of potent and selective ET receptor antagonists and synthesis inhibitors. Over the next few years, clinical studies using these agents should provide more definitive answers to many of the questions that have been raised. It would seem timely, therefore, to review our present understanding regarding the biological properties of ETs in the pulmonary system and the current evidence to implicate these peptides as mediators in asthma and in other respiratory disorders.

2. Endothelin Receptors

The biological effects of the ETs are mediated *via* specific membrane receptors, which are members of the superfamily of G-protein coupled, seven transmembrane spanning receptors. Thus, although in the original report describing the isolation and characterization of porcine ET-1, it was speculated that the mature peptide may be an endogenous ligand for dihydropyridine-sensitive calcium channels [1], the presence of saturable high affinity binding sites, using $[^{125}I]Tyr^{13}$-ET-1, was demonstrated on cultured rat vascular smooth muscle cells. Binding was unaffected by other known vasoconstrictors or by calcium channel blockers, such as nicardipine, verapamil, and diltiazem, suggesting the existence of a novel receptor for ET-1 [2]. Autoradiographic studies indicated that these binding sites are widely distributed, not only to blood vessels, but also in the parenchyma of various organs including the lung [3].

There is evidence from pharmacological, biochemical, and molecular biological studies for ET receptor subtypes (see Chapter 2). For example, the first indication of the existence of more than one type of ET receptor came from the report that the conserved C-terminal hexapeptide ET[16–21] is a full contractile agonist on guinea pig bronchus but is practically devoid of activity on rat thoracic aorta [4]. It was proposed, therefore, that there are two receptor types and these were termed ET_A (for aorta) and ET_B (for bronchus) [4]. Subsequently, the cloning of two ET receptors from several tissues, including lung, was demonstrated [5, 6]; a receptor designated ET_A which preferentially bound ET-1, and an ET_B receptor which bound all three isopeptides with equal affinity.

The issue as to whether additional ET receptor subtypes exist is a controversial one. Thus, although pharmacologically there is considerable evidence for an expansion of the currently ET receptor classification, this is not supported by most of the molecular biological studies. For example, Southern blot analysis of human genomic DNA under conditions of low stringency suggests the existence of only two ET receptor genes [7], corresponding to the cloned ET_A and ET_B receptors. Additional receptor types may therefore represent either novel proteins with low sequence homology to the cloned receptors or alternatively post-translational modifications of these molecules. Support for the latter suggestion derives from a recent study in which the PD142893-sensitive vasodilator response of the aorta and the PD142893-resistant contractile response of the gastric fundus, putative ET_{B1}- and ET_{B2}-mediated responses respectively, were shown both to be completely absent in the ET_B receptor knockout mouse [8].

Evidence has been presented to suggest the existence of an ET-3-selective type of mammalian ET receptor which has been termed ET_C [9, 10]. In 1993, Karne et al. [11] reported the cloning of a receptor from *Xenopus laevis* dermal melanophores, a cell type in which ET-3 induces pigment dispersion with a potency more than 400-fold greater than that of either ET-1 or ET-2. However, when this receptor was expressed in transfected cells it did not display the level of ET-3 selectivity evident in the biological response of pigment dispersion. To date, no mammalian protein corresponding to an ET_C receptor has been cloned.

3. Biological Properties of Endothelin in the Pulmonary System

3.1. Airway Smooth Muscle Contractile Responses

As outlined below, and more extensively in other chapters, ET-1 produces a variety of effects in the pulmonary system, which may be relevant to a pathophysiological role. Shortly after its discovery, ET-1 was identified as a potent airway smooth muscle (ASM) contractile agonist. In an initial report, Uchida et al. [12] found that ET-1 induced a concentration-dependent contractile response in isolated guinea pig tracheal strips with an EC_{50} of 0.53 nM, making it one of – if not the most – potent bronchoconstrictors recognized. The ability of ET-1 to induce ASM contraction was confirmed by other investigators in guinea pigs and other species [16].

Notably, several groups have demonstrated the ability of ET-1 to act as a potent contractile agonist in preparations of isolated *human* large (diameter 3–15 mm) [15, 17–20] and, more recently, small (0.5–1 mm) bronchi [21]. In these studies, EC_{50} values of 10–30 nM have generally been reported. Both ET-2 and ET-3 are also active, although most studies have reported lower potency and/or efficacy than ET-1 [17, 18, 20]. Indomethacin pretreatment has been found not to attenuate ET-1-induced bronchoconstriction [17], suggesting that cyclooxygenase products do not play an important role in this response. In addition, Hay et al. [19] found no significant effect of histamine H1, thromboxane, cysteinyl leukotriene, or PAF antagonists, alone or in combination, on ET-1-induced contractions. Taken together these results indicate that, in contrast to the observations made in some studies using isolated guinea pig airways, ET-1 appears to elicit contraction of human isolated ASM preparations predominantly *via* a direct mechanism with no significant involvement of the release of secondary mediators.

The *in vivo* bronchoconstrictor effects of ETs have also been reported in a variety of species. The ability of intravenously administered ET-1 to induce a bronchoconstrictor response in guinea pigs *in vivo* has been reported by several investigators [22–25]. ET-2 and ET-3 also increase pulmonary inflation pressure *in vivo* with a similar order of potency to ET-1 when administered intravenously to guinea pigs [26], whereas big ET-1 exerts minimal, if any, direct effect [27]. Lagente et al. [28] demonstrated that a nebulized solution of ET-1 (5 or 10 µg/ml) administered to anaesthetized and ventilated guinea pigs produced a dose-dependent bronchoconstrictor response, without effect on systemic blood pressure. Similar effects of ET-1 aerosol challenge on pulmonary resistance were produced in rats [29], dogs [30], and sheep [31, 32]. In the reported studies in humans, aerosolized, ET-1 produced bronchoconstriction; it was significantly more potent in asthmatics than non asthmatic individuals [33].

Studies examining the question of which receptors mediate these ET-induced ASM contractions have demonstrated marked species differences. In guinea pig airways *in vitro*, for example, the ET_B-selective agonists sarafotoxin (SRTX) S6c and IRL 1620 elicit contractions potently whereas ET_A receptor antagonists such as BQ-123 are ineffective against ET ligand-induced responses [34–36], suggesting that these contractile responses are predominantly ET_B receptor-mediated. Similar conclusions have been reached in *in vivo* studies in guinea pigs based on the order of potency of the three isopeptides and on the effects of pharmacological agonists and antagonists [25]. In sheep, on the other hand, the contractile response appears to be mediated by ET_A receptors, as it can be antagonized by BQ-123 both *in vitro* [37] and *in vivo* [32]. In the case of human isolated large [34] and small [20] airways, ET-1-induced contractions are reported to be principally ET_B-mediated.

In addition to causing ASM contraction, ET-1 is able to induce a transient relaxant response under certain circumstances. Using a guinea pig *in vivo* model, White et al. [38] reported that intravenously administered ET-1 (10 nmol/kg) caused a biphasic response consisting of initial tracheal smooth muscle relaxation followed by sustained contraction. Unlike the contractile response, relaxation was unaffected by either indomethacin or by removal of the epithelium.

3.2. Neuromodulation

In addition to their ability to act directly as ASM contractile agonists, low concentrations of ETs may also promote bronchoconstriction by facilitating cholinergic neurotransmission. Early reports using autoradiographic techniques to localize ET binding demonstrated the presence of specific high-density binding sites on parasympathetic nerves in porcine lung [39] and on the cell bodies, axonal processes, and varicosities of cholinergic (and adrenergic) neurons present in primary cultures of guinea-pig tracheal smooth muscle [40]. These receptors were shown to be predominantly of the ET_B type [39, 40], and their activation produced a delayed contraction of adjacent smooth muscle cells [40].

In mouse [41], rat [42], rabbit [43] and human [44, 45] airways, the contractile response to electrical field stimulation (mediated by release of acetylcholine (ACh) from post-ganglionic parasympathetic nerves) is markedly potentiated by ET-1 or SRTX S6c, at agonist concentrations which induce little or no direct spasmogenic action. Increased release of ACh in response to ET-1 and SRTX S6c has been demonstrated directly in rat trachea by measuring efflux of radioactivity from [^{3}H]choline-loaded nerves [42]. In human airways, the potentiation of cholinergic neurotransmission involves both ET_A and ET_B receptors, as it can be inhibited by BQ-123 and BQ-788 in combination, but not by either antagonist alone [45].

3.3. Bronchial Hyperresponsiveness

Relatively few studies have investigated the potential of ETs to increase bronchial reactivity, and there have been conflicting findings. In anaesthetized and ventilated guinea pigs, for example, Kanazawa et al. [46] reported that pretreatment with nebulized ET-1, at a concentration (1 pM) too low itself to induce a measurable change in airway resistance, increased the magnitude and duration of the bronchoconstrictor response to subsequent administration of inhaled histamine, whereas Macquin-Mavier et al. [24] or Lagente et al. [47] were unable to detect any effect on airway responsiveness following an intravenous bolus injection or aerosol administration of ET-1, respectively. However, in conscious sheep, aerosolization of ET-1 (20 µg/ml) was found to induce increased airway responsiveness to carbachol which was evident at 4 h, after the initial increase in basal tone had resolved, and which persisted for 24 h [32]. In this model, the ET-1-induced increase in airway reactivity, like the initial bronchoconstrictor response, was shown to be ET_A-receptor mediated [32].

3.4. Microvascular Permeability

Several investigators have examined the ability of ET-1 to alter microvascular permeability using preparations of isolated perfused guinea pig [48, 49] and rat [50, 51] lungs. In these models, addition of ET-1 to the perfusate results in the rapid development of pulmonary oedema, although aerosol administration has little effect. In part this response appears to result from pulmonary venoconstriction with a consequent rise in capillary hydrostatic pressure [48]. However, an increase in microvascular permeability also occurs in blood-perfused rat lungs, an effect that requires both leukocytes and plasma components [51].

Studies in whole animals have also demonstrated ET-1-induced pulmonary oedema and increased microvascular permeability *in vivo*, although there have been some conflicting findings. Filep et al. [52] reported that intravenous bolus injection of ET-1 in conscious rats resulted in a dose-dependent increase in vascular permeability of the trachea and bronchi, as measured by extravasation of Evans blue dye. This effect appeared to be in part mediated by PAF [52] and thromboxane A_2 [53]. In a subsequent report, the authors showed that the response could be markedly attenuated by BQ-123, indicating the involvement of an ET_A receptor-dependent mechanism [54]. In anaesthetized guinea pigs, Macquin-Mavier et al. [24] reported that bolus intravenous injection of ET-1 did not significantly increase lung permeability. On the other hand, Filep et al. [55] demonstrated marked albumin extravasation in the trachea and bronchi following intravenous injection of ET-1 (0.25–2.50 µg/kg), a response which involved both ET_A and ET_B receptors.

3.5. Mucociliary function

In an early study using isolated ferret trachea, Yurdakos and Webber reported that ET-1 (0.01–100 nM) had no significant effect on basal submucosal gland secretion but inhibited phenylephrine- and methacholine-induced mucus output [56]. Subsequently, however, ET-1 (1 nM–1 µM) was reported to produce a modest increase in glycoconjugate release from isolated feline tracheal submucosal glands [57]. Secretagogue properties of ETs have also been demonstrated in human upper airway mucosa. Using explant cultures from human nasal inferior turbinate tissue, Mullol et al. [58] found that ET-1 and ET-2, but not ET-3, stimulated secretion from submucosal gland serous and mucous cells, assessed by measuring the release of mucous glycoprotein and lactoferrin respectively, although high concentrations (0.1–10 µM) of peptide were required. Topical application of ET-1 to human nasal mucosa *in vivo* has been shown also to induce glandular secretion, assessed by secretion weight and by lysozyme content, and moreover the response elicited was found to be greater in allergic individuals [59].

In addition to these effects on glandular function, ET-1 influences ciliary beat frequency. In cultured canine tracheal epithelial cells, ET-1 produced a rapid and concentration-dependent increase in ciliary beat frequency with an EC_{50} of approximately 3 nM [60]. Similarly, both ET-1 and ET-3 have been reported to increase ciliary beat frequency in cultured rabbit tracheal epithelial cells and also maxillary sinus ciliary beat frequency in rabbits *in vivo* [61].

Mucociliary clearance depends on both the composition and volume of airway secretions and also ciliary beat frequency. Tracheal mucus velocity provides an index of overall function, and can be measured by insufflating radio-opaque Teflon particles into the trachea and using videofluoros-copy to determine the rate at which they travel [62]. Aerosolized ET-1 (0.01–1 µM) induced a significant fall in tracheal mucus velocity in con-scious sheep which was maximal within 30 min [63]. This response was inhibited by BQ-123, but not by BQ-788, indicating that it was mediated through ET_A receptors; neither indomethacin nor the LT receptor antago-nist MK-571 had any influence.

3.6. Proinflammatory Properties

A number of studies have examined the potential of ETs to influence the function of inflammatory cells *in vitro*, sometimes with conflicting results. The first step in the recruitment of circulating leukocytes to sites of inflammation involves interactions between complementary pairs of adhe-sion molecules expressed on the surfaces of leukocytes and vascular endo-thelial cells. This aspect has not been intensively studied in relation to ETs, but one report has described increased expression of intercellular adhesion molecule-1 (ICAM-1) and vascular cell adhesion molecule-1 (VCAM-1) and induction of E-selectin expression on human brain endothelial cells in response to all three ET isopeptides [64]. In addition, ET-1 promotes adhe-sion of neutrophils to endothelial cells by increasing their cell-surface expression of the β_2 integrins LFA-1 and Mac-1, which bind to ICAM-1 [65]. Within tissues, inflammatory cells then migrate under the influence of chemotactic gradients. Chemotaxis of peripheral blood monocytes [66] or neutrophils [67] towards ET-1 has been reported, although other studies did not confirm these findings [68, 69]. Finally, leukocytes undergo a pro-cess of activation to release mediators at sites of inflammation. Although unable to induce activation directly, two reports have suggested that ET-1 (0.1–1 nM) exerts a priming effect on the ability of neutrophils to produce reactive oxygen species in response to other stimuli [69, 70]; however, again other authors have not confirmed this property [71]. ET-1 increases production of superoxide by human [72] and guinea pig [55] macrophages, and induces release of a number of cytokines by human peripheral blood monocytes, including tumour necrosis factor (TNF)-α, granulocyte-macrophage colony-stimulating factor (GM-CSF), interleukin (IL)-1, IL-6, IL-8, and monocyte chemotactic protein (MCP)-1 [73–75]. Finally, ET-1 has been shown to activate guinea pig pulmonary mast cells [76] and murine bone marrow-derived cultured mast cells [77] to release histamine and also, in the latter case, LTC_4. Perhaps surprisingly, published reports have not yet described any possible effects of ETs on eosinophil or T-lymphocyte chemotaxis or activation *in vitro*.

The involvement of ET in pulmonary inflammatory responses *in vivo* has been explored in several animal models. In rats, a single intratracheal instillation of Sephadex G-200 (dextran) beads leads to a rapid influx of inflammatory cells, dominated by eosinophils, neutrophils, and macrophages, intense eosinophilic peribronchitis/panbronchiolitis and, at later time-points, a mononuclear granulomatous alveolitis [78, 79]. In this model, the inflammatory reaction is associated with an increase of up to 60-fold in BAL fluid levels of ET-1 measured 24 h after instillation [80–82] and also, following repeated exposure, with an increase in ET content in lung homogenates [83]. Immunohistochemical analysis has demonstrated increased ET expression by the airway epithelium, but not at other sites, in Sephadex-treated animals, suggesting that bronchial epithelial cells are the principal source of the elevated ET levels [82]. Both the inflammatory changes and the elevation in ET levels can be inhibited by the glucocorticosteroid budesonide, administered either subcutaneously [83] or by inhalation [84]. However, studies of the effect of ET receptor antagonism in this model have yielded conflicting results. Finsnes et al. [82] reported that the mixed ET_A/ET_B receptor antagonist bosentan, administered either intravenously or orally, significantly inhibited the inflammatory response. On the other hand Andersson et al. [81] were unable to demonstrate any significant reduction in inflammatory indices following oral bosentan. An increase in the concentration of immunoreactive ET in BAL fluid has also been reported 24 h, but not 12 h, following challenge with *Ascaris suum* in sensitized guinea pigs, in association with a predominantly eosinophilic inflammatory response [85]. Furthermore, in mice sensitized and challenged with ovalbumin, the ET_A receptor antagonist BQ-123 and the mixed ET_A/ET_B receptor antagonist SB 209670, administered intratracheally shortly before challenge, were found to inhibit partially the BAL eosinophilia and neutrophilia that develop in this model [86]; the ET_B receptor antagonist BQ-788 was ineffective. The mechanism of this effect was shown to involve, at least in part, increased production of interferon (IFN)-γ by pulmonary T-lymphocytes.

Direct administration of ET-1 to guinea pigs, either by intravenous injection [24, 55] or by inhalation [87], has been shown to have no effect on cellular influx into the lungs. However, in some of these studies [24, 55] the time-points examined may have been too early for such changes to occur, and in any case the ability to induce infiltration of cells is likely to be limited by the short *in vivo* half-life of exogenously administered peptide. An alternative approach which is currently under investigation is the use of gene transfer systems, such as that provided by adenoviral vectors, to achieve more sustained overexpression of ET-1 in the lungs of rodents [88].

3.7. Mitogenic and Fibrogenic Properties

ETs have been shown to exert mitogenic properties for several cell types of relevance to pulmonary disease, particularly ASM cells and fibroblasts.

A proliferative response in ASM cells was demonstrated in preparations of rabbit [89], guinea pig [90] and ovine [91] ASM; in the latter case a role for both ET_A and ET_B receptors was reported [92]. In all these studies, however, the magnitude of the ET-1-induced response was small in comparison with that induced by other mitogens such as serum and epidermal growth factor (EGF). Studies of the mitogenic potential of ET-1 for cultured human ASM have yielded conflicting results. Tomlinson et al. [93] reported a modest (approximately 2.5-fold increase in [^{3}H]thymidine incorporation) mitogenic effect of ET-1 on bronchial ASM cells. On the other hand, Panettieri et al. [94] were unable to detect any direct mitogenic effect of ET-1 on human tracheal ASM cells, although there was marked potentiation of the response induced by EGF, *via* an ET_A receptor-dependent mechanism.

In an early study, Takuwa et al. [95] showed that ET-1 stimulated DNA synthesis by murine 3T3 fibroblasts in a concentration-dependent manner with an EC_{50} of approximately 30 pM. This effect was synergistically enhanced by low concentrations of insulin-like growth factor-1 (IGF-1). Other workers confirmed this observation, and demonstrated synergy with other growth factors including transforming growth factor (TGF)-β, basic fibroblast growth factor (bFGF), and platelet-derived growth factor (PDGF) [96]. ET-1 and, to a lesser extent, ET-3 are also mitogenic for human dermal [97], pulmonary artery [98], and foetal lung [99] fibroblasts. In the latter case, ET-1 was active over the concentration range 10 nM–5 µM, and the response was partially inhibited by BQ-123, suggesting a role for ET_A receptors [99]. In addition to stimulating fibroblast proliferation, ET-1 can induce these cells to adopt a myofibroblast-like phenotype, as indicated by increased expression of α-smooth muscle actin and by the development of characteristic morphological features [100]. Finally, ETs can activate fibroblast collagen synthesis. In cultured rat cardiac fibroblasts, for example, both ET-1 and ET-3 (1–100 nM) stimulate synthesis of collagen types I and III, principally *via* an ET_B receptor-dependent mechanism [101]. In addition, ET-1, but not ET-3, was found to decrease the release of collagenase activity by these cells, an effect inhibited by the specific ET_A receptor antagonist, PED-3512-PI [101]. The ability of ET-1 to induce collagen synthesis has been confirmed in human dermal fibroblasts [97].

4. Endothelin as a Putative Mediator in Asthma

4.1. Pathophysiology of Asthma

The fundamental physiological abnormalities in asthma comprise airflow obstruction, which is at least partially reversible, and increased airway reactivity. Airflow obstruction is considered to result from multiple factors including increased ASM tone, increased vagal activity [102], impaired mucociliary function [103–105] resulting in mucus plugging, and oedema [106]. In terms of pathology, numerous bronchoscopy-based studies have demonstrated that even in mild asthma there is an increase in the number and/or activation status of a broad range of inflammatory cells in the airways, including mast cells, eosinophils, T-lymphocytes, and cells of the monocyte/macrophage lineage [107–110]. Particular attention has centred on the role of the T-lymphocyte as a possible orchestrator of this inflammatory response. In asthma, airway T-cells preferentially express Th2-type cytokines such as IL-4 and IL-5 [111]. In addition, recent attention has focused on the presence of structural abnormalities in the airway wall in asthma, changes which are collectively referred to as remodelling [112]. There is a substantial increase in the thickness of the smooth muscle layer in the airways of asthmatics [113], and this change is considered to result from both hypertrophy and hyperplasia of ASM cells [114, 115]. In addition, there is deposition of a thick layer of connective tissue, composed principally of collagen type III and V and fibronectin, beneath the bronchial epithelial basement membrane [116], associated with an increased number of myofibroblasts in this region [117]. These pathological abnormalities in asthma are believed to underlie the physiological and clinical expression of the disease, via the capacity of inflammatory cells to release a range of mediators with bronchospastic and other proinflammatory properties and through the geometric effects of structural changes on airway function.

It is apparent that ETs have the potential to mimic many of the characteristic features of asthma; asthma is the pulmonary disease that has received greatest attention with regard to a possible pathophysiological role for ETs. Most studies have approached this issue by examining whether a disease-related alteration in the *in vitro* or *in vivo* expression of ET or its receptors can be demonstrated.

4.2. Expression of Endothelin in Asthma

4.2.1. Peripheral blood: Circulating ET-1 is normally detectable, and most investigators have reported mean venous plasma levels in the range 1–5 pg/ml in healthy subjects. Several reports have described peripheral blood ET measurements in patients with asthma. Aoki et al. [118] determined plasma ET-1 levels in 62 asthmatic patients with disease of varying severity. Levels were found to increase stepwise in those judged, using a clinical evaluation score, to have mild, moderate, or severe disease. Plasma levels correlated with the degree of airflow obstruction, measured as the FEV_1/FVC ratio. Furthermore, in a subgroup of 9 patients with severe disease who were admitted to hospital and treated with intravenous theophylline therapy, plasma levels were found to decline progressively in parallel with clinical recovery. Another report also indicated that plasma

ET levels were increased in a group of 31 children with atopic asthma compared to a group of 13 age-matched healthy control children, although no indices of disease severity were reported [119]. In a number of other studies [120–122], however, no significant differences could be detected in plasma ET levels between adult subjects with mild stable atopic asthma and healthy control subjects. The principal limitation with measurements of circulating ET is that its origin is unclear, and therefore its relationship to pulmonary pathophysiology can only be considered indirect at best.

4.2.2. Bronchial tissue: In the original report by Yanagisawa et al. [1], preproET-1 mRNA was detected by Northern analysis in porcine aortic endothelial cells but not in porcine pulmonary tissue. However, several reports subsequently demonstrated the presence of immunoreactive ET-1 in homogenates of rat [123] and porcine [124] lung, and of preproET-1 mRNA in homogenates from human foetal [125] and adult [126] lung. Using immunohistochemistry, ET-like immunoreactivity was localized to airway epithelial cells in tissue from rat [127], rabbit [128], and human [129, 130]. Similarly, preproET-1 mRNA was detected in airway epithelial cells, as well as in neuroendocrine cells, in human pulmonary tissue [129]. The predominant localization of ET in the airway epithelium is consistent with evidence from *in vitro* studies indicating the release of immunoreactive ET, either spontaneously or following appropriate stimulation, from cultured porcine [131], canine [131], guinea-pig [132], rabbit [133], and human [134] airway epithelial cells.

A number of investigators have examined ET expression in tissue derived from asthmatic subjects. In an initial report, Springall et al. [135] evaluated ET expression in bronchial biopsy specimens using an indirect immunofluorescence technique (Fig. 1). Expression was detectable in the airway epithelium of 11 of 17 (65%) subjects with asthma but in only 1 of 11 (9%) nonasthmatic control subjects (p < 0.001). Increased expression was also evident in vascular endothelial cells and airways glands in the asthmatic group, although these differences did not reach statistical significance. The asthmatic group in this study was heterogeneous in that it included a mixture of atopic and nonatopic subjects and in addition some subjects were receiving inhaled or oral corticosteroids and other anti-asthma therapies. No clear relation could be established between ET expression and corticosteroid usage.

Recently, we have extended these observations in bronchial biopsy specimens [136]. After immunohistochemical staining, expression of immunoreactive ET by the airway epithelium was quantified using a computer-assisted system of image analysis. There was a highly significant increase in the proportion of the bronchial epithelium immunostained from 14.3 ± 2.0% in nonasthmatic control subjects to 35.4 ± 3.8% in a group of atopic asthmatics receiving treatment only with inhaled β_2-agonists as required (Fig. 2). However, in asthmatic subjects additionally receiving

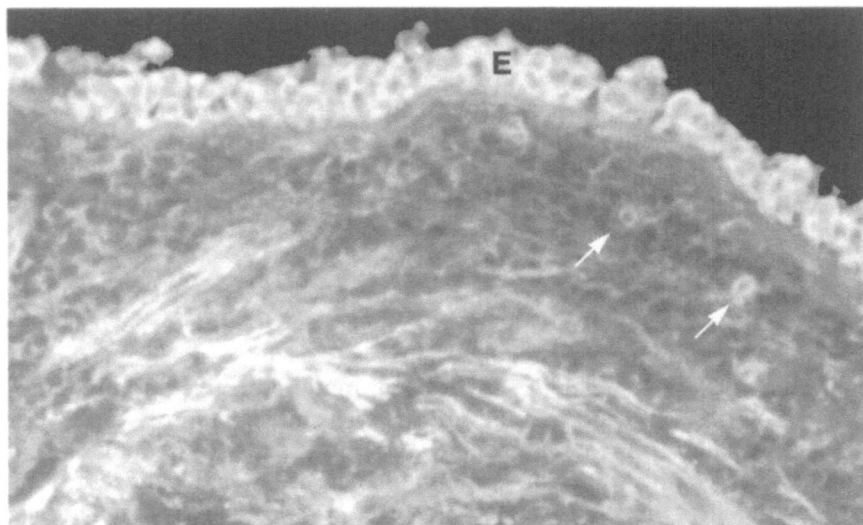

Figure 1. Immunoreactive ET in a bronchial biopsy specimen from an asthmatic subject examined using an indirect immunofluorescence technique. Expression is localized in the airway epithelium (E) and in vascular endothelium (arrows). (Courtesy of Dr. D. R. Springall).

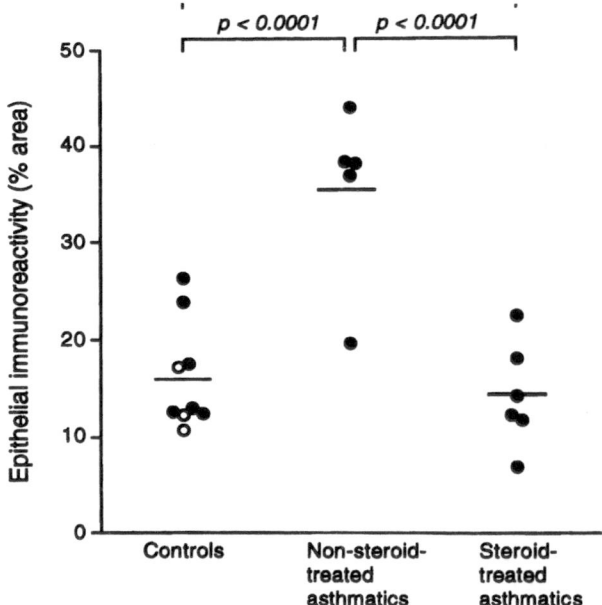

Figure 2. Comparison of airway epithelial expression of immunoreactive ET in bronchial biopsy specimens from control subjects (o: non-atopic; ●: atopic), non-steroid-treated asthmatics, and steroid-treated asthmatics. Horizontal bars represent mean values. (From ref. [136] with permission; Redington et al. *J Allergy Clin Immunol* 1997; 100; [544–552]).

therapy with inhaled and/or oral corticosteroids, levels of ET expression were similar to those in control subjects. In addition to the bronchial epithelium, expression of immunoreactive ET was evident in relation to airway glands and to the vascular endothelium but was not obviously present in relation to macrophages or other inflammatory cells. In another study, Ackerman et al. [137] reported that expression of immunoreactive ET was detectable in bronchial epithelial cells in all of 10 patients with symptomatic asthma receiving treatment with β_2-agonist alone but in only 2 of 10 subjects with asymptomatic asthma. These *in vivo* findings are consistent with *in vitro* reports. For example, Vittori et al. [138] found that cultured bronchial epithelial cells obtained from asthmatic subjects expressed preproET-1 mRNA and released significant quantities of immunoreactive ET into the supernatant, whereas neither mRNA nor peptide expression could be detected in cells from nonasthmatic subjects. Expression of ET-1 by the cells derived from asthmatic subjects was reduced after incubation with hydrocortisone [138]. A reduction in the release of ET-1 in response to corticosteroids has also been reported in a study using the human pulmonary adenocarcinoma cell line A549 [139].

In the biopsy studies discussed above, no relationship could be detected between ET expression and physiological measures of airflow obstruction or bronchial reactivity, nor with indices of airways inflammation and remodelling. However, in a study by Carpi et al. [140], expression of immunoreactive ET was examined in bronchial biopsy specimens from 22 asthmatic subjects, all but four of whom were atopic, who had disease of varying severity as assessed by measurement of lung function and symptom scores. Highly significant inverse correlations were evident in this study between ET expression and both FEV_1 % predicted and airway responsiveness to methacholine, and a relationship was also apparent with responsiveness to nebulized distilled water. Numbers of inflammatory cells in bronchial biopsies, including eosinophils and mast cells, were also correlated with epithelial ET expression.

In all of these immunohistochemical studies, the antisera used have been raised against ET-1 but exhibit significant cross-reactivity with the two other ET isopeptides. Presumably this high degree of cross-reactivity relates to the limited ability of the three peptides to express differing epitopes, although more recently isopeptide-specific antibodies have been described. In a study involving pulmonary tissue obtained from patients with a range of pathologies, expression of the individual ET isopeptides was evaluated using antisera raised against C-terminal sequences of big ET-1, big ET-2, and big ET-3 [130]. It was concluded that big ET-1 and big ET-3 were the principal forms expressed by airway epithelial cells, whereas all three were immunolocalized to submucosal glands. These findings are in agreement with a report that cultured canine tracheal epithelial cells release ET-1 and ET-3 but not ET-2 [131], and also with a study by Mattoli et al. [120] indicating that ET-1 and ET-3, in approximately equal amounts, were detectable by fast protein liquid chromatography in BAL fluid from asthmatic subjects whereas ET-2 could not be measured.

The mechanism of increased epithelial ET expression in asthma may involve exposure to cytokines and other inflammatory mediators present in the airway wall. Release of ET-1 by cultured guinea-pig tracheal epithelial cells is increased by exposure to proinflammatory cytokines, including IL-1, IL-2, IL-6, IL-8 and TNF-α, and to TGF-β [141]. Recent reports have confirmed the ability of IL-1α, IL-1β, and TNF-α to increase preproET-1 mRNA expression and ET-1 release in human airway epithelial cells [137, 142–144]. Histamine has also been shown to exert a similar effect, although its potency was low in comparison with that of IL-1β [137]. In addition, exposure of bronchial epithelial cells to allergen could lead directly to ET release, as suggested by a report indicating that a proportion of bronchial epithelial cells derived from asthmatic subjects express CD23, the low-affinity IgE receptor, and can be activated to release ET *via* an IgE-dependent mechanism [145].

4.2.3. Bronchoalveolar lavage fluid: In an early case report, Nomura et al. [146] found that the level of immunoreactive ET in BAL fluid was six-fold higher in a patient mechanically ventilated for acute severe asthma compared to the amounts measured after recovery. No further reports have appeared describing BAL fluid ET measurements during spontaneous episodes of acute asthma, but several investigators have reported increased levels in stable asthma. In our own study [147], BAL fluid ET levels were increased from a median of 0.08 pM in nonasthmatic control subjects to 0.30 pM in subjects with asthma receiving treatment only with bronchodilators (Fig. 3). In asthmatic subjects receiving long-term treatment with corticosteroids, ET levels were intermediate between those in the other two groups; these relationships were similar whether ET levels were expressed in relation to total protein content in BAL fluid or as concentrations. In the non-steroid-treated asthmatic group, but not in either of the other two groups, an inverse correlation was evident between BAL ET levels and FEV$_1$ % predicted (Fig. 4). No relationship could be detected, however, with bronchial reactivity in either of the asthmatic groups. Similar findings have been described by other investigators. Mattoli et al. [120] reported that the mean BAL fluid ET level in a group of six asthmatics who were not receiving corticosteroid therapy was approximately four-fold greater than in a group of 5 nonasthmatic control subjects. Following a 15-day course of prednisolone, levels in the subjects with asthma were reduced to values similar to those in the control subjects. One other study has also confirmed elevated levels of ET in BAL fluid in asthma, although in this case treatment details were not specified [148].

In our own studies, we have also examined the influence of acute allergen exposure on ET levels in the airways using a segmental bronchoprovocation model [149]. Allergen solution, to which the subject had previously

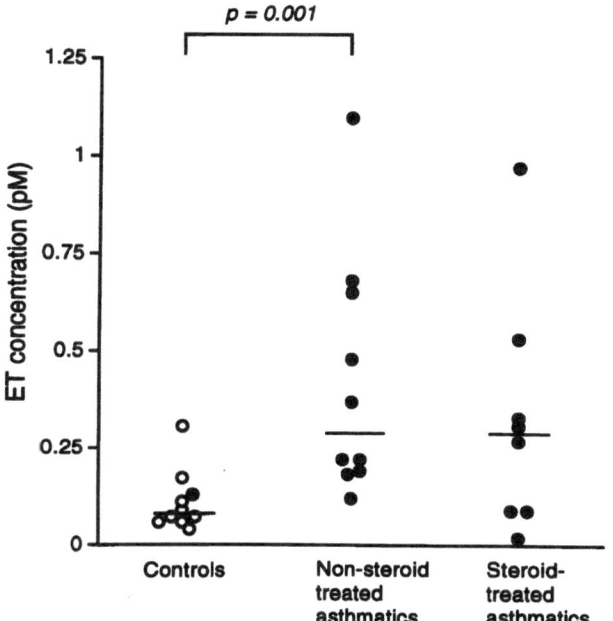

Figure 3. ET levels expressed as concentrations in BAL fluid in control (o: nonatopic; •: atopic), non-steroid-treated asthmatic, and steroid-treated asthmatic subjects. Horizontal bars represent median values. (From ref. [147] with permission; Redington et al. *Am J Respir Crit Care Med* 1995; 151: 1034–1039).

demonstrated skin-prick test sensitivity, and saline as a control were administered to separate bronchopulmonary segments by endobronchial instillation, and BAL was performed during the immediate phase of bronchoconstriction. We hypothesized that allergen exposure would result in the rapid release of ET which would be evident as an increased concentration in BAL fluid. However, in contrast, we found that in patients receiving regular therapy with inhaled corticosteroids no difference between the two sites was evident, whereas in subjects treated only with inhaled β_2-agonists as required, levels of ET were actually significantly lower at the allergen-challenged sites than at the saline-challenged sites (Fig. 5). The failure to detect ET release during the early response in this study is consistent with data from animal studies. For example, Andersson et al. [85] reported that in sensitized guinea pigs challenged with the relevant antigen there was an increase in BAL fluid ET levels which was evident 24 h but not 12 h after exposure. Similarly, Noguchi et al. [32] found that the late phase response in antigen challenged sheep could be partially inhibited by the ET_A receptor antagonist BQ-123 whereas the immediate response was unaffected. On the basis of these studies a role for ET as a mediator of the early asthmatic reaction, therefore, would appear unlikely; this would be consistent with

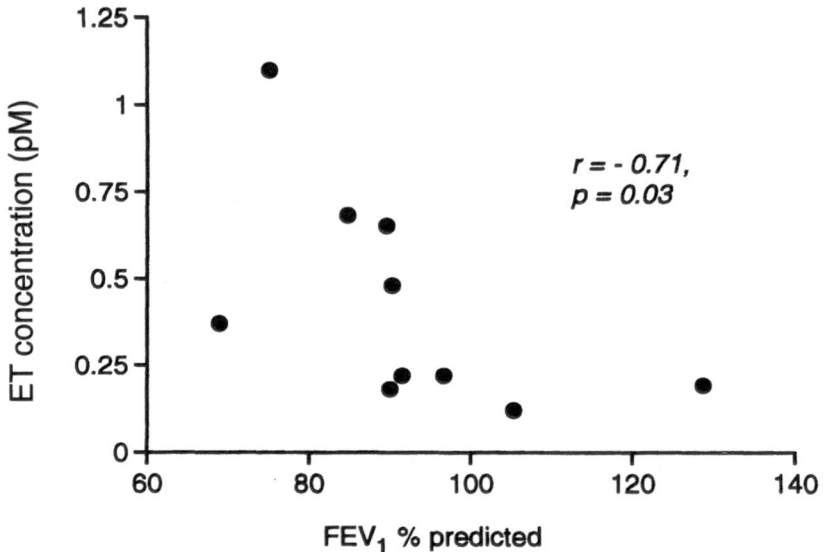

Figure 4. Demonstration of inverse correlation between FEV$_1$ % predicted and BAL fluid ET levels in non-steroid-treated asthmatics. (From ref. [147] with permission; Redington et al. *Am J Respir Crit Care Med* 1995; 151: 1034–1039).

pharmacological evidence implicating mast cell-derived histamine and leukotrienes as the principal spasmogens [150]. Involvement of ET in the late phase response, however, remains possible and this area requires further investigation.

In exercise-induced asthma, it is proposed that bronchoconstriction results from hypertonicity of the airway epithelial lining fluid, which occurs due to loss of water during hyperventilation [151]. However, the precise mechanism of bronchoconstriction in response to an osmotic stimulus is unknown. We have therefore examined the influence of hyper-osmolar saline exposure on ET levels in BAL fluid [152]. This was performed using a similar bronchoprovocation procedure to the one described above – hypertonic and isotonic saline aerosols were delivered to separate bronchopulmonary segments and BAL was performed during the immediate phase of bronchoconstriction. In this study, we found no evidence of ET release in response to the hypertonic stimulus in a group of seven asthmatic subjects, all but one of whom were receiving regular therapy with inhaled corticosteroids in addition to β_2-agonists. In fact, to the contrary, ET levels were significantly lower at the hypertonic saline-exposed sites than the sham isotonic saline-exposed sites. Again previous pharmacological studies have provided fairly strong support for histamine, leukotrienes and, to a lesser extent, prostanoids as important mediators of both hypertonic saline-induced and exercise-induced bronchospasm [153–156].

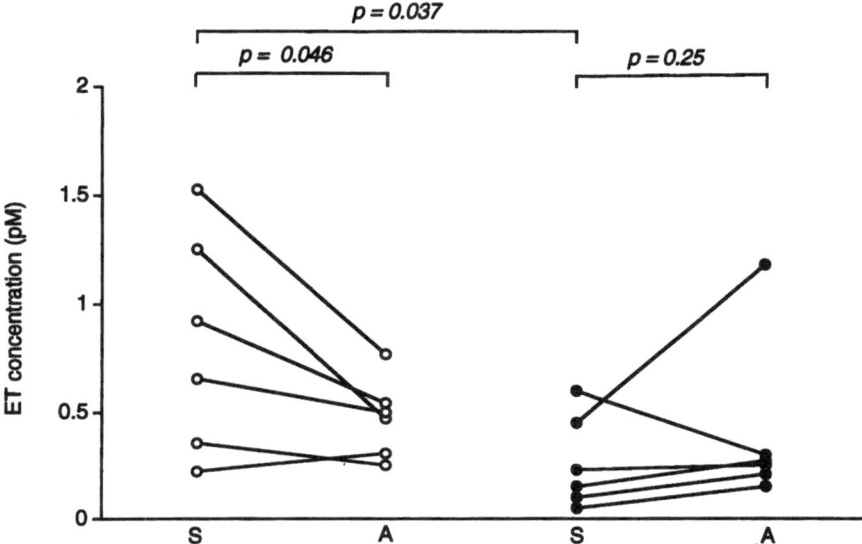

Figure 5. Concentrations of ET in BAL fluid from saline-challenged (S) and allergen-challenged (A) segments in non-steroid-treated (o) and steroid-treated (•) asthmatic subjects. (From ref. [149] with permission; Redington et al. *Eur Respir J* 1997; 10: 1026–1032).

The reason for the unexpected reduction in ET levels observed in the two studies mentioned above is unclear. Perhaps the most straightforward explanation would be that it simply represents a dilutional phenomenon resulting either from plasma extravasation following allergen exposure or from osmotic influx of water from the epithelium and intraepithelial spaces following the hypertonic stimulus. Alternatively, these reductions could result from increased metabolism or binding of ET within the airway lumen. Activated neutrophils, for example, have been shown to release a protease activity, identified as principally cathepsin G, which is able to degrade ET to biologically inactive fragments [157]. The fact that in the allergen challenge study the reduction was only evident in the group of subjects who were not receiving treatment with corticosteroids might suggest that corticosteroid therapy leads to a reduction in the release of ET-degrading enzymes; indeed there was a trend towards a reduction in total numbers of leukocytes in BAL fluid in the steroid-treated compared to the non-steroid-treated group. In both studies, however, the magnitude of the reductions observed was relatively small and it would perhaps be appropriate therefore to interpret these findings with caution at present.

Interestingly, a reduction in BAL fluid ET levels has also been reported in a study of patients with nocturnal asthma [121], none of whom had received recent treatment with corticosteroids. BAL fluid ET, expressed in relation to total protein, was found to be significantly lower in the nocturnal asthma group, measured at 4 a.m., compared to subjects with nocturnal asthma studied during the daytime, nonasthmatic subjects studied during the daytime, and nonasthmatic subjects studied at 4 a.m. Increased binding within tissue or increased metabolism were suggested as potential reasons for this finding.

The principal cellular source of immunoreactive ET in BAL fluid is presumed to be the airway epithelium, in view of the immunohistochemical and *in situ* hybridization findings in pulmonary tissue as discussed above. However, other cell populations may also contribute, although their quantitative importance is uncertain. For example, Chanez et al. [158] reported that expression of immunoreactive ET was significantly increased in alveolar macrophages from patients with asthma compared to normal subjects, although there was no detectable difference between the two groups in spontaneous or lipopolysaccharide-stimulated ET release from cultured macrophages.

4.2.4. Induced sputum: Examination of induced sputum has recently gained popularity as a means to investigate airways inflammation in asthma. This technique provides an alternative to BAL for sampling the airway epithelial lining fluid, and has the advantage that it is less invasive and can be repeated after a short interval if required. A recent report has demonstrated the presence of immunoreactive ET in induced sputum samples at higher concentrations than those measurable in plasma [122]. However, no differences were detected between asthmatic and nonasthmatic subjects, nor between asthmatics treated with inhaled corticosteroids and those treated with β_2-agonists alone. These findings are not, therefore, in agreement with those obtained in studies using BAL [120, 147, 148], and this apparent discrepancy could relate either to differences in the populations of subjects studied or to methodological differences between sputum induction and BAL. In a preliminary report, the same group also examined ET levels in induced sputum following acute allergen-induced bronchoconstriction [159]. As compared to placebo, allergen exposure was found to have no effect on ET levels in sputum obtained 15 min after inhalation. This finding is consistent with our studies using BAL following segmental allergen exposure [149].

4.3. Endothelin Receptor Expression in Asthma

The distribution of specific ET binding sites has been studied in pulmonary tissue derived from asthmatic and nonasthmatic subjects. Goldie et al. [160] reported high levels of [^{125}I]-ET-1 binding associated with bronchial smooth muscle, and identified these sites as principally ET_B receptors but with a small ET_A receptor component. Low levels of specific binding to ET_A, but not ET_B, sites were also detected in bronchial epithelium, lamina propria, submucosal glands, and blood vessels. Quantitative analysis revealed no major differences in receptor density or binding characteristics between asthmatic and nonasthmatic tissue, and similar conclusions were reached in a study of peripheral human lung [161]. However, in functional studies, bronchial smooth muscle from asthmatic subjects was found to exhibit a reduced contractile responsiveness to ET_B receptor stimulation [160], perhaps reflecting chronic exposure to elevated ET levels *in vivo*.

4.4. ECE Expression in Asthma

Although the presence of ECE activity in airways from non-asthmatic individuals has been demonstrated in pharmacological, localization and molecular biological studies [144, 162], comparable analysis of tissue derived from asthmatic subjects has not yet been reported. Airway epithelial cells, endothelial cells, and, to a lesser extent, ASM cells and alveolar macrophages were identified as the principal sites of expression in normal airways [144].

4.5. Conclusions

The studies described above have established a firm framework for viewing ETs as important mediators in asthma. They are able to mimic many of the salient features of the disease, expression in the airways is increased in patients with asthma, and, at least in some studies, correlations have been demonstrated between ET expression and indexes of disease severity. Two directions of research are now required. Many of the relevant biological effects of ETs have so far been demonstrated only *in vitro* or in animal studies, and their relevance to the human pulmonary system *in vivo* needs to be confirmed. Most importantly, the physiological and pathophysiological role of ETs in the airways need to be addressed directly using specific ET receptor antagonists and/or ECE inhibitors. By way of comparison, a role for endogenous ET-1 in the maintenance of vascular tone in normal human subjects has already been unequivocally demonstrated by local intraarterial infusion of phosphoramidon and of BQ-123 [163]. Comparable studies in asthma are now required, and in this regard the development of potent, orally active non-peptidic ET receptor antagonist represents an important advance [164]. Extensive research will be needed, however, to assess the effects of such agents on airway function in healthy subjects and asthmatics, and to compare the results of ET_A, ET_B or combined ET_A/ET_B receptor antagonism and of ECE inhibition. Studies in asthma are likely to focus initially on the experimental paradigm of acute allergen exposure, with subsequent attention turning to possible therapeutic efficacy in clinical disease.

5. Endothelin as a Putative Mediator in Other Respiratory Diseases

5.1. Chronic Obstructive Pulmonary Disease (COPD)

COPD has many clinical and pathophysiological features in common with asthma. Airflow obstruction in COPD results from varying combinations of airways disease and emphysema, and tends to be progressive and to exhibit little reversibility. Hypersecretion of mucus is also a prominent clinical feature, and is associated with goblet cell hyperplasia and submucosal gland enlargement [165]. As with asthma, it is now recognized that in COPD the conducting airways are chronically inflamed, the cellular infiltrate comprising activated T-lymphocytes, macrophages, mast cells, and, at least during exacerbations, eosinophils [166–169]. Subepithelial fibrosis is not a consistent feature in COPD [168], but there may be a striking increase in the thickness of the smooth muscle layer in the small airways [170].

Mattoli et al. [120] measured immunoreactive ET in plasma from five subjects with COPD, all of whom were current or former smokers and who demonstrated airflow obstruction that was not significantly reversed by administration of inhaled salbutamol. No significant differences were detectable in these levels in comparison with a group of five non-smoking control subjects. The same group were also unable to demonstrate any significant difference in BAL fluid ET levels between healthy volunteers and patients with COPD [120]. A subsequent study by Sofia et al. [171] also indicated that plasma ET levels did not differ between COPD patients and healthy control subjects. However, these authors found that, as compared to healthy control subjects, 24-h urinary excretion of ET was significantly increased in a group of 13 COPD patients admitted to hospital with clinical evidence of acute infective exacerbations [171]. Furthermore, a correlation was evident between ET excretion and the degree of arterial hypoxemia. After intervals of approximately 2 weeks, when there was evidence of clinical recovery, ET excretion was reduced but still remained higher than in the control subjects. There is little clearance of injected ^{125}I-ET-1 from the peripheral blood into the urine, however, and urinary ET is therefore believed to be principally of renal origin [172]. These findings in COPD thus probably reflect altered renal production and/or metabolism of ET as a result of chronic hypoxia, rather than any pulmonary abnormality *per se*. In support of this suggestion, Faraone et al. [173], in a recent preliminary report, found that although renal clearance of immunoreactive ET was increased in subjects with COPD compared to healthy control subjects, the ratio of arterial/venous plasma ET, which provides an index of pulmonary extraction/sypthesis, was unchanged.

Shokeir et al. [174] examined expression of immunoreactive ET in resected lung tissue obtained from smokers with evidence of airflow obstruction who were undergoing lobectomy or pneumonectomy for carcinoma.

No differences were detected in the proportion of epithelial cells immunostained either in respiratory or membranous bronchioles compared to a group of non-smoking control subjects. These findings are in agreement with the results of Vittori et al. [138] who were able to detect only sparse expression of preproET-1 mRNA and ET-like immunoreactivity in cultured bronchial epithelial cells derived from subjects with COPD.

Taken together, these studies indicate that, despite the many similarities between COPD and asthma, there is little evidence to suggest a role for ET in the pathophysiology of COPD.

5.2. Idiopathic Pulmonary Fibrosis

Idiopathic pulmonary fibrosis (IPF) is a chronic disorder of unknown aetiology which frequently runs an aggressive clinical course. In the early stages there is a patchy alveolitis, and persistent inflammation leads to the derangement of alveolar structure with loss of type I cells and proliferation of type II cells [175]. Finally, the interstitial matrix becomes expanded with large numbers of fibroblasts, myofibroblasts, smooth muscle cells, and masses of disordered collagen fibres [175]. On account of the mitogenic and fibrogenic properties of ETs, several studies have examined their expression in IPF.

Uguccioni et al. [176] reported that plasma levels of ET-1 were significantly elevated in patients with IPF, irrespective of treatment, compared to control subjects. Furthermore, these levels were found to correlate significantly with duration of disease although not with other parameters of disease activity. One report has also described increased levels of ET-like immunoreactive material in BAL fluid in IPF [148]. Whereas this was below the limit of sensitivity of the assay (< 0.8 pg/ml) in normal subjects, detectable ET was present in all of 10 subjects with IPF with a mean level of 12.4 pg/ml.

Giaid et al. [177] performed a detailed study of lung tissue obtained from patients with IPF, patients with nonspecific focal pulmonary fibrosis, and normal controls. Whereas there was little ET expression in normal lung or in tissue from patients with nonspecific fibrosis, IPF was characterized by striking expression of preproET-1 mRNA and immunoreactive big ET-1 and ET-1 in airway epithelial cells and type II pneumocytes, particularly those adjacent to areas of granulation tissue. Significant correlations were reported between scores for ET-1-like immunoreactivity and histological parameters of disease activity. In a subsequent report, the same authors confirmed these observations and also demonstrated increased expression of ECE-1 in IPF and its colocalization with big ET-1 and ET-1 [144]. One other study has also indicated increased ET expression in tissue from patients with IPF, although in this case small vessel endothelial cells were the principal sites of expression [176].

Further studies will be required to understand the significance of these observations. Initial efforts are likely to focus on defining the role of ETs in animal models of pulmonary fibrosis, such as the well-characterized response to intratracheal instillation of bleomycin in rodents [178, 179]. Expression of several other fibrogenic molecules, such as TGF-β and PDGF, is also increased, both in these animal models [180, 181] and in patients with IPF [182–184], and so the possible interrelationship between ET and these various other putative mediators will need to be addressed.

5.3. Systemic Sclerosis

Systemic sclerosis is frequently associated with pulmonary fibrosis. Although the aetiology of this condition is unknown, vascular endothelial cell injury is believed to occur at an early stage with subsequent inflammatory cell infiltration and fibrosis. Several investigators have reported that plasma levels of ET-1 are elevated in systemic sclerosis [97, 185, 186]. In one of these studies a correlation was reported between plasma ET-1 and the degree of pulmonary involvement, as indicated by impairment of diffusing capacity [186]. Levels of ET-1 in patients with systemic sclerosis are also elevated in BAL fluid, and furthermore ET-1 has been shown to represent a major component of the fibroblast mitogenic activity present in these samples [99]. However, no differences in BAL fluid ET-1 levels could be detected in that study between subjects with and without pulmonary fibrosis as determined by computed tomography (CT) [99]. Plasma ET levels are also apparently unaffected by the presence or absence of pulmonary fibrosis on CT [187], although in both cases these findings may relate to the relative insensitivity of CT as a means of detecting early fibrotic change. Dermal fibroblasts derived from patients with systemic sclerosis are reported to express increased preproET-1 mRNA and to release increased amounts of ET-1 protein [188], but to exhibit an impaired ET-1-induced proliferative response as a result of a reduction in ET_A receptor expression [189].

5.4. Lung Cancer

In view of the localization of ET expression to bronchial epithelial cells and neuroendocrine cells in normal lung tissue [129], it might be expected that these peptides would be expressed by pulmonary tumours derived from these cell types. Using immunohistochemistry and in situ hybridization, Giaid et al. [190] examined expression of ETs by a range of pulmonary neoplasms. The majority of squamous cell carcinomas and adenocarcinomas were found to express both preproET-1 mRNA and ET-like immunoreactivity. In contrast, expression was sparse or absent in the case of small cell carcinomas, large cell carcinomas, and carcinoid tumours. No definite functional implications can be attached to these observations, although it was speculated that ET might act in a paracrine manner to promote tumour growth.

5.5. Acute Lung Injury

Several studies have reported approximately five-fold elevations of mixed venous plasma ET-1 concentrations in patients with acute lung injury/adult respiratory distress syndrome (ARDS) in comparison with those in control subjects [191–193]. Pulmonary clearance of ET-1, calculated from systemic arterial and mixed venous measurements, was also investigated in these studies. Whereas in control subjects a major fraction of circulating ET-1 was cleared during passage through the lung, in ARDS there was a net pulmonary release of immunoreactive ET-1 [191, 192]. In those patients who recovered, these abnormalities of ET metabolism were reversed [192]. A number of the biological properties of ETs are of potential relevance to the pathophysiology of ARDS, particularly their effects on microvascular permeability and on leukocyte function. Further studies will be required, initially in animal models, before the significance of these observational reports becomes clearer.

5.6. Sickle Cell Disease

Pulmonary involvement in sickle cell disease is frequent and may be severe [194]. In the acute chest syndrome, endothelial dysfunction is believed to be a contributory factor in the microvascular occlusive process. A recent report [195] has described increased plasma levels of ET-1 in two patients with sickle cell disease during the initial stages of the acute chest syndrome, followed by a reduction in ET-1 levels with clinical improvement. In addition, exposure to plasma obtained from these two subjects increased synthesis and release of ET-1 by cultured bovine endothelial cells. On the basis of these findings, it was suggested that ET-1 may contribute to the vasoocclusive state in the acute chest syndrome.

5.7. Behçet's Syndrome

Behçet's syndrome is a chronic multisystem disease of unknown aetiology, characterized by relapsing iritis with recurrent buccal and genital ulceration. Pulmonary involvement in this condition is not infrequent, and results at least in part from a vasculitis involving large and small pulmonary arteries [196]. One study has reported that BAL fluid levels of ET were approximately three-fold elevated in a group of subjects with active

Behçet's disease and suspected pulmonary involvement compared to healthy control subjects, although plasma levels were similar in the two groups [197].

5.8. Pulmonary Embolism

One paper has described an increased arterial/venous plasma ET ratio in patients with acute pulmonary embolism as compared to healthy control subjects [193]. It was speculated that abnormal ET metabolism in patients with pulmonary embolism could contribute to several aspects of the pathophysiology of this condition.

References

1 Yanagisawa M, Kurihara H, Kimura S, Tomobe Y, Kobayashi M, Mitsui Y, Yazaki Y, Goto K, Masaki T (1988) A novel potent vasoconstrictor peptide produced by vascular endothelial cells. *Nature* 332: 411–415

2 Hirata Y, Yoshimi H, Takata S, Watanabe TX, Kumagai S, Nakajima K, Sakakibara S (1988) Cellular mechanisms of action by a novel vasoconstrictor endothelin in cultured rat vascular smooth muscle cells. *Biochem Biophys Res Commun* 154: 868–875

3 Koseki C, Imai M, Hirata Y, Yanagisawa M, Masaki T (1989) Autoradiographic distribution in rat tissues of binding sites for endothelin: a neuropeptide? *Am J Physiol* 256: R858–R866

4 Maggi CA, Giuliani S, Pattachini R, Santicioli P, Rovero P, Giachetti A, Meli A (1989) The C-terminal hexapeptide, endothelin-(16-21), discriminates between different endothelin receptors. *Eur J Pharmacol* 166: 121–122

5 Arai H, Hori S, Aramori I, Ohkuba H, Nakanishi S (1990) Cloning and expression of a cDNA encoding an endothelin receptor *Nature* 348: 730–732

6 Sakurai T, Yanagisawa M, Takuwa Y, Miyazaki H, Kimura S, Goto K, Masaki T (1990) Cloning of a cDNA encoding a non-isopeptide-selective subtype of the endothelin receptor. *Nature* 348: 732–735

7 Sakamoto A, Yanagisawa M, Sakurai T, Takuwa Y, Yanagisawa H, Masaki T (1991) Cloning and functional expression of human cDNA for the ET_B endothelin receptor. *Biochem Biophys Res Commun* 178: 656–663

8 Mizuguchi T, Nishiyama M, Moroi K, Tanaka H, Saito T, Masuda Y, Masaki T, de Wit D, Yanagisawa M, Kimura S (1997) Analysis of two pharmacologically predicted endothelin B receptor subtypes by using the endothelin B receptor gene knockout mouse. *Br J Pharmacol* 1427–1430

9 Samson WK, Skala KD, Alexander BD, Huang F-LS (1990) Pituitary site of action of endothelin: selective inhibition of prolactin release *in vitro. Biochem Biophys Res Commun* 169: 737–743

10 Martin ER, Brenner BM, Ballerman BJ (1990) Heterogeneity of cell surface endothelin receptors. *J Biol Chem* 265: 14044–14049

11 Karne S, Yajawickreme CK, Lerner MR (1993) Cloning and characterization of an endothelin-3 specific receptor (ETc receptor) from *Xenopus laevis* dermal melanophores. *J Biol Chem* 268: 19126–19133

12 Uchida Y, Ninomiya H, Saotome M, Nomura A, Ohtsuka M, Yanagisawa M, Goto K, Masaki T, Hasegawa S (1988) Endothelin, a novel vasoconstrictor peptide, as potent bronchoconstrictor. *Eur J Pharmacol* 154: 227–228

13 Maggi CA, Patacchini R, Giuliani S, Meli A (1989) Potent contractile effect of endothelin in isolated guinea-pig airways. *Eur J Pharmacol* 160: 179–182

14 Hay DWP (1990) Mechanism of endothelin-induced contraction in guinea-pig trachea: comparison with rat aorta. *Br J Pharmacol* 100: 383–392

15 Henry PJ, Rigby PJ, Self GJ, Preuss JM, Goldie RG (1990) Relationship between endothelin-1 binding site densities and constrictor activities in human and animal airway smooth muscle. *Br J Pharmacol* 100: 786–792

16 Lee H-K, Leikauf GD, Sperelakis N (1990) Electromechanical effects of endothelin on ferret bronchial and tracheal smooth muscle. *J Appl Physiol* 68: 417–420

17 Advenier C, Sarria B, Naline E, Puybasset L, Lagente V (1990) Contractile activity of three endothelins (ET-1, ET-2 and ET-3) on the human isolated bronchus. *Br J Pharmacol* 100: 168–172

18 McKay KO, Black JL, Armour CL (1991) The mechanism of action of endothelin in human lung. *Br J Pharmacol* 102: 422–428

19 Hay DWP, Hubbard WC, Undem BJ (1993) Endothelin-induced contraction and mediator release in human bronchus. *Br J Pharmacol* 110: 392–398

20 Hemsén A, Franco-Cereceda A, Matran R, Rudehill A, Lundberg JM (1990) Occurrence, specific binding sites and functional effects of endothelin in human cardiopulmonary tissue. *Eur J Pharmacol* 191: 319–328

21 Adner M, Cardell LO, Sjöberg T, Ottosson A, Edvinsson L (1996) Contractile endothelin-B (ET$_B$) receptors in human small bronchi. *Eur Respir J* 9: 351–355

22 Payne AN, Whittle BJR (1990) Potent cyclo-oxygenase-mediated bronchoconstrictor effects of endothelin in the guinea-pig *in vivo*. *Eur J Pharmacol* 158: 303–304

23 Touvay C, Vilain B, Pons F, Chabrier PE, Mencia-Huerta JM, Braquet P (1990) Bronchopulmonary and vascular effects of endothelin in the guinea-pig. *Eur J Pharmacol* 176: 23–33

24 Macquin-Mavier I, Levame M, Istin N, Harf A (1989) Mechanisms of endothelin-mediated bronchoconstriction in the guinea pig. *J Pharmacol Exp Ther* 250: 740–745

25 Noguchi K, Noguchi Y, Hirose H, Nishikibe M, Ihara M, Ishikawa K, Yano M (1993) Role of endothelin ET$_B$ receptors in bronchoconstrictor and vasoconstrictor responses in guinea-pigs. *Eur J Pharmacol* 233: 47–51

26 Pons F, Loquet I, Touvay C, Roubert P, Chabrier P-E, Mencia-Huerta JM, Braquet P (1991) Comparison of the bronchopulmonary and pressor activities of endothelin isoforms ET-1, ET-2, and ET-3 and characterization of their binding sites in guinea pig lung. *Am Rev Respir Dis* 143: 294–300

27 Pons F, Touvay C, Lagente V, Mencia-Huerta J-M, Braquet P (1991) Bronchopulmonary and pressor effects of big-endothelin-1 in the guinea-pig. *Neurochem Int* 18: 481–483

28 Lagente V, Chabrier PE, Mencia-Huerta J-M, Braquet P (1989) Pharmacological modulation of the bronchopulmonary action of the vasoactive peptide, endothelin, administered by aerosol in the guinea-pig. *Biochem Biophys Res Commun* 158: 625–632

29 Di Maria GU, Bellofiore S, Malatino LS, Maggi CA, Torrisi A, Mistretta A (1991) Aerosolized endothelin-1, but not its C-terminal hexapeptide, causes airway narrowing in the rat. *Eur Respir J* 4: 528–531

30 Uchida Y, Hamada M, Kameyama M, Ohse H, Nomura A, Hasegawa S, Hirata F (1992) ET-1 induced bronchoconstriction in the early phase but not late phase of anesthetized dogs is inhibited by indomethacin and ICI 198615. *Biochem Biophys Res Commun* 183: 1197–1202

31 Abraham WM, Ahmed A, Cortes A, Spinella MJ, Malik AB, Andersen TT (1993) A specific endothelin-1 antagonist blocks inhaled endothelin-1-induced bronchoconstriction in sheep. *J Appl Physiol* 74: 2537–2542

32 Noguchi K, Ishikawa K, Yano M, Ahmed A, Cortes A, Abraham WM (1995) Endothelin-1 contributes to antigen-induced airway hyperresponsiveness. *J Appl Physiol* 79: 700–705

33 Chalmers GW, Little SA, Patel KR, Thomson NC (1997) Endothelin-1-induced bronchoconstriction in asthma. *Am J Respir Crit Care Med* 156: 382–388

34 Hay DWP, Luttmann MA, Hubbard WC, Undem BJ (1993) Endothelin receptor subtypes in human and guinea-pig pulmonary tissues. *Br J Pharmacol* 110: 1175–1183

35 Hay DWP (1992) Pharmacological evidence for distinct endothelin receptors in guinea pig bronchus and aorta. *Br J Pharmacol* 106: 759–761

36 Battistini B, Warner TD, Fournier A, Vane JR (1994) Characterization of ET$_B$ receptors mediating contractions induced by endothelin-1 or IRL 1620 in guinea pig isolated airways: effects of BQ-123, FR139317 or PD 145065. *Br J Pharmacol* 111: 1009–1016

37 Goldie RG, Grayson PS, Knott PG, Self GJ, Henry PJ (1994) Predominance of endothelin$_A$ (ET$_A$) receptors in ovine airway smooth muscle and their mediation of ET-1-induced contraction. *Br J Pharmacol* 112: 749–756

38 White SR, Hathaway DP, Umans JG, Tallet J, Abrahams C, Leff AR (1991) Epithelial modulation of airway smooth muscle response to endothelin-1. *Am Rev Respir Dis* 144: 373–378

39 Kobayashi M, Ihara M, Sato N, Saeki T, Ozaki S, Ikemoto F, Yano M (1993) A novel ligand, [^{125}I]BQ-3020, reveals the localization of endothelin ET$_B$ receptors. *Eur J Pharmacol* 235: 95–100

40 Takimoto M, Inui T, Okada T, Urade Y (1993) Contraction of smooth muscle by activation of endothelin receptors on autonomic neurons. *FEBS Lett* 324: 277–282

41 Henry PJ, Goldie RG (1995) Potentiation by endothelin-1 of cholinergic nerve-mediated contractions in mouse trachea via activation of ET$_B$ receptors. *Br J Pharmacol* 114: 563–569

42 Knott PG, Fernandes LB, Henry PJ, Goldie RG (1996) Influence of endothelin-1 on cholinergic nerve-mediated contractions and acetylcholine release in rat isolated tracheal smooth muscle. *J Pharmacol Exp Ther* 279: 1142–1147

43 Yoneyama T, Hori M, Tanaka T, Matsuda Y, Karaki H (1995) Endothelin ET$_A$ and ET$_B$ receptors facilitating parasympathetic neurotransmission in the rabbit trachea. *J Pharmacol Exp Ther* 275: 1084–1089

44 Fernandes LB, Henry PJ, Rigby PJ, Goldie RG (1996) Endothelin$_B$ (ET$_B$) receptor-activated potentiation of cholinergic nerve-mediated contraction in human bronchus. *Br J Pharmacol* 118: 1873–1874

45 Fernandes LB, Henry PJ, Goldie RG (1997) Endothelin-1 potentiates cholinergic nerve-mediated contraction in human bronchus. *Am J Respir Crit Care Med* 155: A573 (Abstract)

46 Kanazawa H, Kurihara N, Kirata K, Fujiwara H, Matsushita H, Takeda T (1992) Low concentration endothelin-1 enhanced histamine-mediated bronchial contractions of guinea pigs *in vivo*. *Biochem Biophys Res Commun* 187: 717–721

47 Lagente V, Boichot E, Mencia-Huerta J, Braquet P (1990) Failure of aerosolized endothelin (ET-1) to induce bronchial hyperreactivity in the guinea pig. *Fundam Clin Pharmacol* 4: 275–280

48 Horgan MJ, Pinheiro JMB, Malik AB (1991) Mechanism of endothelin-1-induced pulmonary vasoconstriction. *Circ Res* 69: 157–164

49 Pons F, Touvay C, Lagente V, Mencia-Huerta JM, Braquet P (1991) Comparison of the effects of intra-arterial and aerosol administration of endothelin-1 (ET-1) in the guinea-pig isolated lung. *Br J Pharmacol* 102: 791–796

50 Bonvallet ST, Oka M, Yano M, Zamora MR, McMurtry IF, Stelzner TJ (1993) BQ123, an ET$_A$ receptor antagonist, attenuates endothelin-1-induced vasoconstriction in rat pulmonary circulation. *J Cardiovasc Pharmacol* 22: 39–43

51 Helset E, Kjaeve J, Hauge A (1993) Endothelin-1-induced increases in microvascular permeability in isolated, perfused rat lungs requires leukocytes and plasma. *Circ Shock* 39: 15–20

52 Filep JG, Sirois MG, Rousseau A, Fournier A, Sirois P (1991) Effects of endothelin-1 on vascular permeability in the conscious rat: interactions with platelet-activating factor. *Br J Pharmacol* 104: 797–804

53 Sirois MG, Filep JG, Roussaeu A, Plante GE, Sirois P (1992) Endothelin-1 enhances vascular permeability in conscious rats: role of thromboxane A$_2$. *Eur J Pharmacol* 214: 119–125

54 Filep JG, Sirois MG, Földes-Filep E, Rousseau A, Plante GE, Fournier A, Yano M, Sirois P (1993) Enhancement by endothelin-1 of microvascular permeability via the activation of ET$_A$ receptors. *Br J Pharmacol* 109: 880–886

55 Filep JG, Fournier A, Földes-Filep E (1995) Acute pro-inflammatory actions of endothelin-1 in the guinea-pig lung: involvement of ET$_A$ and ET$_B$ receptors. *Br J Pharmacol* 115: 227–236

56 Yurdakos E, Webber SE (1991) Endothelin-1 inhibits pre-stimulated tracheal submucosal gland secretion and epithelial albumin transport. *Br J Pharmacol* 104: 1050–1056

57 Shimura S, Ishihara H, Satoh M, Masuda T, Nagaki N, Sasaki H, Takishima T (1992) Endothelin regulation of mucus glycoprotein secretion from feline tracheal submucosal glands. *Am J Physiol* 262: L208–L213

58 Mullol J, Chowdhury BA, White MV, Ohkubo K, Rieves RD, Baraniuk J, Hausfeld JN, Shelhamer JH, Kaliner MA (1993) Endothelin in human nasal mucosa. *Am J Respir Cell Mol Biol* 8: 393–402

59 Riccio MM, Reynolds CJ, Hay DWP, Proud D (1995) Effects of intranasal administration of endothelin-1 to allergic and nonallergic individuals. *Am J Respir Crit Care Med* 152: 1757–1764

60 Tamaoki J, Kanemura T, Sakai N, Isono K, Kobayashi K, Takizawa T (1991) Endothelin stimulates ciliary beat frequency and chloride secretion in canine cultured tracheal epithelium. *Am J Respir Cell Mol Biol* 4: 426–431

61 Amble FR, Lindberg SOH, McCaffrey TV, Runer T (1993) Mucociliary function and endothelins 1, 2, and 3. *Otolaryngol Head Neck Surg* 109: 634–645

62 Friedman M, Stott FD, Poole DO, Dougherty R, Chapman GA, Watson H, Sackner MA (1977) A new roentgenographic method for estimating mucous velocity in airways. *Am Rev Respir Dis* 115: 67–72

63 Sabater JR, Otero R, Abraham WM, Wanner A, O'Riordan TG (1996) Endothelin-1 depresses tracheal mucus velocity in ovine airways via ET-A receptors. *Am J Respir Crit Care Med* 154: 341–345

64 McCarron RM, Wang L, Stanimirovic DB, Spatz M (1993) Endothelin induction of adhesion molecule expression on human brain microvascular endothelial cells. *Neurosci Lett* 156: 31–34

65 López-Farré A, Reisco A, Espinosa G, Digiuni E, Cernadas MR, Alvarez V, Montón M, Rivas F, Gallego MJ, Egido J et al (1993) Effects of endothelin-1 on neutrophil adhesion to endothelial cell and perfused heart. *Circulation* 88: 1166–1171

66 Achmad TH, Rao GS (1992) Chemotaxis of human blood monocytes towards endothelin-1 and the influence of calcium channel blockers. *Biochem Biophys Res Commun* 189: 994–1000

67 Elferink JGR, de Koster BM (1994) Endothelin-induced activation of neutrophil migration. *Biochem Pharmacol* 48: 865–871

68 Bath PMW, Mayston SA, Martin JF (1990) Endothelin and PDGF do not stimulate peripheral blood monocyte chemotaxis, adhesion to endothelium, and superoxide production. *Exp Cell Res* 187: 339–342

69 Hafström I, Ringertz B, Lundeberg T, Palmblad J (1993) The effect of endothelin, neuropeptide Y, calcitonin gene-related peptide and substance P on neutrophil functions. *Acta Physiol Scand* 148: 341–346

70 Ishida K, Takeshige K, Minakami S (1990) Endothelin-1 enhances superoxide generation of human neutrophils stimulated by the chemotactic peptide N-formyl-methionyl-leucyl-phenylalanine. *Biochem Biophys Res Commun* 173: 496–500

71 Kopprasch S, Gatzweiler A, Kohl M, Schröder H-E (1995) Endothelin-1 does not prime polymorphonuclear leukocytes for enhanced production of reactive oxygen metabolites. *Inflammation* 19: 679–687

72 Haller H, Schaberg T, Lindschau C, Lode H, Distler A (1991) Endothelin increases $[Ca^{2+}]_i$, protein phosphorylation, and O_2^--production in human alveolar macrophages. *Am J Physiol* 261: L478–484

73 Helset E, Sildnes T, Seljelid R, Konopski ZS (1993) Endothelin-1 stimulates human monocytes *in vitro* to release TNF-α, IL-1β and IL-6. *Mediators of Inflammation* 2: 374–378

74 Helset E, Sildnes T, Konopski ZS (1994) Endothelin-1 stimulates monocytes *in vitro* to release chemotactic activity identified as interleukin-8 and monocyte chemotactic protein-1. *Mediators of Inflammation* 3: 155–160

75 Cunningham ME, Huribal M, Bala RJ, McMillen MA (1997) Endothelin-1 and endothelin-4 stimulate monocyte production of cytokines. *Crit Care Med* 25: 958–964

76 Uchida Y, Ninomiya H, Sakamoto T, Lee JY, Endo T, Nomura A, Hasegawa S, Hirata F (1992) ET-1 released histamine from guinea pig pulmonary but not peritoneal mast cells. *Biochem Biophys Res Commun* 189: 1196–1201

77 Egger D, Geuenich S, Denzlinger C, Schmitt E, Mailhammer R, Ehrenreich H, Dörmer P, Hültner L (1996) IL-4 renders mast cells functionally responsive to endothelin-1. *J Immunol* 154: 1830–1837

78 Källström L, Brattsand R, Lövgren U, Svensjö E, Roempke K (1985) A rat model for testing anti-inflammatory action in lung and the effect of glucocorticosteroids (GCS) in this model. *Agents Actions* 17: 355–357

79 Bjermer L, Sandström T, Särnstrand B, Brattsand R (1994) Sephadex-induced granulomatous alveolitis in rat: effects of antigen manipulation. *Am J Industrial Med* 25: 73–78

80 Andersson SE, Hemsén A, Zackrisson C, Lundberg JM (1996) Release of endothelin-1 into rat airways following Sephadex-induced inflammation: modulation by enzyme inhibitors and budesonide. *Respiration* 63: 111–116

81 Andersson SE, Hemsén A, Lundberg JM (1996) The effect of endothelin receptor blockade on the development of the Sephadex-induced inflammation in the rat lung. *Acta Physiol Scand* 158: 189–193

82 Finsnes F, Skjonsberg OH, Tonnessen T, Naess O, Lyberg T, Christensen G (1997) Endothelin production and effects of endothelin antagonism during experimental airway inflammation. *Am J Respir Crit Care Med* 155: 1404–1412

83 Andersson SE, Zackrisson C, Hemsén A, Lundberg JM (1992) Regulation of lung endothelin content by the glucocorticosteroid budesonide. *Biochem Biophys Res Commun* 188: 1116–1121

84 Andersson SE, Eirefelt S, Zackrisson C, Hemsén A, Dahlbäck M, Lundberg JM (1995) Regulatory effects of aerosolized budesonide and adrenalectomy on the lung content of endothelin-1 in the rat. *Respiration* 62: 34–39

85 Andersson SE, Zackrisson C, Behrens K, Hemsén A, Forsberg K, Linden M, Lundberg JM (1995) Effect of allergen provocation on inflammatory cell profile and endothelin-like immunoreactivity in guinea-pig airways. *Allergy* 50: 349–358

86 Fujitani Y, Trifilieff A, Tsuyuki S, Coyle AJ, Bertrand C (1997) Endothelin receptor antagonists inhibit antigen-induced lung inflammation in mice. *Am J Respir Crit Care Med* 155: 1890–1894

87 Boichot E, Carre C, Lagente V, Pons F, Mencia-Huerta J-M, Braquet P (1991) Endothelin-1 and bronchial hyperresponsiveness in the guinea pig. *J Cardiovasc Pharmacol* 17 (Suppl. 7): S329–S331

88 Redington AE, Sime PJ, Télémaque S, Yanagisawa M, Gauldie J (1997) Adenovirus-mediated expression of endothelin-1 by pulmonary epithelial cells and fibroblasts. *Am J Respir Crit Care Med* 155: A357 (Abstract)

89 Noveral JP, Rosenberg SM, Anbar RA, Pawlowski NA, Grunstein MM (1992) Role of endothelin-1 in regulating proliferation of cultured rabbit airway smooth muscle cells. *Am J Physiol* 263: L317–324

90 Stewart AG, Grigoriadis G, Harris T (1994) Mitogenic actions of endothelin-1 and epidermal growth factor in cultured airway smooth muscle. *Clin Exp Pharmacol Physiol* 21: 277–285

91 Glassberg MK, Ergul A, Wanner A, Puett D (1994) Endothelin-1 promotes mitogenesis in airway smooth muscle cells. *Am J Respir Cell Mol Biol* 10: 316–321

92 Carratu P, Scuri M, Styblo JL, Wanner A, Glassberg MK (1997) ET-1 induces mitogenesis in ovine airway smooth muscle cells via ET_A and ET_B receptors. *Am J Physiol* 272: L1021–1024

93 Tomlinson PR, Wilson JW, Stewart AG (1994) Inhibition by salbutamol of the proliferation of human airway smooth muscle cells grown in culture. *Br J Pharmacol* 111: 641–647

94 Panettieri Jr RA, Goldie RG, Rigby PJ, Esterhas AJ, Hay DWP (1996) Endothelin-1-induced potentiation of human airway smooth muscle proliferation: an ET_A receptor-mediated phenomenon. *Br J Pharmacol* 118: 191–197

95 Takuwa N, Takuwa Y, Yanagisawa M, Yamashita K, Masaki T (1989) A novel vasoactive peptide endothelin stimulates mitogenesis through inositol lipid turnover in Swiss 3T3 fibroblasts. *J Biol Chem* 264: 7856–7861

96 Brown KD, Littlewood CJ (1989) Endothelin stimulates DNA synthesis in Swiss 3T3 cells: synergy with polypeptide growth factors. *Biochem J* 263: 977–980

97 Kahaleh MB (1991) Endothelin, an endothelial-dependent vasoconstrictor in scleroderma: enhanced production and profibrotic action. *Arthritis Rheum* 34: 978–983

98 Peacock AJ, Dawes KE, Shock A, Gray AJ, Reeves JT, Laurent GJ (1993) Endothelin-1 and endothelin-3 induce chemotaxis and replication of pulmonary artery fibroblasts. *Am J Respir Cell Mol Biol* 7: 492–499

99 Cambrey AD, Harrison NK, Dawes KE, Southcott AM, Black CM, du Bois RM, Laurent GL, McNulty RJ (1994) Increased levels of endothelin-1 in bronchoalveolar lavage fluid from patients with systemic sclerosis contribute to fibroblast mitogenic activity *in vitro*. *Am J Respir Cell Mol Biol* 11: 439–445

100 Villaschi S, Nicosia RF (1994) Paracrine interactions between fibroblasts and endothelial cells in a serum-free coculture model: modulation of angiogenesis and collagen gel contraction. *Lab Invest* 71: 291–299

101 Guarda E, Katwa LC, Myers PR, Tyagi SC, Weber KT (1993) Effects of endothelins on collagen turnover in cardiac fibroblasts. *Cardiovasc Res* 27: 2130–2134

102 Barnes PJ (1986) Neural control of human airways in health and disease. *Am Rev Respir Dis* 134: 1289–1314

103 Bateman JRM, Pavia D, Sheahan NF, Agnew JE, Clarke SW (1983) Impaired tracheobronchial clearance in patients with mild stable asthma. *Thorax* 1983; 38: 463–467

104 Messina MS, O'Riordan TG, Smaldone GC (1991) Changes in mucociliary clearance during acute exacerbations of asthma. *Am Rev Respir Dis* 143: 993–997

105 O'Riordan TG, Zwang J, Smaldone GC (1992) Mucociliary clearance in adult asthma. *Am Rev Respir Dis* 146: 598–603

106 Persson CA (1986) Role of plasma exudation in asthmatic airways. *Lancet* 1126–1129

107 Azzawi M, Bradley B, Jeffery PK, Frew AJ, Wardlaw AJ, Knowles G, Assoufi B, Collins JV, Durham S, Kay AB (1990) Identification of activated T lymphocytes and eosinophils in bronchial biopsies in stable atopic asthma. *Am Rev Respir Dis* 142: 1407–1413

108 Bousquet J, Chanez P, Lacoste JY, Barneon G, Ghavanian N, Enander I, Venge P, Ahlstedt S, Simony-Lafontaine J, Godard P et al (1990) Eosinophilic inflammation in asthma. *N Engl J Med* 323: 1033–1039

109 Djukanovic R, Wilson JW, Britten KM, Wilson SJ, Walls AF, Roche WR, Howarth PH, Holgate ST (1990) Quantitation of mast cells and eosinophils in the bronchial mucosa of symptomatic atopic asthmatics and healthy control subjects using immunohistochemistry. *Am Rev Respir Dis* 142: 863–871

110 Poston RN, Chanez P, Lacoste JY, Lichfield T, Lee TH, Bousquet J (1992) Immunohistochemical characterization of the cellular infiltration in asthmatic bronchi. *Am Rev Respir Dis* 145: 918–921

111 Robinson DS, Hamid Q, Ying S, Tsicopoulos A, Barkans J, Bentley AM, Corrigan C, Durham SR, Kay AB (1992) Predominant T_{H_2}-like bronchoalveolar T-lymphocyte population in atopic asthma. *N Engl J Med* 326: 298–304

112 Redington AE, Howarth PH (1997) Airway wall remodelling in asthma. *Thorax* 52: 310–312

113 Kuwano K, Bosken CH, Paré PD, Bai TR, Wiggs BR, Hogg JC (1993) Small airways dimensions in asthma and in chronic obstructive pulmonary disease. *Am Rev Respir Dis* 148: 1220–1225

114 Heard BE, Hossain S (1972) Hyperplasia of bronchial smooth muscle in asthma. *J Pathol* 110: 319–331

115 Ebina M, Takahashi T, Chiba T, Motomiya M (1993) Cellular hypertrophy and hyperplasia of airway smooth muscles underlying bronchial asthma: a 3-D morphometric study. *Am Rev Respir Dis* 148: 720–726

116 Roche WR, Beasley R, Williams JH, Holgate ST (1989) Subepithelial fibrosis in the bronchi of asthmatics. *Lancet* i: 520–524

117 Brewster CEP, Howarth PH, Djukanovic R, Wilson J, Holgate ST, Roche WR (1990) Myofibroblasts and subepithelial fibrosis in bronchial asthma. *Am J Respir Cell Mol Biol* 3: 507–511

118 Aoki T, Kojima T, Ono A, Unishi G, Yoshijima S, Kameda-Hayashi N, Yamamoto C, Hirata Y, Kobayashi Y (1994) Circulating endothelin-1 levels in patients with bronchial asthma. *Ann Allergy* 73: 365–369

119 Chen W-Y, Yu J, Wang J-Y (1995) Decreased production of endothelin-1 in asthmatic children after immunotherapy. *J Asthma* 32: 29–35

120 Mattoli S, Soloperto M, Marini M, Fasoli A (1991) Levels of endothelin in the bronchoalveolar fluid of patients with symptomatic asthma and reversible airflow obstruction. *J Allergy Clin Immunol* 88: 376–384

121 Kraft M, Beam WR, Wenzel SE, Zamora MR, O'Brien RF, Martin RJ (1994) Blood and bronchoalveolar lavage endothelin-1 levels in nocturnal asthma. *Am J Respir Crit Care Med* 149: 947–952

122 Chalmers GW, Thomson L, Macleod KJ, Dagg KD, McGinn BJ, McSharry C, Patel KR, Thomson NC (1997) Endothelin-1 levels in induced sputum samples from asthmatic and normal subjects. *Thorax* 52: 625–627

123 Matsumoto H, Suzuki N, Onda H, Fujino M (1989) Abundance of endothelin-3 in rat
    intestine, pituitary gland and brain. *Biochem Biophys Res Commun* 164: 74–80
124 Kitamura K, Tanaka T, Kato J, Eto T, Tanaka K (1989) Regional distribution of immuno-
    reactive endothelin in porcine tissues: abundance in inner medulla of kidney. *Biochem
    Biophys Res Commun* 161: 348–352
125 Bloch KD, Eddy RL, Shows TB, Quertermous T (1989) cDNA cloning and chromosomal
    assignment of the gene encoding endothelin 3. *J Biol Chem* 264: 18156–18161
126 Nunez DJR, Brown MJ, Davenport AP, Neylon CB, Schofield JP, Wyse RK (1990) Endo-
    thelin-1 mRNA is widely expressed in porcine and human tissues. *J Clin Invest* 85:
    1537–1541
127 Rozengurt N, Springall DR, Polak JM (1990) Localization of endothelin-like immunore-
    activity in airway epithelium of rats and mice. *J Pathol* 160: 5–8
128 Rennick RE, Loesch A, Burnstock G (1992) Endothelin, vasopressin, and substance P like
    immunoreactivity in cultured and intact epithelium from rabbit trachea. *Thorax* 47:
    1044–1049
129 Giaid A, Polak JM, Gaitonde V, Hamid QA, Moscoso G, Legon S, Uwanogho D, Ron-
    calli M, Shinmi O, Sawamura T et al (1991) Distribution of endothelin-like immuno-
    activity and mRNA in the developing and adult human lung. *Am J Respir Cell Mol Biol*
    4: 50–58
130 Marciniak SJ, Plumpton C, Barker PJ, Huskisson NS, Davenport AP (1992) Localization
    of immunoreactive endothelin and proendothelin in the human lung. *Pulm Pharmacol* 5:
    175–182
131 Black PN, Ghatei MA, Takahashi K, Bretherton-Watt D, Krausz T, Dollery CT, Bloom SR
    (1989) Formation of endothelin by cultured airway epithelial cells. *FEBS Lett* 255:
    129–132
132 Ninomiya H, Uchida Y, Ishii Y, Nomura A, Kameyama M, Saotome M, Endo T, Hasegawa
    S (1991) Endotoxin stimulates endothelin release from cultured epithelial cells of guinea-
    pig airways. *Eur J Pharmacol* 203: 299–302
133 Rennick RE, Milner P, Burnstock G (1993) Thrombin stimulates release of endothelin and
    vasopressin, but not substance P, from isolated rabbit tracheal epithelial cells. *Eur J Phar-
    macol* 230: 367–370
134 Mattoli S, Mezzetti M, Riva G, Allegra L, Fasoli A (1990) Specific binding of endothelin
    on human bronchial smooth muscle cells in culture and secretion of endothelin-like ma-
    terial from bronchial epithelial cells. *Am J Respir Cell Mol Biol* 3: 145–151
135 Springall DR, Howarth PH, Counihan H, Djukanovic R, Holgate ST, Polak JM (1991)
    Endothelin immunoreactivity of airway epithelium in asthmatic patients. *Lancet* 337:
    697–701
136 Redington AE, Springall DR, Meng Q· H, Tuck AB, Holgate ST, Polak JM, Howarth PH
    (1997) Immunoreactive endothelin in bronchial biopsy specimens: increased expression in
    asthma and modulation by corticosteroid therapy. *J Allergy Clin Immunol* 100: 544–552
137 Ackerman V, Carpi S, Bellini A, Vassalli G, Marini M, Mattoli S (1995) Constitutive
    expression of endothelin in bronchial epithelial cells of patients with symptomatic and
    asymptomatic asthma and modulation by histamine and interleukin-1. *J Allergy Clin
    Immunol* 96: 618–627
138 Vittori E, Marini M, Fasoli A, de Franchis R, Mattoli S (1992) Increased expression of
    endothelin in bronchial epithelial cells of asthmatic patients and effect of corticosteroids.
    Am Rev Respir Dis 146: 1320–1325
139 Calderón E, Gómez-Sánchez CE, Cozza EN, Zhou M, Coffey RG, Lockey RF, Prockop
    LD, Szentivanyi A (1994) Modulation of endothelin-1 production by a pulmonary epithe-
    lial cell line. I. Regulation by glucocorticoids. *Biochem Pharmacol* 48: 2065–2071
140 Carpi S, Marini M, Vittori E, Vassalli G, Mattoli S (1993) Bronchoconstrictive responses
    to inhaled ultrasonically nebulized distilled water and airway inflammation in asthma.
    Chest 104: 1346–1351
141 Endo T, Uchida Y, Matsumoto H, Suzuki N, Nomura A, Hirata F, Hasegawa S (1992) Regu-
    lation of endothalin-1 synthesis in cultured guinea pig airway epithelial cells by various
    cytokines. *Biochem Biophys Res Commun* 186: 1594–1599
142 Nakano J, Takizawa H, Ohtoshi T, Shoji S, Yamaguchi M, Ishii A, Yanagisawa M, Ito K
    (1994) Endotoxin and pro-inflammatory cytokines stimulate endothelin-1 expression and
    release by airway epithelial cells. *Clin Exp Allergy* 24: 330–336

143 Aubert J-D, Juillerat-Jeanneret L, Leuenberger P (1997) Expression of endothelin-1 in human broncho-epithelial and monocyte cell lines: influence of tumor necrosis factor-α and dexamethasone. *Biochem Pharmacol* 53: 547–552

144 Saleh D, Furukawa K, Tsao M-S, Maghazachi A, Corrin B, Yanagisawa M, Barnes PJ, Giaid A (1997) Elevated expression of endothelin-1 and endothelin-converting enzyme-1 in idiopathic pulmonary fibrosis: possible involvement of proinflammatory cytokines. *Am J Respir Cell Mol Biol* 16: 187–193

145 Campbell AM, Vignola AM, Chanez P, Godard P, Bousquet J (1994) Low-affinity receptor for IgE on human bronchial epithelial cells. *Immunol* 82: 506–508

146 Nomura A, Uchida Y, Kameyama M, Saotome M, Oki K, Hasegawa S (1989) Endothelin and bronchial asthma. *Lancet* ii: 747–748 (Letter)

147 Redington AE, Springall DR, Ghatei MA, Lau LCK, Holgate ST, Polak JM, Howarth PH (1995) Endothelin in bronchoalveolar lavage fluid and its relation to airflow obstruction in asthma. *Am J Respir Crit Care Med* 151: 1034–1039

148 Sofia M, Mormile M, Faraone S, Alifano M, Zofra S, Romano L, Carratù L (1993) Increased endothelin-like immunoreactive material on bronchoalveolar lavage fluid from patients with bronchial asthma and patients with interstitial lung disease. *Respiration* 60: 89–95

149 Redington AE, Springall DR, Ghatei MA, Madden J, Bloom SR, Frew AJ, Polak JM, Holgate ST, Howarth PH (1997) Airway endothelin levels in asthma: influence of endobronchial allergen challenge and maintenance corticosteroid therapy. *Eur Respir J* 10: 1026–1032

150 Redington AE, Polosa R, Walls AF, Howarth PH, Holgate ST (1995) Role of mast cells and basophils in asthma. *Chem Immunol* 62: 22–59

151 Anderson SD (1984) Is there a unifying hypothesis for exercise-induced asthma? *J Allergy Clin Immunol* 73: 660–665

152 Makker H, Springall DR, Redington AE, Ghatei MA, Bloom SR, Polak JM, Howarth PH, Holgate ST (1997) Airway endothelin levels in asthma: influence of endobronchial hypertonic saline challenge. *Clin Exp Allergy;* in press

153 Finnerty JP, Wilmot C, Holgate ST (1989) Inhibition of hypertonic saline-induced bronchoconstriction by terfenadine and flurbiprofen: evidence for the predominant role of histamine. *Am Rev Respir Dis* 140: 593–597

154 Finnerty JP, Holgate ST (1990) Evidence for the roles of histamine and prostaglandins as mediators in exercise-induced asthma: the inhibitory effects of terfenadine and flurbiprofen alone and in combination. *Eur Respir J* 3: 540–547

155 Manning PJ, Watson RM, Margolskee DJ, Williams VC, Schwartz JI, O'Byrne PM (1990) Inhibition of exercise-induced bronchoconstriction by MK-571, a potent leukotriene D4-receptor antagonist. *N Engl J Med* 323: 1736–1739

156 Finnerty JP, Wood-Baker R, Thomson H, Holgate ST (1992) Role of leukotrienes in exercise-induced asthma: inhibitory effect of ICI 204219, a potent leukotriene D4 receptor antagonist. *Am Rev Respir Dis* 145: 746–749

157 Fagny C, Michel A, Nortier J, Deschodt-Lanckman M (1992) Enzymatic degradation of endothelin-1 by activated human polymorphonuclear neutrophils. *Regul Peptides* 42: 27–37

158 Chanez P, Vignola AM, Albat B, Springall DR, Polak JM, Godard P, Bousquet J (1996) Involvement of endothelin in mononuclear phagocyte inflammation in asthma. *J Allergy Clin Immunol* 98: 412–420

159 Chalmers GW, Macleod KM, Thomson LJ, McSharry C, Patel KR, Thomson NC (1997) The influence of the immediate allergic reaction in asthma on endothelin-1 (ET-1) in induced sputum and plasma. *Am J Respir Crit Care Med* 155: A820 (Abstract)

160 Goldie RG, Henry PJ, Knott PG, Self GJ, Luttmann MA, Hay DWP (1995) Endothelin-1 receptor density, distribution, and function in human isolated asthmatic airways. *Am J Respir Crit Care Med* 152: 1653–1658

161 Knott PG, D'Aprile AC, Henry PJ, Hay DWP, Goldie RG (1995) Receptors for endothelin-1 in asthmatic human peripheral lung. *Br J Pharmacol* 114: 1–3

162 McKay KO, Battistini B, Jeng AY, Sirois P, Black JL (1997) Endothelin-converting enzyme activity in human airways. *Am J Respir Crit Care Med* 155: A127 (abstract)

163 Haynes WG, Webb DJ (1994) Contribution of endogenous generation of endothelin-1 to basal vascular tone. *Lancet* 344: 852–854

164 Clozel M, Breu V, Burri K, Cassal J-M, Fischli W, Gray GA, Hirth G, Löffler B-M, Müller M, Neidhart W et al (1993) Pathophysiological role of endothelin revealed by the first orally active endothelin antagonist. *Nature* 365: 759–761

165 Reid L (1954) Pathology of chronic bronchitis. *Lancet* i: 275–279

166 Ollerenshaw SL, Woolcock AJ (1992) Characteristics of the inflammation in biopsies from large airways of subjects with asthma and subjects with chronic airflow limitation. *Am Rev Respir Dis* 145: 922–927

167 Saetta M, DiStefano A, Maestrelli P, Ferraresso A, Drigo R, Potena A, Ciaccia A, Fabbri LM (1993) Activated T-lymphocytes and macrophages in bronchial mucosa of subjects with chronic bronchitis. *Am Rev Respir Dis* 147: 301–306

168 Pesci A, Rossi GA, Bertorelli G, Aufiero A, Zanon P, Olivieri D (1994) Mast cells in the airways lumen and bronchial mucosa of patients with chronic bronchitis. *Am J Respir Crit Care Med* 149: 1311–1316

169 Saetta M, di Stefano A, Maestrelli P, Turato G, Ruggieri MP, Roggeri A, Calcagni P, Mapp CE, Ciaccia A, Fabbri LM (1994) Airway eosinophilia in chronic bronchitis during exacerbations. *Am J Respir Crit Care Med* 150: 1646–1652

170 Cosio MG, Hale KA (1980) Morphologic and morphometric effects of prolonged cigarette smoking on the small airways. *Am Rev Respir Dis* 122: 265–271

171 Sofia M, Mormile M, Faraone S, Carratù P, Alifano M, Di Benedetto G, Carratù L (1994) Increased 24-hour endothelin-1 urinary excretion in patients with chronic obstructive pulmonary disease. *Respiration* 61: 263–268

172 Abassi ZA, Tate JE, Golomb E, Keiser HR (1992) Role of neutral endopeptidase in the metabolism of endothelin. *Hypertension* 20: 89–95

173 Faraone S, Sofia M, Maniscalco M, Alifano M, Battiloro R, Micco A. Mormile M, Carratù P (1997) Endogenous endothelin production/extraction in patients with COPD. *Am J Respir Crit Care Med* 155: A594

174 Shokeir MO, Paré P, Wright JL (1994) Relation of smoking to immunoreactive endothelin in the bronchiolar epithelial cells. *Thorax* 49: 786–788

175 Crystal RG, Bitterman PB, Rennard SI, Hance AJ, Keogh BA (1984) Interstitial lung diseases of unknown cause: disorders characterized by chronic inflammation of the lower respiratory tract. *N Engl J Med* 310: 154–166

176 Uguccioni M, Pulsatelli L, Grigolo B, Facchini A, Fasano L, Cinti C, Fabbri M, Gasbarrini G, Meliconi R (1995) Endothelin-1 in idiopathic pulmonary fibrosis. *J Clin Pathol* 48: 330–334

177 Giaid A, Michel RP, Stewart DJ, Sheppard M, Corrin B, Hamid Q (1993) Expression of endothelin-1 in lungs of patients with cryptogenic fibrosing alveolitis. *Lancet* 341: 1550–1554

178 Snider GL, Hayes JA, Korthy AL (1978) Chronic interstitial pulmonary fibrosis produced in hamsters by endotracheal bleomycin: pathology and stereology. *Am Rev Respir Dis* 117: 1099–1108

179 Thrall RS, McCormick JR, Jack RM, McReynolds RA, Ward PA (1979) Bleomycin-induced pulmonary fibrosis in the rat: inhibition by indomethacin. *Am J Pathol* 95: 117–130

180 Khalil N, Bereznay O, Sporn M, Greenberg AH (1989) Macrophage production of transforming growth factor β and fibroblast collagen synthesis in chronic pulmonary inflammation. *J Exp Med* 170: 727–737

181 Santana A, Saxena B, Noble NA, Gold LI, Marshall BC (1995) Increased expression of transforming growth factor β isoforms (β1, β2, β3) in bleomycin-induced pulmonary fibrosis. *Am J Respir Cell Mol Biol* 13: 34–44

182 Antoniades HN, Bravo MA, Avila RE, Galanopoulos T, Neville-Golden J, Maxwell M, Seiman M (1990) Platelet-derived growth factor in idiopathic pulmonary fibrosis. *J Clin Invest* 86: 1055–1064

183 Broekelmann TJ, Limper AH, Colby TV, McDonald JA (1991) Transforming growth factor β1 is present at sites of extracellular matrix gene expression in human pulmonary fibrosis. *Proc Natl Acad Sci USA* 88: 6642–6646

184 Khalil N, O'Connor RN, Unruh HW, Warren PW, Flanders KC, Kemp A, Bereznay OH, Greenberg AH (1991) Increased production and immunohistochemical localization of transforming growth factor-β in idiopathic pulmonary fibrosis. *Am J Respir Cell Mol Biol* 5: 155–162

185 Yamane K, Kashiwagi H, Suzuki N, Miyauchi T, Yanagisawa M, Goto K, Masaki T (1991) Elevated plasma levels of endothelin-1 in systemic sclerosis. *Arthritis Rheum* 34: 243–244 (Letter)

186 Yamane K, Miyauchi T, Suzuki N, Yuhara T, Akama T, Suzuki H, Kashiwagi H (1992) Significance of plasma endothelin-1 levels in patients with systemic sclerosis. *J Rheumatol* 19: 1566–1571

187 Morelli S, Ferri C, Polettini E, Bellini C, Gualdi GF, Pittoni V, Valesini G, Santucci A (1995) Plasma endothelin-1 levels, pulmonary hypertension, and lung fibrosis in patients with systemic sclerosis. *Am J Med* 99: 255–260

188 Kawaguchi Y, Suzuki K, Hara M, Hidaka T, Ishizuka T, Kawagoe M, Nakamura H (1994) Increased endothelin-1 production in fibroblasts derived from patients with systemic sclerosis. *Ann Rheum Dis* 53: 506–510

189 Kikuchi K, Kadono T, Sato S, Tamaki K, Takehara K (1995) Impaired growth response to endothelin-1 in scleroderma fibroblasts. *Biochem Biophys Res Commun* 207: 829–838

190 Giaid A, Hamid QA, Springall DR, Yanagisawa M, Shinmi O, Sawamura T, Masaki T, Kimura S, Corrin B, Polak JM (1990) Detection of endothelin immunoreactivity and mRNA in pulmonary tumours. *J Pathol* 162: 15–22

191 Druml W, Steltzer H, Waldhäusl W, Lenz K, Hammerle A, Vierhapper H, Gasic S, Wagner OF (1993) Endothelin-1 in adult respiratory distress syndrome. *Am Rev Respir Dis* 148: 1169–1173

192 Langleben D, Demarchie M, Laporta D, Spanier AH, Sclesinger RD, Stewart DJ (1993) Endothelin-1 in acute lung injury and the adult respiratory distress syndrome. *Am Rev Respir Dis* 148: 1646–1650

193 Sofia M, Faraone S, Alifano M, Micco A, Albisinni R, Maniscalco M, Di Minno G (1997) Endothelin abnormalities in patients with pulmonary embolism. *Chest* 111: 544–549

194 Johnson CS, Verdigem TD (1988) Pulmonary complications of sickle cell disease. *Semin Respir Med* 9: 287–296

195 Hammerman SI, Kourembanas S, Conca TJ, Tucci M, Brauer M, Farber HW (1997) Endothelin-1 production during the acute chest syndrome in sickle cell disease. *Am J Respir Crit Care Med* 156: 280–285

196 Raz I, Okon E, Chajek-Shaul T (1989) Pulmonary manifestations in Behçet's syndrome. *Chest* 95: 585–589

197 Hamzaoui A, Hamzaoui K, Chabbou A, Ayed K (1996) Endothelin-1 expression in serum and bronchoalveolar lavage from patients with active Behçet's disease. *Br J Rheumatol* 35: 357–358

Index

adenylyl cyclase   95
adhesion molecule   237
adult respiratory distress syndrome (ARDS)
    136, 178, 253
airway   51
airway epithelium   110, 185, 238, 241, 242,
    248
airway hyperresponsiveness   132
airway resistance   126
airway smooth muscle (ASM) cell   233,
    234, 238–240, 439,
airway smooth muscle (ASM), contraction
    of   84, 108, 233, 234
airway smooth muscle (ASM), relaxation of
    92, 234
airway smooth muscle (ASM)
    hypertrophy/hyperplasia   185
airway smooth muscle (ASM), mitogenesis
    of   93
airway smooth muscle (ASM), proliferation
    of   132
albuterol, β-adrenoceptor agonist
    130
allergen   244, 245, 247–249
allergic, intranasally to symptomatic and
    non-allergic individuals   128
allergic sheep   132
alveoli   127
anti-aggregatory   182
antiserum, anti-ET-1   133
arachidonic acid metabolism   161
arachidonic acid product   162
Ascaris suum antigen   132
asthma   36, 51, 77, 185, 188, 189,
    239–241, 243, 244, 249
asthma, bronchial   166
asthma, exercise-induced   246
asthma, nocturnal   247, 248
asthma, symptomatic   167
asthmatic patient   130
autoradiography   199

Behçet's disease   254
Behçet's syndrome   253
binding site, in asthmatic and non-asthmatic
    airway   130
biopsy   243
bronchial biopsy   241–243
bronchial circulation   97
bronchial hyperresponsiveness   235
bronchoalveolar lavage (BAL) fluid   64,
    238, 244–248, 250, 252, 253
bronchoconstriction, ET-1 induced
    125
bronchoconstrictor   163, 164

calcium channel, L-type   88
calcium, cytosolic   84
calcium, extracellular   87
calcium, intracellular   85
calcium, intracellular mobilization   113
cardiovascular development   144
chemotaxis   237
chloride channel   91
chloride secretion   163, 164
cholinergic nerve   98
chronic heart failure   14
chronic obstructive pulmonary disease
    (COPD)   178, 250, 251
ciliary beat   164
Clara cell   70
clearance receptor   74
collagen synthesis, stimulate   132
collagen   239, 240, 251
congenital heart disease   214
corticosteroid   163, 164, 238, 241,
    243–248
cross-overspilling   77
cryptogenic fibrosis alveolitis (CFA)   76
cyclooxygenase (COX)   69, 161, 162,
    164
cyclooxygenase product   163
cystic fibrosis   73, 185
cytokine   55, 156, 157, 161

development   144
diacylglycerol (DAG)   89, 147

ECE   249, 251
ECE-1, knockout of   76
ECE-1α 65
ECE-1β 65
ECE-2   65
edema   188, 240
edema formation   184
endothelial cell   183
endothelin   4, 117
endothelin, "big-"   4, 64
endothelin-1   64, 107, 198
endothelin-1, proinflammatory actions of
    187, 178
endothelin-3   198
endothelin converting enzyme   9
endothelin peptide   50
endothelin, proinflammatory actions of
    178
endothelin receptor   11, 158, 159
endothelin receptor agonist   13, 25
endothelin receptor antagonist   223
endothelin release   157
endothelin synthesis   156

endothelium-derived contracting factor
    (EDCF)   2
endothelium-derived relaxing factor (EDRF)
    2
eosinophil   237, 238, 240, 243, 250
eosinophil infiltration   182
eosinophilic inflammation   182
epithelial cell   96
epithelium   243
epithelium denudation   164
epithelium, removing the   165
ET-1   132
ET-1, aerosol of   128
ET-1 analog [diaminoproprionic acid1-
    Asp15] ET-1   128
ET-1, mitogenic response to   186
ET-1 null mutation mice   50
ET-1 receptor   160
ET-3, aerosol of   128
ET_A receptor antagonist, BQ-123 and
    FR 139317   126
ET_A receptor   22, 114, 125, 235
ET_A receptor antagonist, BQ-610   137
ET_A receptor-selective antagonist   26
ET_A/ET_B receptor antagonist
    (PD-156707-0015)   137
ET_A/ET_B receptor antagonist, bosentan
    127
ET_A/ET_B receptor antagonist, mixed
    30
ET_B receptor   22, 114, 125
ET_B receptor agonist, IRL 1620   126
ET_B receptor antagonist, BQ-788   126
ET_B receptor-selective antagonist   30
ET_C receptor   23
ET_C receptor, putative   125

fibroblast   238, 239, 251, 252
fibroblast, proliferation of   132, 239
fibrosis   186, 188
furin   9

gene targeting   14
gland, airway   241
gland, basal submucosal   236
gland product release   166
gland, submucosal   243, 249
gland, submucosal serous and mucous cell
    236
gland, trachael submucosal   236
glucocorticoid   163
GMP, cyclic   77
G-protein   147
gross pulmonary edema   137

heart disease   37
hemangioendothelioma   14
Hirschsprung disease   145
hypoxia   56

hypoxic pulmonary hypertension   213, 216,
    220, 222, 223

idiopathic pulmonary fibrosis (IPF)   67,
    251, 252
inflammation   237, 251
inflammation, airway   243, 248
inflammation, pulmonary   177, 178
inflammatory action, pro-   127
inflammatory cell   179, 187
inflammatory cell, influx into airway   127
inflammatory cell, stimulate   189
inflammatory process   187
iNOS   166
intranasal, to sympotomatic allergic and
    non-allergic   128

kidney   37

leukocyte   237, 247, 253
leukotriene   128
LTD_4 antagonist, MK-571   128
lung, adult   146
lung, fetal   146
lung, postnatal   146
lung cancer   252
lung development   50, 144
lung edema   184
lung elastase   126
lung injury, acute   187, 189
lung resistance   126

macrophage   179, 181, 237, 238, 243, 248,
    249, 250
mast cell   237, 240, 243, 250
mast cell activation   182
metalloprotease   10
microvascular leakage   97
microvascular permeability   184, 235, 236,
    253
mitogen   148
mitogen-activated protein kinase (MAPK)
    12, 93, 149
mitogenic effect of ET-1   221
monocrotaline-induced pulmonary hyperten-
    sion   217, 224
monocyte   179, 181, 237
mononuclear cell, activation of   188
mucociliary clearance   163
mucosal inflammation   185
mucus   236, 237, 240, 250
myofibroblast   239, 240, 251

neonatal pulmonary hypertension   214,
    218, 221, 222, 225
nerves, non-adrenergic non-cholinergic
    197
nerves, parasympathetic   197
nerves, sensory   197

nerves, sympathetic   197
neuromodulator   197, 234
neurotransmission   99, 197
neurotransmission, central   197
neurotransmission, peripheral autonomic
   197
neutral endopeptidase (NEP)   158
neutrophil   178, 179, 183, 237, 238, 247
neutrophil, accumulation   178, 181, 188
neutrophil, activation   178
neutrophil, intrapulmonary accumulation of
   179
neutrophil influx   188
nitric oxide (NO)   164
nitric oxide (NO)/endothelin-1 balance   73
nitric oxide synthase   64
nose   129

oedema, see edema

parenchyma   54
peripheral airway   126
phosphoinositide pathway   113
phosphoinositide turnover   85
phospholipase A2   91
phospholipase C   147
phosphoramidon   67
phosphoramidon, inhibitor of enzyme neutral endopeptidase (NEP)   126
platelet, activation   181
platelet aggregation, inhibition of   182
porcine aortic, originally isolated   156
postjunctional site   200
prejunctional site   200
preproendothelin   4
prostaglandin (PGI$_2$)   148
protein kinase C   89, 113
proto-oncogenes c-fos and c-myc, expression
   of   186
provocative concentration of ET-1 that
   caused 35% decrease in specific airway
   conductance (PC$_{35Sgaw}$)   130
pulmonary blood vessel   116
pulmonary circulation   55
pulmonary development   144
pulmonary edema   184
pulmonary embolism   254
pulmonary fibrosis   54, 189, 252
pulmonary hypertension   36, 56, 136, 189
pulmonary hypertension, animal model of
   216, 219, 221–223
pulmonary hypertension, human   213, 223

pulmonary hypertension, primary   64
pulmonary hypertension, treatment of   215
pulmonary insufflation pressure   69
pulmonary vascular remodeling   147
pulmonary vascular resistance   136
pulmonary vasoconstriction   126
pulmonary vasodilatation   117
pulmonary vasoreactivity to endothelin
   219

receptor   232–235, 237–239, 244, 245,
   248, 249
receptor subtypes   126, 159
relaxation   112
remodelling   240, 243
restenosis   38
rhinitis, allergic   168
ryanodine receptor   86

sarafotoxin, S6a-S6d   6
sarafotoxin S6b   200
sarafotoxin S6c   202
sarafotoxin S6c, specific ET$_B$ receptor
   agonist   132
secretion and nasal symptoms, indices of
   128
sepsis   137
sickle cell disease   253
signal transduction process   83
smooth muscle cell, contraction   235
smooth muscle cell, proliferation   148
sodium-hydrogen antiporter   90
sputum, induced   248
steroids   161
subjects, normal   130
submucosal gland   97, 164, 165
systemic sclerosis   252

thromboxane A$_2$ (TxA$_2$) inhibitor, OKY-046
   126
tissue resistance   126
T-lymphocyte   238, 240, 250
T-lymphocyte chemotaxis   237
trachael mucus velocity (TMV), measure of
   mucociliary clearance   128
trisphosphate (IP$_3$)   147
TXA$_2$ 148

vascular permeability   184
vascular smooth muscle, proliferation of
   132
vasoconstriction   166

**RPP – Respiratory Pharmacology
and Pharmacotherapy**

Stockley R. A., University Hospital Birmingham,
UK (Ed.)

Molecular Biology of the Lung

Vol. I: Emphysema and Infection
Vol. II: Asthma and Cancer

Stockley R. A. (Ed.)
Molecular Biology of the Lung
Vol. I: Emphysema and Infection
1999. 230 pages. Hardcover
ISBN 3-7643-5857-2

The techniques and knowledge in molecular biology are
advancing rapidly. The basic mechanisms involved in the
pathogenesis of various lung conditions including asthma,
COPD and lung cancer are being clarified by the application of this powerful technology. Genetic causes of
single gene disorders such as cystic fibrosis and Alpha-1
antitrypsin deficiency have now been well delineated. This
progress leads to an understanding of the pathogenic
process that result in disease.

The purpose of Volumes I and II of Molecular Biology of
the Lung is to provide an update on the use of this powerful technology. As such, the books provide an insight into
the techniques for the non-specialist. Scientists embarking
upon studies in chronic lung disease will find a review of
the current status of research. At the same time, the books
are useful to clinicians, both specialist and academic, and
to scientists already involved in the basic aspects of the
pathogenesis of lung disease. Both volumes deal with
basic mechanisms of cell biology, receptors and cell activation and provides an insight as to how the technology
influences our concepts of pathogenesis and vice versa.

Stockley R. A. (Ed.)
Molecular Biology of the Lung
Vol. II: Asthma and Cancer
1999. 222 pages. Hardcover
ISBN 3-7643-5968-4

Volume I focuses on Emphysema and Infection, Volume II
focuses on Asthma and Cancer.

Stockley R. A. (Ed.)
Molecular Biology of the Lung
Two Vols. Set
1999. 452 pages. Hardcover
ISBN 3-7643-5969-2

Please order through your bookseller or directly from:

**Birkhäuser Verlag AG, P.O. Box 133
CH-4010 Basel / Switzerland
Tel.: +41 / (0)61 / 205 07 07
Fax: +41 / (0)61 / 205 07 92
e-mail: orders@birkhauser.ch**

Birkhäuser

RPP – Respiratory Pharmacology and Pharmacotherapy

Martinet Y., Brabois Hospital, Vandoeuvre, France / **Hirsch F.R.,** Copenhagen University Hospital, Denmark /
Martinet N., INSERM U14, Vandoeuvre, France / **Vignaud J.-M.,** Nancy Central Hospital, France /
Mulshine J.L., National Cancer Institute, Rockville, MD, USA (Ed.)

Clinical and Biological Basis of Lung Cancer Prevention

Lung cancer is a disease with pandemic public health implications as it is now the leading cause of cancer mortality throughout the world. This book results from two recent International Association for the Study of Lung Cancer (IASLC) Workshops on lung cancer prevention. It strikes a balance between considering public health approaches to tobacco control and population-based screening, advances in clinical evaluation of chemoprevention approaches, and the biology of lung carcinogenesis.

Indeed, while the science of smoking cessation is evolving as new pharmacological tools are moving into clinical evaluation, the current impact of molecular diagnostics is profound. The rapidly-evolving diagnostic technologies are revolutionizing basic scientific investigation of cancer, and this trend is expected to soon spill over into the clinical practice of medicine. The evolution of economical diagnostic platforms to allow for direct bronchial epithelial evaluation in high-risk populations promises to improve the diagnostic lead-time for this disease. The hope is that enough progress will occur to permit lung cancer detection in advance of clinical cancer so that the disease can be addressed early on, while it is still confined to the site of origin.

Chemoprevention, which is designed to intervene in the early phase of carcinogenesis prior to any subjective clinical manifestation of a cancer, is also generating greater research interest. Moreover, the benefit of aero-solized administration of chemoprevention agents over conventional oral administration has strong appeal and may result in the reduction of the incidence of cancer when combined with new diagnostic technologies.

Martinet Y. et al. (Ed.)
Clinical and Biological Basis of Lung Cancer Prevention
1997. 344 pages. Hardcover
ISBN 3-7643-5778-9

BioSciences with Birkhäuser

Please order through your bookseller or directly from:

**Birkhäuser Verlag AG, P.O. Box 133
CH-4010 Basel / Switzerland
Tel.: +41 / (0)61 / 205 07 07
Fax: +41 / (0)61 / 205 07 92
e-mail: orders@birkhauser.ch**

Birkhäuser

RPP – Respiratory Pharmacology and Pharmacotherapy

Isenberg D. A., University College, London, UK / **Spiro S. G.**, Middlesex Hospital, London, UK (Ed.)

Autoimmune Aspects of Lung Disease

The lung forms an integral part of the body's immune system
and is subject to a range of diseases which are either autoimmune in nature or have clear-cut immunological abnormalities. Autoimmune Aspects of Lung Disease provides a concise
review of the lung's role in the immune system and a detailed
account of both primary and secondary lung diseases which
are characterised by immunological perturbation or frank auto-
immunity.

The volume presents a detailed, up-to-date account of disorders
ranging from infection to neoplasia and is written in both an
informative and stimulating style by a prestigious group of
authors. The chapters are extensively referenced and provide
numerous insights into the aetiopathogenesis and clinical
features and treatment of immunologically-linked pulmonary
disease.

The book is intended as both an overview for physicians and
scientists with an established interest in diseases of the lung,
immunologists seeking to learn more about relevant disorders
in the lung and general physicians, whether specialists or in
training, seeking to enrich their knowledge of the links between the pulmonary and immune systems.

Isenberg D. A., Spiro S.G. (Ed.)
**Autoimmune Aspects
of Lung Disease**
1997. 288 pages. Hardcover.
ISBN 3-7643-5719-3

BioSciences with Birkhäuser

Please order through your bookseller or directly from:

Birkhäuser Verlag AG, P.O. Box 133
CH-4010 Basel / Switzerland
Tel.: +41 / (0)61 / 205 07 07
Fax: +41 / (0)61 / 205 07 92
e-mail: orders@birkhauser.ch

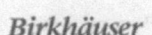

Birkhäuser